Urban Sustainability

The Urban Sustainability Book Series is a valuable resource for sustainability and urban-related education and research. It offers an inter-disciplinary platform covering all four areas of practice, policy, education, research, and their nexus. The publications in this series are related to critical areas of sustainability, urban studies, planning, and urban geography.

This book series aims to put together cutting-edge research findings linked to the overarching field of urban sustainability. The scope and nature of the topic are broad and interdisciplinary and bring together various associated disciplines from sustainable development, environmental sciences, urbanism, etc. With many advanced research findings in the field, there is a need to put together various discussions and contributions on specific sustainability fields, covering a good range of topics on sustainable development, sustainable urbanism, and urban sustainability. Despite the broad range of issues, we note the importance of practical and policy-oriented directions, extending the literature and directions and pathways towards achieving urban sustainability.

The series will appeal to urbanists, geographers, planners, engineers, architects, governmental authorities, policymakers, researchers of all levels, and to all of those interested in a wide-ranging overview of urban sustainability and its associated fields. The series includes monographs and edited volumes, covering a range of topics under the urban sustainability topic, which can also be used for teaching materials.

Ali Cheshmehzangi · Sara Alidoust ·
Wendy Y. Chen · Richard Fuller
Editors

Nature-based Urbanism

Springer

Editors
Ali Cheshmehzangi
School of Architecture, Design & Planning
University of Queensland
Brisbane, QLD, Australia

Wendy Y. Chen
Department of Geography
University of Hong Kong
Hong Kong, Hong Kong

Sara Alidoust
School of Architecture, Design
and Planning
University of Queensland
Brisbane, QLD, Australia

Richard Fuller
School of the Environment
University of Queensland
Brisbane, QLD, Australia

ISSN 2731-6483 ISSN 2731-6491 (electronic)
Urban Sustainability
ISBN 978-981-95-6863-5 ISBN 978-981-95-6864-2 (eBook)
https://doi.org/10.1007/978-981-95-6864-2

This Springer imprint is published by the registered company Springer Nature Singapore Pte Ltd.
The registered company address is: 152 Beach Road, #21-01/04 Gateway East, Singapore 189721, Singapore

We collectively dedicate this book to the thinkers and actors who are committed to shaping cities as living systems—ecologists, planners, architects, and citizens alike—whose dedication to balance and beauty turns the idea of ***nature-based urbanism*** *into a lived reality.*

Preface

"Look deep into nature, and then you will understand everything better".

—Albert Einstein

Cities all around the world are dealing with a confluence of issues that go right to the heart of how urban growth has been conceived, rationalised, and institutionalised for over a century. The prevailing paradigms of urban growth are no longer appropriate, as evidenced by rising temperatures, failing drainage systems, declining biodiversity, growing social injustices, and persistent public health stressors. Many cities continue to grow using approaches that prioritise short-term profits over long-term resilience, externalise environmental costs, and consider nature as disposable. The result is an urbanisation pattern that is increasingly unhealthy, environmentally brittle, and socially fragmented. It is now impossible to sidestep the need for a new logic of city-making.

In this context, *nature-based urbanism* emerges as a genuine rethinking of urban systems rather than as another planning notion or a superficial "greening" program. Its fundamental idea is simple: cities that are built as extensions of the natural systems that support them perform better ecologically, socially, and economically. Adding parks or trees as compensatory gestures is not enough to achieve this. It demands that cities include ecological thought into their identity, performance, and structure. This entails creating roadways that function as blue-green corridors, planning stormwater systems that behave like wetlands, reshaping public areas to accommodate both human and non-human users, and rehabilitating degraded landscapes as active urban assets. These are not aesthetic goals. They are structural conditions for survival in an era of climate volatility.

This volume takes that challenge seriously. This book, which is the second in a trilogy about "nature in cities", focuses on case studies from around the world to examine how *nature-based urbanism* might progress from ideation to reality. Every chapter shows how cities—big and small, wealthy and resource-constrained, quickly expanding and long-standing—are experimenting with novel connections with the natural world. These are not all-encompassing solutions; rather, they represent various cultures, climates, and political structures. However, when taken as a whole, they show a distinct pattern: cities that include ecological systems into their planning and design frameworks are typically more robust, more liveable, and better equipped to adjust to stress.

A recurring theme in the ensuing chapters is that cities cannot continue to solve structural urban issues with disjointed solutions or symbolic "green touches". Heat stress, flooding, pollution, declining biodiversity, and socioeconomic injustice are all signs of a structural breakdown between cities and the ecological systems that support them. By reintroducing ecological purpose into the architecture of urban decision-making, *nature-based urbanism* provides a remedy rather than visualising nature. The global examples in this collection demonstrate how urban landscapes become more equitable, adaptable, and ultimately future-ready when planning, design, and governance are in line with ecological processes. What distinguishes these examples is not scale or sophistication but the willingness to shift away from outdated assumptions about how cities should grow. They show that, more than ever, ecological thinking has become essential to urban competency and is no longer optional.

This book is a plea for evidence-based, realistic change rather than an affirmation of idealism. The case studies show that *nature-based urbanism* is real and is already changing cities in quantifiable ways. If the last century of urbanisation was defined by the dominance of engineering logics, the current one ought to be shaped by the maturity of ecological logics. The change is in progress. Whether cities will accept it in due course is the question.

Brisbane and Hong Kong

Ali Cheshmehzangi
Sara Alidoust
Wendy Y. Chen
Richard Fuller

Acknowledgements

We acknowledge everyone's contributions to this exciting book project. We extend our gratitude to all the authors, co-authors, chapter contributors, and collaborators who have shared their expertise, passion, and creativity with us. This sentiment applies to our earlier and forthcoming work as well, where we deeply value the dedication and vision of all contributors who helped create this book. Such collaborative efforts foster new dialogues, networking opportunities, and a valuable research exchange platform for scholars worldwide.

About This Book

As the second volume of this trilogy focused on nature in cities, this book focuses on global case study examples relevant to "nature-based urbanism". As a relatively newer field of research, nature-based urbanism encompasses the integration of ecological systems into the planning, design, and management of urban environments, ensuring that cities function not just as human habitats but as active components of broader natural systems. The book uses ecological restoration, biophilic design, and nature-based solutions to build urban forms and environments that improve biodiversity, lessen the effects of climate change, and advance human well-being.

By integrating ecological thought into the fundamental fabric of cities—street networks that serve as blue-green corridors, stormwater systems that resemble wetlands, and public areas that welcome both human and non-human life—nature-based urbanism goes beyond token greening or discrete interventions. It reinterprets urban growth as a chance for mutual flourishing rather than as an invasion of nature.

Nature-based urbanism is positioned as a revolutionary method of urban development in the Anthropocene by integrating ecosystem services, cultural values, and climate adaptability. In addition to improving liveability and creating a revitalised sense of place, it tackles urgent global issues, including heat islands, flooding, and air pollution. This book makes the case that understanding nature as a co-creator of urban life, rather than just a source of inspiration, is essential to the future of cities. In doing so, it presents a picture of cities as living, breathing landscapes that are resilient, adaptive, and intricately linked to the ecological systems that support them.

This book:

1. Showcases innovative global case studies redefining cities through ecological integration.
2. Introduces nature-based urbanism from multiple practical and theoretical perspectives
3. Bridges climate resilience, biodiversity, and liveability in a unified urban vision.
4. Offers practical, scalable strategies for embedding nature at every city scale.

Contents

Editors and Contributors

About the Editors

Ali Cheshmehzangi is Head of the School of Architecture, Design and Planning and a Professor at The University of Queensland, Australia. He is among the top 20 global scholars in the urban sustainability research area. He is the Editor-in-Chief of the Urban Sustainability Book Series with Springer Nature. So far, Ali has published over 600 journal papers, articles, conference papers, book chapters, and reports. He has published more than 55 academic books, some of which have received awards at the international, national, provincial, and municipal levels. He has also received international awards and recognition for his research on urban resilience studies and sustainability research, as well as the 2018 Vice-Chancellor's award for his impactful contribution to higher education.

Sara Alidoust is a Senior Lecturer in Urban Planning at The School of Architecture, Design and Planning at The University of Queensland, Australia. Her research centres on the intersections of Urban Planning and Public Health, with a strong emphasis on interdisciplinary collaboration and practical application. Sara delivers transdisciplinary solutions to some of the most compelling challenges of our time, "housing equity" and "healthy cities". Her work stands out through the application of Systems Thinking to complex planning issues, and she explores the connections between planning and the physical, mental, and social health of individuals and communities.

Wendy Y. Chen is a professor in the Department of Geography at The University of Hong Kong. She serves as the Editor-in-Chief for Urban Forestry & Urban Greening, a top international journal in the field of urban greening design and management. Her research agenda has focused on key knowledge gaps in evaluating and modelling urban green-blue spaces (GBS) that are reserved, modified, and deployed in urban/peri-urban areas. She is among a small group of scholars who have begun to experiment and validate the utilisation of classical non-market approaches in China's transitional context and facilitate the quantitative assessment of GBS' contribution to the quality of life. She is ranked as a top 1% scholar worldwide by Clarivate Analytics.

Richard Fuller is a leading expert in urban ecology research. He is a Professor at the School of the Environment at The University of Queensland, Australia. He works on pure and applied topics in biodiversity and conservation. Much of his work is interdisciplinary, focusing on the interactions between people and nature, how these can be enhanced, and how these relationships can be shaped to converge on coherent solutions to the biodiversity crisis. His current research topics include the ecology of urban environments, understanding what drives some people to show stronger environmental concern than others, and strategies for designing efficient conservation plans.

Contributors

Sara Alidoust School of Architecture, Design and Planning, The University of Queensland, Brisbane, Australia

Diana M. Benjumea Mejia College of Interdisciplinary and Experiential Learning, Singapore University of Social Sciences, Bukit Timah, Singapore;
The National University of Colombia, Caldas, Colombia

Richard Fuller School of the Environment, The University of Queensland, Brisbane, Australia;
Centre for Biodiversity and Conservation Science, The University of Queensland, Brisbane, Australia

Ahmad Fariz Al Hazmi School of Architecture, Design and Planning, The University of Sydney, Sydney, NSW, Australia

Ammon Beyerle Here Studio, Melbourne, Australia

Federico Camerin Department of Urbanismo y Representación de la Arquitectura, Universidad de Valladolid, Valladolid, Spain

Fernando Carmona Mateos Department of Urbanística y Ordenación del Territorio, Universidad Politécnica de Madrid, Madrid, Spain

Padmaja Chakravarty Department of Planning and Architecture, National Institute of Technology Rourkela, Rourkela, Odisha, India

Wendy Y. Chen Department of Geography, The University of Hong Kong, Hong Kong, China

Ali Cheshmehzangi School of Architecture, Design and Planning, The University of Queensland, Brisbane, Australia;
School of Architecture and Urban Planning, Hunan University, Changsha, China

Rafael Córdoba Hernández Department of Urbanística y Ordenación del Territorio, Universidad Politécnica de Madrid, Madrid, Spain

Thanh Hung Dang University of Architecture, Ho Chi Minh City, Viet Nam

Nigel K. Downes College of Environment and Natural Resources, Can Tho city, Viet Nam

Tania Giraldo Ospina The National University of Colombia, Caldas, Colombia

Perrine Hamel Asian School of the Environment, Nanyang Technological University, Nanyang, Singapore;
Earth Observatory of Singapore, Nanyang Technological University, Nanyang, Singapore

Jennifer L. Kent School of Architecture, Design and Planning, The University of Sydney, Sydney, NSW, Australia

Azadeh Khaki Ghasr Department of Architecture, Faculty of Architecture and Urban Planning, Shahid Beheshti University, Tehran, Iran

Minh Trang Khong University of Architecture, Ho Chi Minh City, Viet Nam

Janice S. H. Lee Asian School of the Environment, Nanyang Technological University, Nanyang, Singapore;
Earth Observatory of Singapore, Nanyang Technological University, Nanyang, Singapore

Adrian Loo School of Biological Sciences, National University of Singapore, Kent Ridge, Singapore

Isidoro R. Malaque III University of the Philippines Mindanao, Davao, Philippines

Javier Morán Uriel Department of Urbanística y Ordenación del Territorio, Universidad Politécnica de Madrid, Madrid, Spain

Rachel R. Y. Oh Department of Geography, National University of Singapore, Kent Ridge, Singapore

Ali Rad Yousefnia School of Architecture, Design and Planning, University of Queensland, Brisbane, Australia

Nabanita Saha Department of Planning and Architecture, National Institute of Technology Rourkela, Rourkela, Odisha, India

Basudatta Sarkar Department of Planning and Architecture, National Institute of Technology Rourkela, Rourkela, Odisha, India

Rachel Mae P. Sitcharon Provincial Government of Agusan del Sur, Agusan del Sur, Philippines

Amir Tayyebi Architecture and Planning Department, Jundi-Shapur University of Technology, Dezful, Iran

Yiwen Zeng Asian School of the Environment, Nanyang Technological University, Nanyang, Singapore;
School of Social Sciences, Nanyang Technological University, Nanyang, Singapore

Viet Anh Vu University of Architecture, Ho Chi Minh City, Viet Nam

Reframing Urban Futures: The Promise of Nature-Based Urbanism

Ali Cheshmehzangi, Sara Alidoust, Wendy Y. Chen, and Richard Fuller

Abstract *Nature-Based Urbanism* is the second volume of the *Nature and Cities* trilogy, advancing the case for embedding ecology into the core of urban planning, design, and governance. Building on the foundation established in *Nature in Cities, Nurturing Cities*, this book shifts from theory to practice, exploring how cities worldwide are integrating ecological systems as infrastructure for resilience, equity, and liveability. The volume brings together a wide spectrum of cases, from Southeast Asian green-space transitions and vernacular riverine landscapes to multispecies planning, nature-driven placemaking, and regenerative urban governance across India and Europe, to show how cities are re-imagining their relationship with nature. Throughout the case study chapters, ecological thinking transforms planning, design, and daily urban living as a theoretical lens and useful toolset. In addition to providing new avenues for participation, governance, and community agency, the contributions show how nature-based initiatives can address climatic volatility, biodiversity loss, social disintegration, and the deterioration of human-nature relationship. When taken as a whole, they show that incorporating nature into urban design is not just an environmental need but also a cultural and civic revolution that is now occurring in places all over the world. By uniting ecological restoration, biophilic design, and nature-based solutions under the framework of urbanism, this book positions cities as living ecologies rather than static built environments. Its intellectual ambition is to redefine urbanism as regenerative practice; its practical aim is to provide scalable models for professionals, policymakers, and communities. More importantly, *Nature-Based*

A. Cheshmehzangi (✉) · S. Alidoust
School of Architecture, Design and Planning, The University of Queensland, Brisbane, Australia
e-mail: a.chesh@uq.edu.au

A. Cheshmehzangi
School of Architecture and Urban Planning, Hunan University, Changsha, China

W. Y. Chen
Department of Geography, The University of Hong Kong, Hong Kong, China

R. Fuller
School of the Environment, The University of Queensland, Brisbane, Australia

Centre for Biodiversity and Conservation Science, The University of Queensland, Brisbane, Australia

A. Cheshmehzangi et al. (eds.), *Nature-based Urbanism*, Urban Sustainability,
https://doi.org/10.1007/978-981-95-6864-2_1

Urbanism contributes to global debates on sustainability by making a bold claim: the future of urban life depends on building cities with nature, not against it.

Keywords Nature-based urbanism · Regenerative design · Urban ecology · Climate resilience · Biodiversity · Liveability · Sustainability · Nature-based solutions · NBS

1 Introducing Nature-Based Urbanism: A New Paradigm for Cities

Cities today stand at a threshold. More than half of the world's population lives there, and they are hubs for innovative culture and economic activity. However, they are also the epicentre of biodiversity loss, climate change, and growing social inequality. These convergent crises are beyond the scope of conventional theories of urban growth, which are motivated by extraction, expansion, and technological solutions. Rethinking what cities are and how they interact with the natural systems that support them is necessary, not making small changes.

In this environment, nature-based urbanism becomes both a worldview and a practice (Scott et al., 2016). It sees cities as living ecologies that are inextricably linked to rivers, soil, forests, and species rather than as isolated human constructions. This method reframes urbanism as a regenerative process by utilising biophilic design, ecological restoration, and nature-based solutions (Cheshmehzangi, 2025). Rooftops become meadows, stormwater systems resemble wetlands, street networks serve as blue-green corridors, and public areas welcome both human and non-human life (Stuiver, 2023). By doing this, nature-based urbanism transforms the city into a catalyst for ecological repair rather than a place of ecological stress.

This volume is the second in a trilogy on *Nature and Cities*. The first, *Nature in Cities, Nurturing Cities*, laid the theoretical groundwork for the idea that we must integrate nature into cities in order to nurture them. This book expands on that idea by focusing on global illustration and applied practice. It illustrates how nature-based urbanism is being implemented in a variety of settings, from community-driven farms in Latin America to rewilded riverbanks in European towns, from vertical forests in densely populated Asian cities to mangrove buffers safeguarding coastal populations. These examples show a broader shift in the conception and construction of cities rather than being isolated experiments.

The urgency of this shift cannot be overstated. Heat islands that threaten health, floods that overwhelm infrastructure, air pollution that shortens life, and biodiversity decline that weakens resilience are just a few of the overlapping threats that characterise the Anthropocene. Conventional engineering is not enough to solve these problems. Nature-based urbanism shows that natural systems, which cool neighbourhoods, absorb rainfall, filter air, store carbon, and promote community well-being, are essential infrastructure rather than luxury.

By situating nature as a co-creator of urban life, this volume advances a paradigm where urban futures are no longer measured solely by growth or efficiency but by resilience, adaptability, and reciprocity. It insists that the future of cities and the environmental systems they occupy are inextricably linked. Adopting nature-based urbanism means acknowledging the interdependence of the health of our cities, communities, and planet.

1.1 Strengthening the Narrative: From Greening to Living Ecologies

In recent decades, the phrase "greening the city" has been popular in planning documents, legislative agendas, and even marketing efforts (Branz & Dümpelmann, 2011). Pocket parks, tree-planting campaigns, and green roofs are often pushed as outward manifestations of environmental stewardship. Even though these programs are beneficial, they frequently continue to be tokenistic, or perhaps, appears to be as small-scale fixes applied to essentially unsustainable urban systems. There is a chance that "greening" will turn into a flimsy gesture, i.e., a patch of greenery covering more serious ecological flaws.

Nature-based urbanism advances a different proposition. It shifts the emphasis from individual actions to transformative routes, from dispersed greenery to systemic integration. In this case, nature is a co-structuring principle of the urban environment rather than a cosmetic addition. Nature-based urbanism starts with the idea that cities are already a part of larger ecosystems rather than asking how to incorporate nature into current urban structures. Therefore, rather than trying to dominate or marginalise natural logics—water flows, species habitats, soil systems, and atmospheric dynamics—urban planning and design must be in line with them.

This narrative change is crucial because the problems cities face are existential rather than aesthetic. Spreading vegetation throughout the terrain won't solve issues like heat waves, flooding, biodiversity loss, and air pollution. They call for systemic fixes. A network of urban forests and blue-green corridors that control temperature, absorb runoff, and offer pathways for species movement is more revolutionary than a single street tree. While a rain garden mitigates the effects of runoff, resilience is essentially redefined by a citywide water-sensitive urban design framework that emulates natural hydrological cycles.

Importantly, this route also redefines policy and governance. Token greening is frequently carried out through pilot programs or short-term projects that are susceptible to changing political objectives. In contrast, ecological concepts must be included in long-term planning frameworks, building codes, and investment plans to achieve nature-based urbanism. It advocates for participatory techniques where communities co-design and manage local ecologies, as well as interdepartmental collaboration that integrates planning, health, infrastructure, and the environment.

By doing this, it establishes ecological integration as a prerequisite for sustainable urban futures rather than as an optional policy add-on.

This approach also carries a cultural and philosophical dimension. Greening typically emphasises aesthetic enhancement—making areas appear appealing or natural. On the other hand, living ecologies emphasise reciprocity and interactions. They encourage locals to view their city as a networked environment where non-human and human life coexist. Urban identity itself is altered by this viewpoint; a city is now characterised by the resilience of its ecosystems, the diversity of its animals, and the health of its rivers rather than just its skyline.

This kind of narrative strengthening is important because it changes both action and imagination. Nature-based urbanism is a step toward regenerative urbanism, i.e., cities that repair ecological harm, improve biodiversity, and foster resilience as inherent qualities, rather than a list of green measures. It presents a picture of reciprocal flourishing rather than conflict between urban and ecological life by going beyond tokenistic greening.

This serves as the basis for this collection, which documents and analyses how cities all over the world are transitioning from green gestures to systemic, transformative tactics of nature-based urbanism.

2 Nature-Based Urbanism in Practice: Redefining Planning, Design, and Policy

The relevance of nature-based urbanism comes from how it transforms practice rather than from abstract aspirations (Dushkova & Haase, 2020). It requires that ecological systems be treated as core infrastructure in planning, design, and policy, influencing the structure and operation of cities in ways that improve resilience, equality, and liveability. With this approach, practitioners must transition from isolated programs to integrated frameworks that acknowledge the city as a socio-ecological system.

In planning practice, nature-based urbanism reframes land-use decisions through ecological logics (Li et al., 2021; Tousi et al., 2025). While grey infrastructure continues to dominate the rest of the city, traditional zoning frequently divides green areas into separate parks or reserves. Nature-based planning, on the other hand, envisions networks of interconnected habitats, i.e., blue-green corridors that connect forests, marshes, and rivers to plazas, streetscapes, and transportation networks. These networks have several functions, including improving biodiversity, controlling urban microclimates, and offering leisure activities. For example, planning laws in European cities like Copenhagen and Rotterdam now incorporate climate adaptation (van der Berg, 2023), requiring green infrastructure and water-sensitive design at all levels, from individual developments to neighbourhood masterplans.

The consequences are similarly revolutionary in urban design. Beyond aesthetics, nature-based design reimagines the relationship between built form and ecological processes. Rooftops serve as ecosystems, public plazas incorporate shade, water, and biodiversity, and streets are built as sponges rather than drains. Singapore's Gardens by the Bay shows how large-scale ecological architecture can redefine national identity and tourism (McNeill, 2022), while Milan's Bosco Verticale (or Vertical Forest) shows how vertical trees may reimagine high-density living (Al-Kodmany, 2024). These are examples of how nature can become an important part of design DNA, concurrently influencing social space, ventilation, cooling, and stormwater retention.

Policy frameworks are critical in moving from demonstration to systemic change. Nature-based initiatives run the danger of staying boutique projects that are susceptible to funding cuts or political turnover in the absence of supportive legislation. Ecological measurements are incorporated into resilience plans, building codes, and infrastructure budgets under progressive governance. For instance, China's "sponge city" effort integrates ecological thinking at the level of national policy by requiring water-sensitive urban design in dozens of metropolitan districts (Chen, 2023; Li et al., 2025). In a similar vein, New York City's Green Infrastructure Plan connects environmental results to legal requirements by directly tying ecological investments to stormwater compliance (Bloomberg & Strickland, 2012). These examples demonstrate how policy may change urban nature from a luxury to a necessary service.

Inclusive engagement is also necessary for the governance of nature-based urbanism. When communities participate in the creation and upkeep of ecological solutions, they are most successful. For example, community-led urban farms in Latin America promote social cohesion and local stewardship in addition to producing food and enhancing biodiversity. Cities can guarantee that nature-based approaches are egalitarian, culturally grounded, and long-lasting by incorporating co-production into governance frameworks.

The practice of nature-based urbanism ultimately requires a synthesis: designers incorporating living systems into built form, planners incorporating ecological corridors into growth strategies, and legislators establishing financial and regulatory frameworks that make ecological integration commonplace rather than exceptional. When taken as a whole, these changes redefine city-making itself, moving from reactive problem-solving to proactive development of inclusive, resilient, and adaptable urban futures. Thus, to reflect on these key factors, this book, through a variety of case studies, illustrates these changes, demonstrating that nature-based urbanism is a practice that is already changing cities all over the world rather than a far-off goal.

3 Purpose and Ambition of This Volume: Building Cities with Nature

The purpose of this second volume in the *Nature and Cities* trilogy is to extend and deepen the narrative introduced in *Nature in Cities, Nurturing Cities*. While the first book established the intellectual foundation, i.e., arguing that nurturing cities requires embedding nature at the heart of their identity, this volume advances the conversation by focusing on practice. *Nature-based Urbanism* seeks to demonstrate how ecological integration can be designed, implemented, and scaled through real-world strategies and case studies, making the idea of nurturing cities tangible across diverse urban contexts.

The intellectual aim of this book is to clarify nature-based urbanism as both a paradigm and a practical pathway. Urbanism, in this framing, is not just about the spatial arrangement of buildings and infrastructure but about the relationships between ecological processes, human communities, and cultural identities. By treating nature as a co-creator of urban life, the book challenges the long-standing dichotomy between human settlement and natural systems. It aims to establish nature-based urbanism as a coherent field of scholarship and practice that unites ecological restoration, biophilic design, and nature-based solutions under a single framework of regenerative city-making.

Practically speaking, this volume aims to give professionals, decision-makers, and communities models that may be modified and used. The combined challenges of social inequality, biodiversity loss, and climate change are plaguing cities all over the world. Conventional methods, which are based on compartmentalised governance and grey infrastructure, are failing. Urban trees that reduce temperatures, wetlands that lessen floods, mangroves that safeguard coastlines, and community gardens that provide food and empower marginalised communities are just a few of the scalable techniques highlighted in this book. The book shows that nature-based urbanism is a workable framework for mainstream urban development rather than a specialised experiment by tying these tactics to larger systems of planning, design, and government.

Another core ambition is to bridge global diversity. High-density Asian megacities, European towns rewilding industrial rivers, Latin American neighbourhoods developing urban farms, and coastal settlements enhancing resilience through ecosystem restoration are just a few of the diverse regions and cultures represented in the volume's case studies. The book emphasises that nature-based urbanism is a flexible, adaptable strategy rather than a single model by combining these examples. The common understanding that cities flourish when they operate as components of wider ecological systems is what binds these contexts together rather than uniformity.

This book offers an important viewpoint in the broader international discussion on "urban sustainability", the central focus of this book series, too. Economic frameworks or technological solutions have frequently dominated discussions on sustainability. These are significant, but they run the risk of ignoring the ecological underpinnings of urban life. According to nature-based urbanism, it is impossible to envision cities in the future without nature at its core; hence, nature is understood as resilience, infrastructure, culture, and identity—all at the same time.

By clarifying both the intellectual vision and the practical goals, this volume seeks to advance a global conversation: building cities with nature is no longer optional; it is the defining challenge of our time and the next generations. As the second volume in a trilogy, it provides the practical grounding for a journey that will continue in the third book, which will explore speculative futures and emergent possibilities for cities fully conceived as living ecologies. It then paves the way for the final volume, which focuses on *designing nature into cities.*

3.1 Structure of the Book: From Concepts to Global Case Studies

In order to help readers comprehend nature-based urbanism as a growing worldview rather than a singular approach, the book is structured as a purposeful journey from fundamental concepts to grounded practice and ultimately to lived experiences. The chapters develop gradually, allowing theory to be shaped by experience and practice to be enhanced by examples from many ecological situations, cultures, and geographical areas. Although each component presents a unique perspective, the three work together to demonstrate how cities rethink, regulate, and eventually coexist with nature in ways that build more resilient and significant urban futures.

The conceptual framework is established in ***Part I, Rethinking Urban Theory Through Nature***. It begins by re-examining well-known urban theories from a nature-centred perspective, placing traditional concepts—like Lynch's imageability—within frameworks with an ecological foundation. The chapters show how nature-driven placemaking transforms both spatial experience and urban identity by moving from conceptual articulation to situated examples. Southeast Asian viewpoints add complexity to these concepts by illuminating how densely populated, quickly urbanising environments manage nature as a developmental obstacle as well as a cultural anchor. An extensive examination of Ho Chi Minh City's green space transformation, which demonstrates how nature-based thinking becomes crucial in cities under extreme growth pressure, concludes this section.

In ***Part II, Governance, Participation, and Multispecies Planning***, the emphasis is shifted from theory to the institutional and social systems that facilitate the adoption of nature-based solutions. Governance is seen as an active area where priorities are debated, contested, and altered rather than as a bureaucratic formality. The chapters look at multispecies planning strategies that defy anthropocentric conventions, participatory design procedures that uncover unanticipated possibilities, and urban environmental governance in India. The geographical perspective is expanded with a comparative chapter on regenerative urbanism in Europe, which highlights planning ideas and policy frameworks that promote long-term ecological restoration. This section reaffirms that governing cultures that emphasise community agency and acknowledge complexity are essential to nature-based urbanism.

In ***Part III, Nature as Everyday Practice and Place-Making***, the topic is expanded to include everyday life's textures. Here, nature is an everyday companion rather than a grand strategy, interwoven with neighbourhood design, childhood mobility, and traditional methods of living with landscapes. The chapters cover everything from the socio-ecological rhythms of riverine communities in the Philippines to the construction of ecocentric precincts that foster human-nature connections. Nature-based urbanism is most effective when it feels lived rather than forced, as this last theme segment demonstrates.

The book concludes with Chapter 13, which ties the threads together by reflecting on how vision translates into practice. It provides a forward-looking explanation of how nature-based urbanism may lead cities toward more inclusive, adaptable, and ecologically conscious futures by synthesising lessons from all the chapters.

4 Toward Regenerative Urban Futures: Concluding Reflections

Future cities involve more than just skylines, infrastructure, and economic engines. It is to imagine living landscapes that are resilient, adaptive, and intricately linked to the ecological processes that enable life. According to nature-based urbanism, regenerative activities that mend past harm, revitalise ecological processes, and support communities are more important for the future of urban life than token acts of greening.

This book shows that nature-based urbanism is not an abstract ideal but a pathway already unfolding across the globe. The case studies included here represent indicators of change, ranging from community farms to river rewilding, from vertical forests to coastal mangrove restoration. They show that cities can be planned as ecologies, where biodiversity and constructed form coexist peacefully, and resilience is ingrained in day-to-day existence. These narratives serve as a reminder that the Anthropocene is both a time of crisis and opportunity, with the lines separating natural and human systems being redefined.

Regenerative urban futures are synonymous with reciprocity (Camrass, 2020; Warden, 2021). Cities cannot prosper by continuously taking resources from rivers, soils, and atmospheres. Instead, they must give back by rehabilitating wetlands, boosting biodiversity, cooling microclimates, and establishing areas of justice and inclusivity. Beyond sustainability, regeneration aims to heal and rejuvenate rather than just preserve. In this way, nature-based urbanism necessitates humility, patience, and creativity and is as much a philosophy of caring as a design method.

The volume is also placed within the larger arc of the Nature and Cities trilogy in its final reflection. The first book, *Nature in Cities, Nurturing Cities* (Cheshmehzangi et al., 2026), laid the groundwork for the idea that ecological care is necessary to nurture cities. This second volume has followed the implementation of that idea, demonstrating how "nature-based urbanism" can transform planning, design, and policy in a variety of cultural and geographical contexts. In the third book, cities will be fully envisioned as living ecologies, i.e., adaptive, creative, and interdependent, in speculative and emergent futures. The trilogy, as a whole, presents a manifesto for change in addition to a critique of contemporary urbanism.

Although urgent, the work at hand is not overwhelming. Building regenerative urban futures means seeing cities as gardens of possibilities where nature is fundamental rather than peripheral, formative rather than brittle. It is to envision urbanism as cooperation rather than dominance. This reorientation is essential to the future of cities and the planet. To hope is to act, and to regenerate is to hope.

References

Al-Kodmany, K. (2024). Promoting health in dense cities through vertical greeneries: The case of plant-and tree-covered tall buildings. *The Routledge Handbook on Greening High-Density Cities* (pp. 509–529). Routledge.

Bloomberg, M. R., & Strickland, C. H. (2012). NYC green infrastructure. *2012 Annual report.* https://www.uni-groupusa.org/PDF/NYCGreenannual_report_2013.pdf

Brantz, D., & Dümpelmann, S. (Eds.). (2011). *Greening the city: Urban landscapes in the twentieth century.* University of Virginia Press.

Camrass, K. (2020). Regenerative futures. *Foresight, 22* (4), 401–415.

Chen, S. (2023). *Research on the design of Guangzhou Tianhe Wetland Park based on the reflection of Sponge city construction* (Doctoral dissertation). Politecnico di Torino.

Cheshmehzangi, A. (2025). *Resourceful urbanism: Designing regenerative cities through adaptive reuse and circular innovation.* Springer Nature.

Cheshmehzangi, A., Alidoust, S., Chen, W. Y., & Fuller, R. (eds.). (2026). *Nature in cities, nurturing cities.* Springer Nature.

Dushkova, D., & Haase, D. (2020). Not simply green: Nature-based solutions as a concept and practical approach for sustainability studies and planning agendas in cities. *Land, 9*(1), 19.

Li, L., Cheshmehzangi, A., Chan, F. K. S., & Ives, C. D. (2021). Mapping the research landscape of nature-based solutions in urbanism. *Sustainability, 13*(7), 3876.

Li, P., Zhuang, L., Lin, K., She, D., Chen, Q., Wang, Q., & Xia, J. (2025). New perspectives on urban stormwater management in China, with a focus on extreme rainfall events. *Natural Hazards, 121*(4), 3745–3774.

McNeill, D. (2022). Botanic urbanism: The technopolitics of controlled environments in Singapore's gardens by the bay. *International Journal of Urban and Regional Research, 46*(2), 220–234.

Scott, M., Lennon, M., Haase, D., Kazmierczak, A., Clabby, G., & Beatley, T. (2016). Nature-based solutions for the contemporary city/Re-naturing the city/Reflections on urban landscapes, ecosystems services and nature-based solutions in cities/Multifunctional green infrastructure and climate change adaptation: Brownfield greening as an adaptation strategy for vulnerable communities?/Delivering green infrastructure through planning: Insights from practice in Fingal, Ireland/Planning for biophilic cities: From theory to practice. *Planning Theory & Practice, 17*(2), 267–300.

Stuiver, M. (Ed.). (2023). *The symbiotic city: Voices of nature in urban transformations.* Brill Publishing.

Tousi, E., Mela, A., & Tseliou, A. (2025). Nature-based urbanism for enhancing senior citizens' outdoor thermal comfort in high-density Mediterranean cities: ENVI-met findings. *Urban Science, 9*(5), 152.

van der Berg, A. (2023). Climate adaptation planning for resilient and sustainable cities: Perspectives from the city of Rotterdam (Netherlands) and the city of Antwerp (Belgium). *European Journal of Risk Regulation, 14*(3), 564–582.

Warden, J. (2021). Regenerative Futures. *From sustaining to thriving together. London, published by royal society for arts, manufacturing and commerce.* https://www.thersa.org/wp-content/uploads/2021/10/from-sustaining-to-thriving-together-final.pdf

Rethinking Urban Theory Through Nature

The Image of the Nature-Based City: Rethinking Lynch's Theory Via Case Studies

Azadeh Khaki Ghasr

Abstract Kevin A. Lynch's significant urban design theory, outlined in The Image of the City, focuses on how people perceive, navigate, and make sense of urban surroundings. The framework consists of five essential elements—paths, edges, districts, nodes, and landmarks—that together contribute to the imaginability of cities. According to the etymology of "Image," a couple of concepts arise to rethink what the means of city image and if the Lynch theory covers other kinds of cities, especially nature-based cities? This chapter applies Lynch's theoretical model to nature-based cities, in which ecological processes are fundamental parts of the urban fabric, actively impacting spatial practices, perceptual experiences, and socio-environmental interactions. By studying some case studies in a global scope, the chapter critically rethinks Lynch's five elements through the lenses of natural sensitivities, such as water, sunlight, heat, wind, moisture, soil, vegetation, and safety. According to the study, nature-based infrastructures frequently shape the birth and development of some of Lynchian features in addition to supporting them. Most often, these infrastructures manifest themselves within the urban environment; however, sometimes they are tangible, but not necessarily visible. Among the most salient examples are the qanat systems in the desert cities of Iran, whose hidden hydrological networks have historically shaped patterns of settlement and spatial organization, and some Lynchian elements are a by product of them. To recognize both man-made and nature-made infrastructures as support of the image of the nature-based city, this chapter defines a conceptual enhancement of Lynch's original theory through an evolved version of it. This deepening highlights the need to progress beyond a merely visual interpretation of urban form toward a more systemic and experiential understanding of the city as a socio-natural construct. What is understood, felt, and experienced via ever-changing interactions with natural processes also contributes to shaping the urban image, in addition to what is immediately apparent.

Keywords Lynch's theory · Nature-based city · Urban design · Urban perception · Nature-based infrastructures

---

A. Khaki Ghasr (✉)
Department of Architecture, Faculty of Architecture and Urban Planning, Shahid Beheshti University, Tehran, Iran
e-mail: a_khakighasr@sbu.ac.ir

A. Cheshmehzangi et al. (eds.), *Nature-based Urbanism*, Urban Sustainability,
https://doi.org/10.1007/978-981-95-6864-2_2

1 Introduction

Kevin A. Lynch, an influential American urban planner and author, significantly contributed to urban planning and environmental psychology. Lynch (1960), analyzed how observers perceive cities using Boston, Jersey City, and Los Angeles as case studies. Lynch's research marked a shift in urban planning, moving away from a top-down approach to focusing on the "image in the city dweller's mind."

Lynch's theory emphasizes the importance of imageability and legibility in a city's mental image. It suggests that a city with easily identifiable districts, landmarks, and pathways can be easily organized into a coherent mental map. This mental image comprises three components: identity, structure, and meaning. Imageability is essential for orientation, wayfinding, emotional security, and a sense of belonging. It allows people to move through the city efficiently and confidently, fostering a positive and engaging experience. Lynch argued that enhancing city imageability could lead to more psychologically satisfying and vibrant urban spaces. This mental image is formed by the uniqueness, memorability, spatial relationships, and emotional associations evoked by objects.

To understand how people form these mental images, Lynch identified five basic elements that consistently appear in individuals' cognitive maps of the city:

Paths, including streets, sidewalks, transit lines, canals, and railroads, are crucial in shaping people's perception of cities. Their continuity, spatial qualities, facades, and proximity to special features significantly influence their imageability. Paths with clear origins and destinations have stronger identities, tying the city together in the observer's mind.

Edges are linear elements that represent boundaries between phases or breaks in continuity, such as shores, railroad cuts, development edges, walls, or riversides. They can function as barriers, separating regions, or as seams, connecting regions. Strong edges contribute to urban space structuring and enhance the identity of adjacent areas, contributing to the overall structure of urban spaces.

Districts are two-dimensional sections of a city with common characteristics, recognizable from both inside and outside. They are determined by physical features such as texture, space, form, detail, symbol, building type, use, activity, inhabitants, maintenance degree, and topography. Districts provide a sense of place and contribute to the overall organization of the city in an individual's mental map.

Nodes are strategic points in a city where an observer can enter and travel from. They can be junctions, transportation breaks, or concentrations of characteristic or activity. Successful nodes often have a unique character and intensify surrounding city characteristics, such as a square, park, or transportation hub. They can also serve as focal points for movement and movement.

Landmarks are localized physical objects that serve as a point of reference for observers, ranging from small signs to large buildings, towers, mountains, or mobile points like the sun. Their singularity is a key characteristic, helping people navigate the city and providing external cues for wayfinding.

Employing the term Image in Lynch's seminal work is the point of departure in this rethinking. Refer to the etymology of Image; some questions arise in the generalization of Lynch's theory. In the entry of Image as noun, there are some definitions: 1) a visual representation of something, 2) a mental picture or impression of something, 3) a popular conception, 4) exact likeness, 5) a tangible or visible representation, 6) a vivid or graphic representation or description, 7) a reproduction or imitation of the form of a person or thing, 8) figure of speech, 9) a set of values given by a mathematical function (Merriam-Webster, 2025). American heritage dictionary (2025) entries some other definitions: 1) A representation of the form of a person or object, 2) a—The opinion or concept of something that is held by the public, b—The concept or character projected to the public, 3) A vivid description or representation in words, ... 4) A mental picture of something not real or present, 5) Obsolete An apparition. In addition, the etymology of the word "image" has been said to have the same base as imitārī, "to follow as a pattern" (Merriam-Webster, 2025).

Based on the properties and dimensions defined for the term "Image" parallel to Lynchian five key elements, some couple concepts drive the rethinking the image of the city theory (Table 2.1).

Revisiting Lynch's theory, notice the Image Couple concepts, five key elements are interpreted through the lens of visual cognition and experiential clarity that leans toward category 2 as the aim and scope of the theory. This chapter proposes to examine the theory in another kind of cities that meet category 1 as the aim scope of the image concepts. The question that drives this examination is: if category 1 represents the image concepts, do the five elements cover all the images of the city? The case studies are nature-based cities, wherein nature is not an incidental element but an essential and formative component of urban experience, shaping patterns of movement, social interaction, and spatial perception.

Table 2.1 Some outlines of Image Couple concepts extracted from etymologies

	Image Couple concepts	
	Category 1	Category 2
1	Tangible	Visible
2	Vivid	Real or not real
3	Form the people mentality	Formed by people
4	To be followed as a pattern	Just a sample
5	Need background knowledge to remember	Easily recalling

2 Defining the Nature-Based City: Principles and Integration

Before the specialized combination of "Nature-based cities" was coined, it had ancient roots. Throughout history, architects and city planners have worked on this basis all over the world. For instance, Nova (2006) examines how classical architectural theorists—Vitruvius, Alberti, and Scamozzi—considered the role of wind in urban planning, thereby highlighting the historical continuity of environmental awareness in city formation (Table 2.2).

The concept of nature-based cities as a scientific term evolved, with early contributions from various thinkers. While there isn't one single "first theorist," Ebenezer (1965) is widely recognized for his influential "Garden City" concept (1898–1902), which laid the groundwork for modern ecological city theory. His publication "Garden Cities of To-morrow" was the description of a utopian city in which people live harmoniously together with nature. Patrick Geddes (1915) applied ecological principles to urban planning, developing a more systematic approach to regional planning based on ecological considerations. Lewis Mumford (1961) developed the idea of urban cycles, acknowledging that cities grow, develop, and eventually decline. Ian McHarg (1969) wrote a text in the field of ecological planning and design, emphasizing the importance of understanding natural systems when designing urban areas. More recently, the concept of NBS (nature-based solutions) has gained prominence, focusing on using natural systems to address societal challenges like climate change, water management, and biodiversity loss.

The nature-based city model is a significant shift in urban development, transforming the conventional approach of placing nature as a decorative or marginal element. The emphasis is not merely on increasing the quantity of green spaces, but on rethinking their role as critical infrastructure for achieving sustainable urban futures (Frantzeskaki et al., 2018). This integrated model supports the reconceptualization of cities as social-ecological systems, where natural processes are embedded within urban planning, governance, and infrastructure design (Nesshöver et al., 2017). The

Table 2.2 Some outlines of the effect of winds in the urban design approach (Nova, 2006)

The scholars		
Vitruvius	Alberti	Scamozzi
- A set of urban planning ideas considering the eight main winds - Emphasizing the structural influence of the wind: The urban design strategy considers the avoidance of direct wind impingement on buildings	- The significance of orientation: Consideration of the geographical location of the city and wind flow patterns - In addition to the wind's influence on the broader urban environment, it highlights the specifics of fenestration in individual structures within the city	- The directional layout of urban thoroughfares is contingent upon the vectors of salubrious and adverse wind patterns - A standardized global paradigm for urban wind regimes is not a universally valid concept

evolution of this framework is influenced by a range of disciplines, including urban planning, landscape ecology, and environmental psychology. The primary goals of developing nature-based cities are to foster environmental sustainability, enhance resilience against climate-related risks, and improve the livability and psychological well-being of urban dwellers (Kabisch et al., 2016). Nature-based urbanism promotes a systems-thinking approach where urban infrastructure is planned to synergize with natural ecosystems, thereby delivering co-benefits that are social, ecological, and economic in nature.

Several fundamental ideas guide nature-based cities. Multifunctionality, in which green infrastructure fulfills many functions at the same time, is an essential component. For example, urban wetlands or green roofs may simultaneously operate in stormwater management, biodiversity support, microclimate adjustment, and recreational provisioning (Raymond et al., 2017). Multifunctionality necessitates that planners adopt a comprehensive approach, balancing natural processes with human-centered requirements. A second premise is systemic integration, which requires understanding and planning for the interdependence of urban systems and natural networks (Eggermont et al., 2015). Rather than seeing green spaces as separate parts, the city is understood as a continuous socio-ecological environment. Integrative planning requires a knowledge base that includes ecology, hydrology, urban sociology, and environmental psychology in order to successfully manage interactions between the built and natural environments.

Equity and inclusivity represent another pivotal dimension. NbS must be implemented in ways that ensure equal access to ecological benefits across diverse socioeconomic groups. Additionally, context specificity is vital. One-size-fits-all solutions may be ineffective or even counterproductive if they fail to account for local constraints or capacities. Adaptive planning frameworks that evolve through iterative engagement and learning processes are increasingly advocated (Davies & Lafortezza, 2019). Finally, cross-sectoral collaboration is indispensable for integrating natural systems into urban development. Nature-based urbanism necessitates partnerships between architects, planners, ecologists, public health officials, and citizens. The integration of NbS into urban planning yields a wide array of benefits. Foremost among these is climate resilience. NbS can mitigate and adapt to the impacts of climate change by regulating temperature, absorbing stormwater, and enhancing urban flood resilience (Depietri & McPhearson, 2017). Table 2.3 summarizes specific components of nature-based cities.

3 Methodology

This section explores a range of traditional cities whose spatial organizations and identities have been historically shaped by their natural environments. By conducting comprehensive field studies that include observations, map-based investigations, archival analysis, and photographic documentation, the study can effectively categorize cities based on the predominant natural elements that have influenced their

Table 2.3 Aspects of nature-based cities

Aspect	Description
Core Concept	The nature-based city redefines urban development by systematically integrating natural elements into the urban fabric as essential infrastructure
Paradigm Shift	Moves away from decorative use of nature to viewing it as central to sustainable urban systems
Theoretical Foundations	Rooted in urban planning, landscape ecology, and environmental psychology; cities are viewed as socio-ecological systems
Primary Goals	Environmental sustainability, climate resilience, enhanced livability, and psychological well-being
Planning Approach	Systems thinking that synergizes urban infrastructure with natural ecosystems to produce co-benefits
Key Principles	Multifunctionality, systemic integration, equity and inclusivity, context specificity, cross-sectoral collaboration

development. This approach not only deepens our understanding of urban forms but also highlights the vital relationship between nature and city planning. These typologies reflect integration and responses oriented to the natural context, and will be reexamined through Lynch's five elements. Furthermore, at the end of this section, in each case, the image of the city is assessed as if it could be covered by Lynch's theoretical framework.

3.1 Case Study

(a) **River-Based Cities**

Isfahan, Iran, with its Zayandeh-Rud River, historically sustained agriculture, supplied water to gardens and palaces, and structured public life along its banks. Bridges serve as both nodes and landmarks, integrating movement and rest, function and form. In addition to the main course of the river, there exists a network of "Madi." "Madi" network constitutes a system of rivulets and water channels within Isfahan, branching off from the river and playing a vital role in the development and prosperity of the city. Similarly, Florence, Italy, along the Arno River, uses the waterway as a dominant axis. The Ponte Vecchio functions as both a bridge (path) and a vibrant commercial node. The river not only facilitates orientation but also frames the aesthetic identity of the Renaissance city, which thrives on visibility, reflection, and visual depth (Rossi, 1984). Cologne (Köln), Germany, presents another compelling example of a riverine city where the Rhine plays a formative role in urban experience. The Rhine is not only a major geographical feature but also a defining element of the city's spatial and cultural identity.

(b) **Sea-Based Cities**

In Bushehr, Iran, street layouts perpendicular to the sea serve not only navigational clarity but also provide natural ventilation, reduce humidity, and facilitate passive cooling. This orientation toward the sea turns wind into an urban design parameter, creating path systems aligned with climatic flows. In Venice, Italy, the lagoon is the primary infrastructure. Waterways act as paths, while thresholds between solid and liquid terrain define edges. Districts correspond with historical functions—mercantile, religious, or residential—while nodes concentrate around plazas and bridges. Here, imageability is not only formed by architecture but by the rhythm and flow of water and boats, making fluid motion a part of cognitive mapping. The other case is Tunis, Tunisia, known for its waterfront setting, which has been used for strategic and morphological purposes. Tunis has a linear pattern, with major arteries serving as navigational spines linking the maritime edge with civic and religious landmarks. The traditional urban core is oriented toward prevailing coastal winds, influencing street direction, courtyard placement, and building massing. The city's narrow alleyways and dense built forms create shaded microclimates, while openings capture cooling winds, reflecting its vernacular adaptation to maritime humidity and heat. The climatic wind flow shapes spatial perception and environmental performance. In addition, the shadow layers that cover the paths are not counted in Lynchian elements. It is such a ceiling for different parts of the sub-city. Tunis' lagoonal geography, including canals and marshlands, significantly influences the urban imaginary.

(c) **Terrain-Based Cities**

Terrain-based cities refer to urban environments that are significantly shaped by the underlying topography. These cities often exhibit unique characteristics due to their adaptation to the natural landscape, including variations in building layouts, street patterns, and even infrastructure placement. Masuleh, Iran, is a prime example where buildings are constructed into a steep mountain slope such that the roof of one structure serves as the courtyard of another. Another ancient case is Kyoto, Japan, which is nestled between mountains and features a unique street pattern adapted to the natural contours. A similar case can be observed in Chefchaouen, Morocco, which exemplifies how steep terrain can shape a distinctive urban fabric that enhances both spatial legibility and cultural identity. Chefchaouen's morphology is characterized by terraced housing and cascading urban layers, where structures appear to emerge from the rock face itself. Much like in similar cases, many rooftops double as public or semi-private terraces, contributing to a vertically integrated system of circulation and gathering spaces.

(d) **Subterranean-Based Cities**

Iran's Nushabad serves as an underground city, which was first constructed to stave off invaders and harsh weather, and has an intricate network of tunnels, water storage, and ventilation control. Although they are concealed from a casual observer, these systems have an impact on the direction and form of settlements. Another case is Coober Pedy, which lies in South Australia's dry outback, and is another example.

Coober Pedy is well-known for its dugouts, which are underground homes dug out to avoid the intense heat of the day. The underground plane lacks or has dispersed landmarks, districts, and even nodes. However, non-visual signals such as temperature changes, air pressure variations, auditory resonance, and tactile input from surfaces are still used for spatial navigation. Orientation based on ocular perception is replaced with multisensory perception, necessitating an extension of Lynch's theory to account for experience components that are frequently ignored in urban studies. Likewise, Derinkuyu, located in Turkey, offers an ancient multi-level underground city. The vast underground network was constructed both as a defensive and adaptable urban structure to deal with both natural factors and frequent external threats. The spatial logic of Derinkuyu is what sets it apart.

(e) **Natural Kinetics-Based Cities**

Kinderdijk windmills and supporting canals in the Netherlands are strategically positioned for wind capture and are embedded in a system of dikes and canals (Fig. 2.1).

The Shushtar Historical Hydraulic System is an ancient complex of interconnected water management structures in Shushtar, Iran. It includes dams, bridges, canals, and watermills, all designed to control the Karun River's flow for irrigation and urban water supply (Fig. 2.2). Water use based on climatic conditions by creating an underground network beneath the old city is considered one of the most intelligent forms of water urban architecture (Ayatollahi et al., 2016).

The Asbads of Nashtifan in Iran exemplify another historical instance of how wind energy was harnessed for sustainable and identified settlement (Fig. 2.3). Nashtifan is subjected to strong seasonal winds known as the "120-day wind," which shaped the

Fig. 2.1 Kinderdijk windmills and supporting canals (*Source* https://www.jordicarrio.com/lang/en/blog/a-tour-of-the-kinderdijk-windmills)

Fig. 2.2 Shushtar Hydraulic System (*Source* https://en.wikipedia.org/wiki/Shushtar_Historical_Hydraulic_System)

form and orientation of its urban structures and infrastructures. These windmills are integrated within the edge of the town, forming part of its ecological infrastructure.

One notable example of qanat-integrated urban design can be observed in Yazd, Iran (see Fig. 2.4). These subterranean infrastructures play a crucial role in defining land values, neighborhood identities, and street alignments. Although the qanats themselves are hidden from view, their presence is marked by visible elements at nodal points (refer to Fig. 2.5). While the water system remains largely concealed, it emerges at the edges of the city, where irrigation becomes apparent. This creates a network of water connections by linking public and private nodes for water usage, which can be seen in various forms such as Payab (Fig. 2.6), water reservoirs (Fig. 2.7), and other manifestations of qanat within the city (Fig. 2.8), which serve as landmarks.

3.2 Rethinking Lynchian Elements in the Cases

The case studies suggest that in nature-based cities, five elements often emerge from the interplay between human needs and environmental conditions. Results of the case study summary are in Table 2.4.

Fig. 2.3 Asbads (*Source* photos by author)

Fig. 2.4 Two qanat strings near Yazd, Iran (*Source* Yazdi & Khaneiki, 2017, p V)

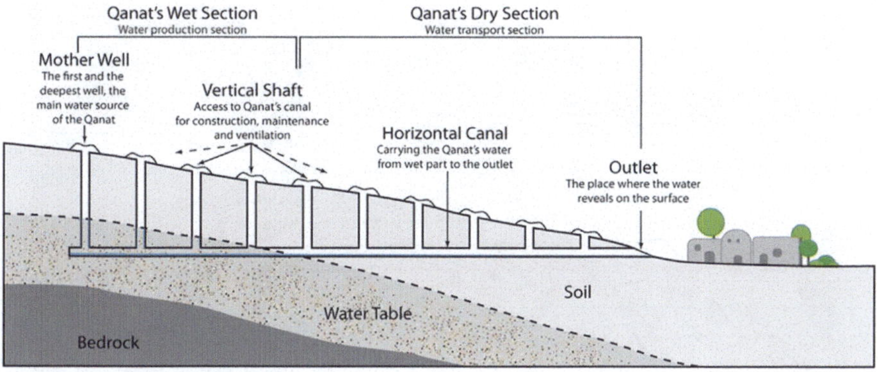

Fig. 2.5 Profile of a typical qanat (*Source*: Omidvari & Golzari, 2016)

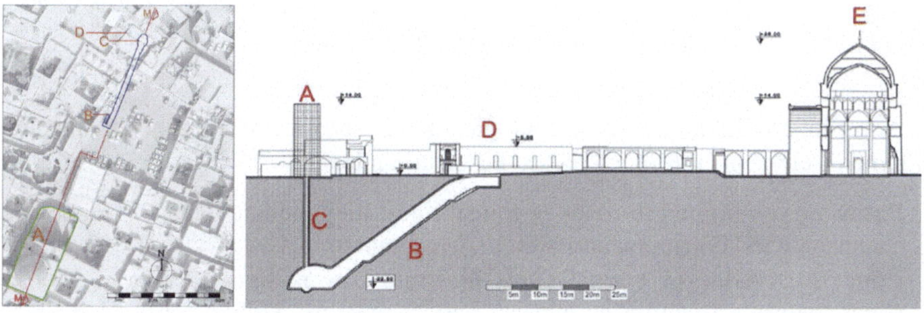

Fig. 2.6 A sample of Payab: The extent and boundaries of the Payab of Qanat-e Zarch in the "Vaght o Sa'at square in Yazd; A: Shrine of Seyed Rokn al-Din, B: Entry point to the Payab after the reorganization of the entrance in 2023, C: Well shaft at the end of the Payab, D: Path of water flow in the gallery of the Shirin branch of the qanat in the past, M-M: Section line. (*Source*: Shahabinejad & Lookzadeh, 2024)

Fig. 2.7 Water reservoir (*Source*: Yazdi & Khaneiki, 2017, p 149)

Fig. 2.8 Irrigation pool (*Source* Yazdi & Khaneiki, 2017, p 150)

According to Table 2.4, some ecological bases necessitate a reconceptualization of Lynch's elements through a nature-based urban lens:

- **Paths** are not limited to roads or transit lines; they include qanats, rivers, water channels, wind corridors, and even underground tunnels.
- **Edges** in riverine or coastal cities illustrate how natural boundaries such as water margin's function as barriers and connectors. Unlike artificial walls or infrastructure, natural edges fluctuate and interact dynamically with the urban form.
- **Districts** in cities arise from environmental differentiation—variations in wind orientation, topographic level, or proximity to qanats—rather than mere land use zoning. Thus, districts can be understood as ecological zones as much as cultural or economic entities.
- **Nodes** such as plazas near rivers, or the appearance places of qanats represent not only spatial convergence but also resource convergence, where environmental infrastructure becomes a focal point of social life.
- **Landmarks**, while traditionally architectural, in nature-based cities are often eco-technical artifacts—wind catchers, water towers, bridges, and mills are imbued with both functional necessity and symbolic significance.

4 Discussion and Implications

The case study demonstrates that when examining Lynch's theory in terms of environmental responsiveness, the concept of image extends beyond category 2, which pertains to the origins of images. For instance, subterranean cities present a significant challenge to Lynch's framework: how can we account for systems that are

Table 2.4 Results of the case study

	Categories	Cases, countries	The results
a	River-based cities	Isfahan, Iran	The cases exemplify how natural elements, when intertwined with historical layers and urban design, can structure perception, orientation, and meaning in the city. The rivers and their channels are more than a path or edge. They organize the spatial imagination of the entire city
		Florence, Italy	
		Cologne, Germany	
b	Sea-based cities	Bushehr, Iran	The hydrological features define the urban edge and have historically influenced district differentiation. The visual connection to the sea/lake provides orientation anchors, helping residents and visitors construct a mental map based on natural and built landmarks
		Venice, Italy	
		Tunis, Tunisia	
c	Terrain-based cities	Kyoto, Japan	The cases demonstrate how topography becomes both constraint and compositional opportunity, influencing the arrangement of paths, nodes, and districts. They underscore the power of terrain not merely as a physical condition, but as a formative agent in the creation of legible, resilient, and character-rich urban environments
		Masuleh, Iran	
		Chefchaouen, Morocco	
d	Subterranean-based cities	Nushabad, Iran	In contrast to aboveground cities, vertical shafts serve as ventilation and escape routes, access points are concealed, and corridors are small and simple to shut. The urban experience is more about embodied memory, stored social information passed down through the generations, and sensory navigation than visual legibility
		Coober Pedy, South Australia	

(continued)

Table 2.4 (continued)

	Categories	Cases, countries	The results
		Derinkuyu, Turkey	
e	Natural kinetics-based cities	Kinderdijk, Netherlands	The cases serve as archetypes of nature-based urban infrastructure: rooted in natural kinetics, embedded in local ecologies, and expressive of deep cultural identities. They reinforce the place-based nature-based city is not only a novel invention but a rediscovery and reconnection with long-standing traditions of living in harmony with nature. In such cities, the seamless integration of the elements forms a sophisticated spatial network where the natural environment is as vital and impactful as the built surroundings. This design demonstrates how deeply water/wind can shape urban living
		Shushtar, Iran	
		Nashtifan, Iran	
		Yazd, Iran	

hidden, non-visual, or understood solely through experience or expert knowledge? The images associated with these systems are tangible and belong to category 1. This situation necessitates the conceptualization of a foundational idea for city imagery: infrastructural substrata. According to the case study, infrastructural substrata are formed through the interactions between humans and their environment. Therefore, these elements are a combination of both man-made and natural environments. We can term this as a "man-nature"-made environment, which can be abbreviated to "manature-made environment." This characteristic is particularly relevant in category E, where nature has been harnessed by humans and its energy has been utilized as part of urban infrastructure. In summary, there are conceptual models available to define and compare conditions in relation to Lynch's theory (see Diagram 2.1, 2.2, 2.3).

Such infrastructures, for example, Qanat, Wind corridors/catchers/exhausts, subterranean tunnels/ passages, may not necessarily seem as overt spatial components, yet they impact shape, function, defense, comfort, and experience. Significantly, they frequently surpass mere visibility, necessitating cognitive and cultural awareness for their identification—thereby broadening Lynch's initial concept of imageability from a purely visual framework to a multi-sensory and knowledge-oriented one. This root advocates for an epistemic broadening of urban legibility that incorporates ecological literacy and people's mentality.

Diagram 2.1 Original Lynch theory

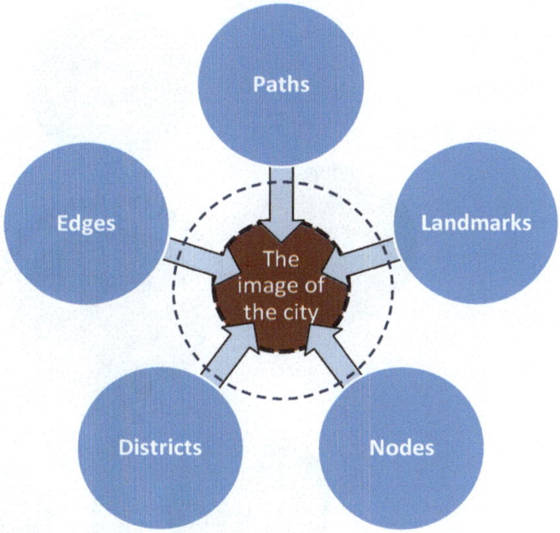

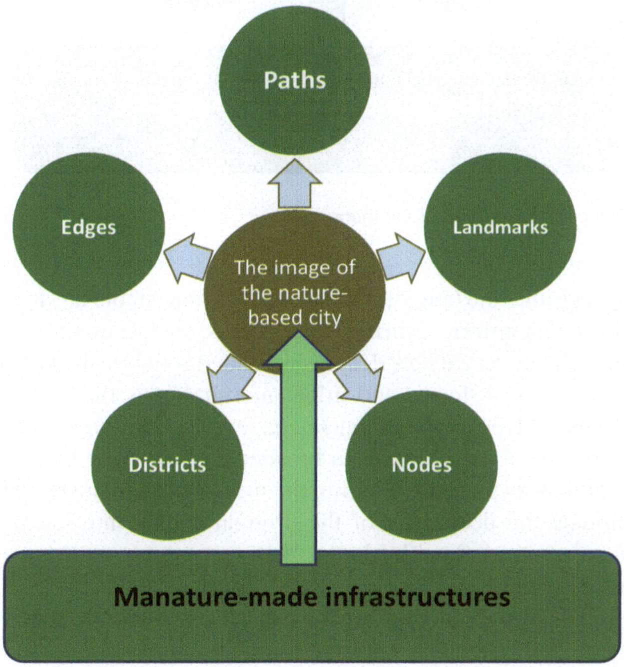

Diagram 2.2 A conceptual model for the image of the nature-based city

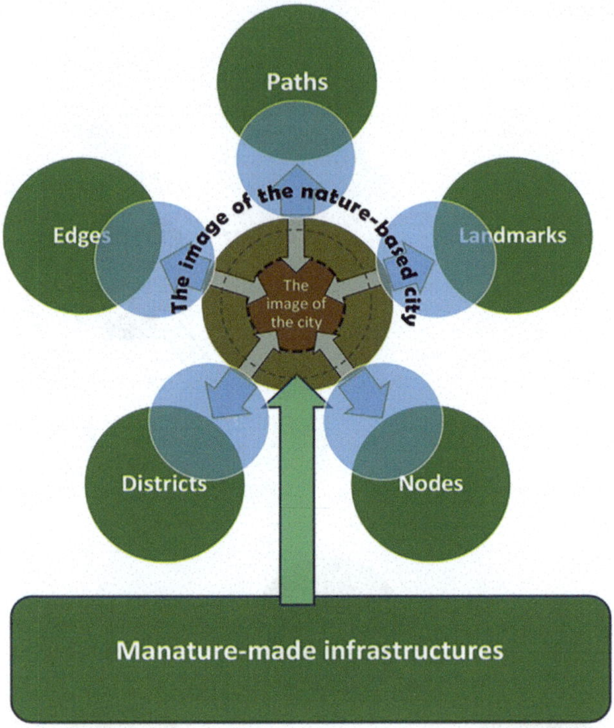

Diagram 2.3 The evolved version of the theory

The cases foreground the tangible infrastructures that enable urban life in extreme conditions—thermal comfort, security, water access, and air quality—without overt physical presence. They reveal how deeply embedded ecological and climatic factors can generate architectural forms and urban morphologies that defy conventional visual categorization. This tactile urbanism, experienced through bodily immersion and environmental feedback, suggests a broader epistemology of city-making that is non-iconic but no less imageable within the mental map of the residents. In doing so, we can validate the deepening of the original theory into the background of infrastructural substrata defined by logics that support life in the manature-made environment as a nature-based city.

According to Diagram 3, in adapting the two conceptual models derived from the original theory and adding depth to it, the following points are made: The original theory and evolved version share a core similarity: both utilize the same five categories of city image elements. However, despite their shared names, these categories have both commonalities and distinctions. The overlap between the two versions, for example, in the "nodes" category, indicates that some nodes are identified in both city types (nature-based and non-nature-based) by both the original and evolved versions. In contrast, the differences suggest that certain nodes are only recognizable and extractable through either the original theory or the evolved version. Consequently, the specific examples within the five categories are not necessarily identical, even if they share some commonalities. For instance, a node might be identified by the original theory but not by the evolved version in nature-based cities, as it may lack the necessary characteristics. Therefore, when analyzing nature-based cities, it's essential to use the evolved theory version for identifying the city image, with the original theory serving a supplementary role to reinforce and clarify the commonalities. A fundamental difference between them is that the evolved version is supported by foundational elements deeply rooted in human-nature interaction. The city image concepts from the original theory are subsumed under the broader scope of the evolved version. In other words, the nature-based city image covers a wider range of concepts within the "image" couple concepts (categories 1 and 2 in Table 2.1), and as a result, the original theory is considered a subset of it. In Diagram 2.3, there are two-way arrows between the formation of the image of the city and the elements. It provides a synergistic path between what the infrastructure creates in the minds of people and what is built by the minds of people, and this evolutionary role in the formation of the mental map based on nature.

5 Conclusion

The concept of the nature-based city emphasizes the deliberate integration of ecological systems into the urban fabric as a means of addressing contemporary societal challenges, enhancing the quality of urban life. In this paradigm, natural elements are no longer regarded merely as decorative or supplementary features, but rather as fundamental strategies for improving human well-being, supporting environmental health, and counteracting the negative impacts of urbanization. The traditional nature-based cities studied offer valuable insights for contemporary urban planning, especially in the context of nature-based urban design. Several lessons emerge: contextual adaptation, infrastructure as identity celebrated, and integrated into the urban image, ecological cognition by environmental awareness, and multi-sensory urbanism in creating legible and livable cities.

This chapter has examined the intersection of Lynch's theory and the environmental logic of traditional, nature-based cities, through global case studies. By reinterpreting Lynch's five core elements—paths, edges, districts, nodes, and landmarks—through the lens of natural infrastructural responsiveness, the study offers an evolved version of the original theory for recognizing the image of the nature-based city. The chapter advocates for a paradigm shift in urban design and environmental psychology—from a predominantly visual and formal understanding of urban space to one that is ecologically immersive, culturally grounded, and functionally integrative. In conclusion, it is recommended to argue for a holistic, environmentally grounded urbanism for planning new cities or the development of existing cities.

References

American heritage dictionary. (2025). https://ahdictionary.com/word/search.html?q=image

Ayatollahi, S. M. H., Mlekzadeh, F. & Pakseresht, A. S. (2016, December). The Role of Wind as a Generator of Cultural Landscape in Desert Climate of Iran. In *International Conference kerpiç* (Vol. 16, pp. 16-23).

Davies, C., & Lafortezza, R. (2019). Transitional path to the adoption of nature-based solutions. *Land Use Policy, 80*, 406–409.

Depietri, Y., & McPhearson, T. (2017). Integrating the grey, green, and blue in cities: nature-based solutions for climate change adaptation and risk reduction. *Nature-based solutions to climate change adaptation in urban areas: Linkages between science, policy and practice*, 91-109.

Ebenezer, H. (1965). *Garden cities of to-morrow*. Рипол Классик.

Eggermont, H., Balian, E., Azevedo, J. M. N., Beumer, V., Brodin, T., Claudet, J., et al. (2015). Nature-based solutions: new influence for environmental management and research in Europe. *GAIA-Ecological perspectives for science and society, 24*(4), 243–248.

Frantzeskaki, N., Van Steenbergen, F., & Stedman, R. C. (2018). Sense of place and experimentation in urban sustainability transitions: The Resilience Lab in Carnisse, Rotterdam, The Netherlands. *Sustainability Science, 13*, 1045–1059.

Geddes, P. (1915). *Cities in evolution: an introduction to the town planning movement and to the study of civics*. Williams.

Kabisch, N., Frantzeskaki, N., Pauleit, S., Naumann, S., Davis, M., Artmann, M., et al. (2016). Nature-based solutions to climate change mitigation and adaptation in urban areas: perspectives on indicators, knowledge gaps, barriers, and opportunities for action. *Ecology and Society, 21*(2).

Lynch, K. (1960). *The image of the city*. MIT Press.

Merriam-Webster. (2025). https://www.merriam-webster.com/dictionary/image#dictionary-entry-1

McHarg, I. L. (1969). *Design with nature* (pp. 7–17). American Museum of Natural History.

Mumford, L. (1961). *The city in history: Its origins, its transformations, and its prospects* (Vol. 67). Houghton Mifflin Harcourt.

Nesshöver, C., Assmuth, T., Irvine, K. N., Rusch, G. M., Waylen, K. A., Delbaere, B., et al. (2017). The science, policy and practice of nature-based solutions: An interdisciplinary perspective. *Science of the Total Environment, 579*, 1215–1227.

Nova, A. (2006). The role of the winds in architectural theory from Vitruvius to Scamozzi 1. In *Aeolian winds and the spirit in Renaissance architecture* (pp. 70–86). Routledge.

Omidvari, S., & Golzari, E. (2016). Tourism of qanat: renewal after drought. *E-ISBN, 9789752438019*, 24.

Raymond, C. M., Frantzeskaki, N., Kabisch, N., Berry, P., Breil, M., Nita, M. R., et al. (2017). A framework for assessing and implementing the co-benefits of nature-based solutions in urban areas. *Environmental Science & Policy, 77*, 15–24.

Rossi, A. (1984). *The architecture of the city*. MIT Press.

Shahabinejad, A., & Lookzadeh, H. (2024). Reidentification of the Time Measuring Device of Roknieh Complex of Yazd Based on the Jame Al-Khairat Endowment Text. *Athar, 45*(105), 61–84.

Yazdi, A. S., & Khaneiki, M. L. (2017). *Qanat knowledge*. Springer Press.

Nature-Based Urban Design for Climate-Resilient Cities: The Case of Dezful, Iran

Amir Tayyebi

Abstract Climate change and rapid urbanization have amplified environmental and social vulnerabilities in semi-arid regions, making resilience-oriented approaches to urban design increasingly urgent. Dezful, a historic city in southwestern Iran, represents a unique context where centuries-old cultural traditions coexist with severe climatic pressures. The city experiences prolonged hot seasons, irregular rainfall patterns, and frequent flash floods, while also facing growing demands on its limited water resources. These conditions present both challenges and opportunities for the integration of Nature-Based Urban Design (NBUD) strategies into its planning and development processes.

Dezful's natural assets—including extensive gardens, productive agricultural lands, and the riparian habitats of the Dez River—constitute a valuable ecological foundation for embedding ecosystem services into urban spaces. Complementing these environmental features are locally adapted architectural elements such as Shavadan (underground chambers designed for thermal comfort), Qomesh (ventilation and water-access shafts), and Kat (stone structures engineered for microclimate regulation). Together, these systems embody a long-standing synergy between human ingenuity and the natural environment, offering inspiration for contemporary climate-adaptive urban design.

The proposed NBUD framework for Dezful emphasizes the restoration and enhancement of green–blue infrastructures, including riparian buffers and interconnected ecological corridors, to mitigate flood hazards and reduce the urban heat island effect. Strategies also include the revitalization of historic gardens into multifunctional public spaces that simultaneously preserve cultural heritage and provide recreational, ecological, and social benefits. Integrating urban agriculture within vacant or underutilized spaces is identified as a pathway to enhance food security while strengthening community engagement with the land.

Methodologically, the chapter combines spatial analysis, participatory design workshops, and climate-responsive design principles to develop an adaptable planning model. Quantitative assessments—such as thermal mapping and flood modeling—are paired with qualitative insights from local stakeholders to ensure

A. Tayyebi (✉)
Architecture and Planning Department, Jundi-Shapur University of Technology, Dezful, Iran
e-mail: a_tayyebi@jsu.ac.ir

A. Cheshmehzangi et al. (eds.), *Nature-based Urbanism*, Urban Sustainability,
https://doi.org/10.1007/978-981-95-6864-2_3

cultural and ecological relevance. This approach underscores the necessity of integrating traditional knowledge systems with modern sustainability practices.

The findings suggest that implementing the Nature-Based Urban Development (NBUD) framework in Dezful can yield multiple co-benefits, from mitigating climate risks such as flooding and heat stress to fostering social cohesion through equitable, nature-rich public spaces. The innovative contribution of this chapter lies in integrating green–blue infrastructure planning with climate-responsive urban design specifically tailored to semi-arid cities. By developing a context-sensitive NBUD framework that connects riparian restoration, ecological corridors, and public space networks, the study demonstrates how nature-based strategies can function as essential urban infrastructure rather than ornamental elements. This approach not only reframes urban planning toward resilience and sustainability but also provides a transferable framework for other semi-arid cities facing similar climatic, ecological, and socio-cultural conditions. In doing so, the chapter contributes to advancing global sustainable urban development agendas, particularly SDG 11, SDG 13, and SDG 15.

Keywords Nature-based urban design · Dezful · Climate adaptation · Green–blue infrastructure · Ecosystem services · Semi-arid cities · Shavadan · Qomesh · Kat

1 Introduction

The accelerating pressures of climate change have turned modern cities into complex arenas of vulnerability. Recurrent heat waves, unpredictable rain patterns, and severe flooding now expose the fragility of conventional development philosophies. In semi-arid regions, such disruptions extend beyond environmental damage—they erode the social and cultural continuities that once allowed communities to live in harmony with their surroundings. Many cities, in their haste to modernize, abandon their adaptive intelligence, severing the ties that linked people and place. Nature-Based Urban Design (NBUD) emerges as a counter-narrative to this rupture, re-establishing nature as a participant in urban design—a collaborator that shapes resilience, identity, and emotional well-being rather than a passive backdrop.

NBUD regards ecological processes not as barriers but as building blocks of spatial reasoning. It blends environmental performance with the cultural meaning of place, drawing its ideas from ecological urbanism (Mostafavi & Doherty, 2016), biophilic design (Kellert & Calabrese, 2015), and the vision of cities that think like planets (Alberti, 2016). Yet its true maturity is visible when the theory encounters practice—particularly in cities whose survival has always depended on ecological wisdom. Among these, research in Iran's semi-arid regions has demonstrated that when vernacular experience is merged with new climatic design, the outcome is not simply environmental adaptation but also the renewal of social bonds (Tayyebi, 2025a; Tayyebi & Momeni, 2024).

Within this wider frame, Dezful stands out as a living laboratory where NBUD's philosophy is inseparable from everyday life. Its long history of environmental adaptation—through *Shavadan* (thermal underground chambers), *Qomesh* (ventilation and water shafts), and *Kat* (stone structures devised for climate regulation)—shows the depth of climatic intelligence that grows from local observation and collective ingenuity. As Tayyebi (2025b) notes, these architectural systems translate environmental empathy into physical form, uniting material aesthetics with ecological pragmatism. They are neither relics of the past nor romantic souvenirs—they are continuous threads of resilience woven through centuries of coexistence between people and place.

In recent decades, however, this organic balance has weakened. The rapid urban growth and infrastructural transformations have fragmented the once-integrated Dez River ecosystem, compromising its role as a natural corridor linking agriculture, habitation, and biodiversity. Traditional gardens that moderated temperature and fostered social interaction are being lost to land conversion and water scarcity. Studies on healthy building and resilient cities (Tayyebi & Momeni, 2024) indicate that this disappearance diminishes not only physical comfort but also a city's cultural and psychological harmony. To restore those relationships, the philosophical essence of ecological design must be re-imagined in contemporary form.

NBUD offers precisely that interpretive framework. It approaches climate adaptation as more than a technical modification—it considers it a cultural and ethical act that reclaims natural intelligence as part of urban logic. Rather than treating resilience as a defense mechanism, NBUD defines it as a process of co-evolution: an ongoing conversation between human creativity and the regenerative rhythms of nature. In Dezful, this translates into reconnecting riverside buffers, rebuilding green–blue networks, and respecting the continuity of vernacular patterns that bind dwellings with landscape and memory.

A parallel conceptual shift involves the role of community. NBUD moves away from expert-centered design, advocating instead for participatory frameworks where residents collaborate as co-designers. Tayyebi (2025a) demonstrates that when planning integrates personal narratives and local experience, its outcomes gain authenticity and endurance. Dezful's communal traditions—rooted in cooperation and shared environmental care—provide fertile ground for this participatory ethos, ensuring that the design is not imposed from above but nurtured from within everyday urban culture.

Viewed through this lens, Dezful becomes more than an example—it transforms into an epistemological model, a city whose wisdom lies in dialogue rather than dominance. Its resilience arises from adaptive patterns built through necessity, empathy, and collective attention to the environment. Reviving these traditions within NBUD's theoretical discourse means reinstating nature as a value system, not a commodity. Such a reframing aligns directly with the United Nations Sustainable Development Goals (SDG 11, SDG 13, SDG 15), emphasizing sustainable growth through cultural continuity, inclusive knowledge, and ecological responsibility (IUCN, 2021; Raymond et al., 2017).

Ultimately, integrating Dezful's vernacular intelligence into NBUD thinking expands our idea of resilience into one that is relational and ethical. It suggests that robust cities are not those armored against nature, but those porous enough to let its rhythms guide them. The goal is not to harden physical boundaries but to soften the interface between built and natural systems, between expertise and tradition, between human intention and ecological response. Through such reconciliation, Dezful reflects a broader global aspiration—to design in partnership with nature, ensuring that the urban future remains both ecologically wise and culturally alive.

2 Philosophical and Theoretical Foundations

From the earliest phases of urban thought, the meaning of *nature* has undergone a profound transformation—from an external landscape to an active force shaping urban existence. Classical philosophers such as Aristotle envisioned the *polis* as a manifestation of human rational order, detached from the raw spontaneity of wilderness. Civilization, in that worldview, emerged through taming and restructuring the natural world. But the ascent of ecological science and the birth of environmental ethics in the twentieth century upended this anthropocentric logic.

Visionaries like Hans Jonas (1984) and Arne Naess (1990) articulated new moral and spiritual relationships between humanity and the biosphere. Jonas framed technological society as morally accountable for the life systems it alters, while Naess advanced *deep ecology*—an understanding of humans as inseparable participants within nature's self-renewing processes. Their ethics later informed environmental aesthetics and design philosophy, crystallizing ideas of *ecological consciousness* and *place-based ethics* (Seixas & Lobner, 2024).

Within this intellectual lineage, Nature-Based Urban Design (NBUD) can be seen as a late-modern synthesis—a paradigm that redefines the city not as nature's opposite but as one of its evolutionary continuities. NBUD treats nature as a creative infrastructure and an epistemic collaborator in design thought, aligning conceptually with ecological urbanism (Mostafavi & Doherty, 2016), landscape urbanism (Waldheim, 2012), and adaptive design approaches that emphasize resilience and ecological processes in urban systems (Ahern, 2011; Alberti, 2016). This perspective also resonates with the broader discourse on nature-based solutions and ecosystem-based planning in cities (Kabisch et al., 2017). Instead of imposing control over ecological systems, NBUD derives spatial form and urban logic from ecological feedback and dynamic landscape processes.

1. Genealogy of "Nature in Design"

The philosophical lineage of "nature within design" stems from phenomenology and environmental hermeneutics. In *Building Dwelling Thinki* (1951), Heidegger reimagined dwelling as a state of being attuned to earth and sky—a foundation for genuine place-making. Following this, landscape theorists (Corner, 1999; Sanders, 1985) argued that design should illuminate rather than conceal ecological processes. Nature thereby shifted from object to *medium*: a bearer of meaning and continuity.

This evolution led to ecological aesthetics, where beauty arises from adaptability and complexity rather than geometric perfection. Through this lens, NBUD inherits both the contemplative depth of phenomenology and the analytical rigor of ecology. Nature is conceived as *processual*—a self-organizing system whose interactions shape the social and spatial metabolism of a city. The designer becomes not an engineer seeking dominance, but a thinker working in dialogue with natural systems.

2. Reassessing Development in Semi-Arid Regions

Semi-arid landscapes—from Iran's plateau to the broader Mediterranean littorals—illustrate the consequences of planning that ignored ecological reciprocity. Throughout the twentieth century, development policies treated climatic extremes as technical malfunctions to be corrected by expanding infrastructure—via dams, pipelines, and sealed architecture—rather than integrating local ecological intelligence. This mindset of environmental conquest, inherited from colonial and industrial paradigms, dismantled vernacular harmony between settlement and landscape.

In Iran, modern zoning schemes replaced fluid networks of shaded alleys and water-based public domains with rigid asphalt corridors. Vernacular adaptations—like the cooling geometry of Dezful's historic quarters—were dismissed as obsolete. Yet, as Tayyebi and Momeni (2024) demonstrate, resilience was once achieved not by mechanization but by *empathetic environmental design*: a lived understanding of seasonal rhythm. NBUD revives that sensibility, arguing that the semi-arid city must rediscover the wisdom embedded in its ecological ancestry instead of silencing it through technology.

Over-centralized water controls and impermeable surfaces have, paradoxically, heightened flood risks—where drought and deluge now coexist. The logic of adaptive design, articulated by Ahern 2011) and Alberti (2016), rejects such linear solutions and advocates feedback-based resilience. Here, infrastructure transforms into ecological tissue—porous, cyclical, and symbiotic.

3. NBUD Among Contemporary Resilience Paradigms

In contemporary urban theory, Nature-Based Urban Design (NBUD) contributes to the evolving discourse on resilient and climate-adaptive cities by integrating ecological processes into urban spatial planning and design. This perspective aligns with recent scholarship on nature-based solutions, urban resilience, and climate-responsive urbanism, which emphasizes the role of green–blue infrastructure in mitigating environmental risks and enhancing urban livability (2019; Kabisch et al., 2017; Raymond et al., 2017; Seddon et al., 2020). By framing nature as a core component of urban infrastructure rather than an auxiliary element, NBUD strengthens theoretical and practical approaches that seek to reconnect urban development with ecological systems and long-term climate resilience. Whereas resilience theory (Holling, 1973) measures a system's ability to reorganize under stress, ecological urbanism extends that idea to urban form, imagining cities as active ecosystems. Adaptive design operationalizes it through participatory governance. NBUD unites these threads under a common philosophical stance: *co-evolution.*

Resilience, from this viewpoint, is transformative rather than restorative—it evolves alongside environmental dynamics. In places like Dezful, built patterns that follow water and heat cycles—evident in Shavadan and Qomesh structures—embody that co-evolution. Tayyebi (2025a) emphasizes that participatory planning deepens this adaptive process, granting citizens agency within ecological governance. Resilience thus becomes a collective cultural act, not a technical protocol.

Nature-Based Urban Design (NBUD) introduces a moral and philosophical dimension into ecological urbanism, positioning nature as a *co-author* rather than merely a resource. It reframes urban design through reciprocity, care, and interdependence—values increasingly central to posthumanist and more-than-human design discourses (Bennett, 2020; Fry, 2017; Latour, 2017). In this view, design decisions emerge from processes of ecological communication and mutual adaptation instead of purely aesthetic or technocratic prescription, resonating with contemporary theories of ecological ethics and relational urbanism. As illustrated in Fig. 1, the paradigm crystallizes through three interconnected integrities:

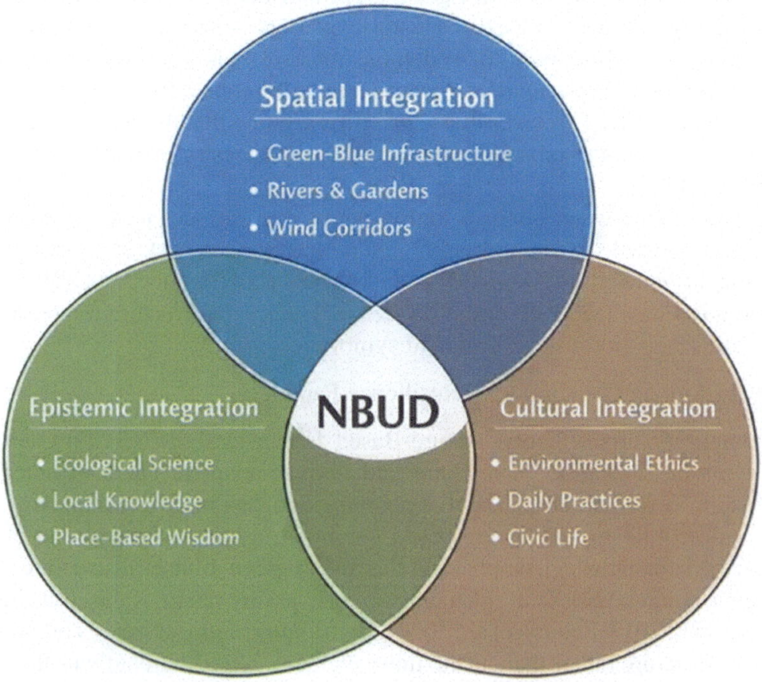

Fig. 1 Nature-Based Urban Design (NBUD) framework (Photo made by the authors)

i. Epistemic Integration: Bridging formal ecological science with traditional, place-based wisdom.
ii. Spatial Integration: Embedding natural infrastructures—rivers, gardens, wind corridors—into civic structure.
iii. Cultural Integration: Sustaining environmental ethics through the rhythm of daily urban life.

By harmonizing these planes, NBUD transcends the managerial tone of "sustainability" and approaches an ontological understanding of the city as part of life's continuum.

4. Humanity and the Cultural Ecosystem in Iranian Cities

Iranian philosophical cosmology conceives human existence as an element within a continuous web of creation, not an occupying force above it. Historically, cities like Dezful, Isfahan, and Kashan expressed this balance through spatial metaphors where architecture linked heaven and earth (*āsmān o zamīn*). Courtyards, windcatchers (badgirs), subterranean spaces, and gardens formed an integrated environmental system within traditional Iranian urbanism. These architectural typologies were carefully designed to regulate microclimate by maximizing the use of natural elements such as airflow, shading, evaporation, and thermal mass. Courtyards functioned as climatic moderators, enabling passive cooling through vegetation, water features, and controlled solar exposure, while also organizing domestic and social life around a protected inner landscape (Edwards et al., 2006; Memarian & Sadoughi, 2011). Windcatchers captured prevailing winds and directed them into interior spaces, often in combination with subterranean rooms or water reservoirs to enhance evaporative cooling and reduce indoor temperatures in hot-arid climates (Bahadori, 2018; Roaf et al., 2009). Gardens, particularly in the Persian *chaharbagh* tradition, introduced shade, humidity, and biodiversity into the urban fabric while symbolically expressing ecological harmony (Moynihan, 1979; Pirnia & Memarian, 2005). Together, these elements orchestrated an environment attuned to cycles of light, air, and water—demonstrating how traditional urban systems embedded environmental ethics and climatic intelligence into everyday spatial design.

These structures embody what contemporary theorists call the cultural ecosystem—a system of memory and morality rather than mere material exchange. In Dezful, it reveals itself in river stewardship, seasonal agriculture, and artisanal techniques such as *Kat* stonework and *Shavadan* cooling chambers. They represent not only functional adaptations but also social ethics of care.

As Tayyebi (2025b) argues, conserving and fostering vernacular wisdom—rather than reducing it to a touristic or development commodity—constitutes a form of philosophical continuity. NBUD inherits this moral stance by positioning humans not as consumers of space, but as custodians who maintain and actively cultivate ecological relationships across time. Seen through this prism, Iranian urbanism exemplifies symbiotic resilience—nature has never been peripheral to civic life but its measure of justice and balance. Reimagining NBUD for Dezful, therefore, is less an innovation than a cultural recovery, reviving dialogue between ancient ecological consciousness and scientific rationality.

3 Methodology: Constructivist–Interpretive Framework

This study develops its conceptual model through constructivist and interpretive reasoning rather than empirical data collection. The main objective is not to measure phenomena but to understand *how meaning itself is constructed* between nature, community, and urban form. According to Lincoln (1985), constructivist inquiry rests on multiple realities shaped by human interaction; here, those realities emerge through the dialogue of ecological processes and cultural expressions within Dezful.

The interpretive approach, inspired by Schutz (1967) and later ecological phenomenology (Naess & Næss, 1990), views architecture and urban space as textual narratives through which society negotiates its relationship with nature. Thus, the framework translates historical patterns of adaptation into abstract theoretical propositions for Nature-Based Urban Design (NBUD).

Methodologically, the first step integrates vernacular intelligence with modern ecological thought (Alberti, 2016). Dezful's architectural repertoire—*Shavadan*, *Qomesh*, and *Kat*—is interpreted not as nostalgia but as *metaphorical prototypes* of environmental design. Each element conveys an ecological principle: *Shavadan* embodies thermal harmony, *Qomesh* represents ventilation and water dialogue, and *Kat* symbolizes material endurance. These become conceptual metaphors guiding the NBUD framework (cf. Tayyebi & Momeni, 2024) (Fig. 2).

4 Stages of Conceptual Modeling

1. Ecological Derivation: Extracting adaptive principles from Dezful's traditional morphology to reveal how local practices encode climate reasoning.
2. Cultural Abstraction: Translating these practices into universal patterns of *resilient culture*—community self-organization and sustainable material use (similar to Ahern, 2011).
3. Model Synthesis: Framing the NBUD triad—nature, community, morphology—as co-evolutionary layers that form the city's living ecosystem (Tayyebi, 2025a).

This process embodies what Mostafavi and Doherty (2016) call *ecological urbanism*: design learning from nature's feedback instead of controlling it.

Through conceptual modeling, the framework maps relations between ecosystem services (cooling, water regulation, vegetation buffers) and cultural space (gathering, identity, ritual). The alignment clarifies how resilience operates simultaneously as ecological and social functionality.

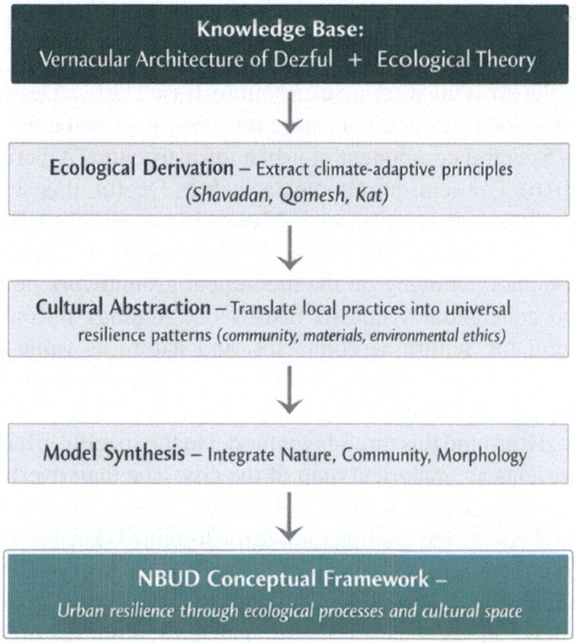

Fig. 2 Methodological framework for developing the NBUD conceptual model (Photo made by the authors)

Metaphorical analysis deepens interpretation: each spatial form is read as moral dialogue with nature—*Shavadan* as humility, *Qomesh* as breath, *Kat* as persistence (Tayyebi, 2025b). Together, they articulate an ethics of coexistence, grounding NBUD in Iran's cultural ecosystem rather than external sustainability templates.

By adopting this constructivist–interpretive methodology, the chapter develops NBUD as a theoretical architecture of *meaning and moral ecology*. Knowledge here is not discovered but constructed through reflection, allowing Dezful's vernacular wisdom to reshape modern urban resilience theory.

A key innovation of this methodology lies in its explicit treatment of urban resilience not merely as a technical or ecological attribute, but as a deeply cultural phenomenon rooted in long-standing human–environment relationships (Tayyebi & Momeni, 2024). By systematically translating the city's vernacular systems—Shavadan, Qomesh, and Kat—into conceptual metaphors and co-evolutionary design principles, the constructivist–interpretive framework integrates traditional collective wisdom with contemporary ecological science. This approach makes the resulting NBUD model both forward-looking (scientifically adaptive) and memory-sensitive (culturally continuous), thereby bridging the gap between historical adaptation patterns and future-oriented urban design in semi-arid contexts.

5 Conceptual Synthesis: Nature as Urban Infrastructure

In the evolving discourse on Nature-Based Urban Design (NBUD), a growing body of scholarship argues that nature must be integrated into cities not as an afterthought but as a central component of urban infrastructure (Alberti, 2016; Mostafavi & Doherty, 2016). For semi-arid contexts such as Dezful, this shift in perspective is more than aesthetic—it is existential. Nature is not merely what surrounds the city; it is the system that regulates its survival, connects its communities, and shapes its moral compass. Drawing on the theoretical groundwork developed earlier in this chapter, the conceptual synthesis distills our findings into a three-layer model: nature as regulator, nature as connector, and nature as value system. Each layer is deeply rooted in both ecological science and vernacular wisdom (Tayyebi, 2025a, 2025b; Tayyebi & Momei2024).

To ground this three-layer model in the specific urban morphology of Dezful, Fig. 3 presents an analytical map of the city. The map overlays the primary natural infrastructures (Dez River and its riparian zones, historic gardens, and productive agricultural lands) and architectural infrastructures (key locations of Shavadan underground chambers, Qomesh ventilation and water-access shafts, and Kat stone structures within the historic quarters) onto the contemporary urban fabric. Proposed NBUD interventions—such as interconnected ecological corridors and riparian buffers—are shown with dashed lines. Color-coding directly corresponds to the three conceptual layers: blue for regulatory functions (climate and hydrological regulation), green for connective functions (biodiversity and social linkage), and ochre for value-system functions (cultural heritage and ethical stewardship). This visualization demonstrates how vernacular elements and modern ecosystem services can be spatially integrated to operationalize NBUD in a semi-arid historic city (Fig. 4).

A. Nature as Regulator

At its foundational level, nature acts as a regulator of climate and micro-environment, buffering the extremes that define semi-arid living (Ahern, 2011). Dezful's historic architecture offers compelling evidence of how human ingenuity has harnessed natural processes to achieve thermal comfort (Tayyebi & Momeni, 2024). *Shavadan*—subterranean chambers—leverage the thermal mass of the earth to create cool sleeping and working spaces during prolonged summer heat (Babaee & Heidari, 2025; Dadfar et al., 2014; Sadoughi et al., 2019). *Qomesh* structures channel air and water through vertical shafts, using simple but effective convection principles to refresh indoor and semi-outdoor spaces (Tayyebi, 2025b). *Kat* stone façades, with their particular material density, slow down heat transfer, allowing interiors to remain habitable despite high surface temperatures (Tayyebi, 2025b; Tayyebi & Momeni, 2024).

These mechanisms embody what Ahern (2011) describes as safe-to-fail systems: designs inherently able to absorb environmental stresses without catastrophic breakdown. Translating this vernacular logic into NBUD means embedding green–blue

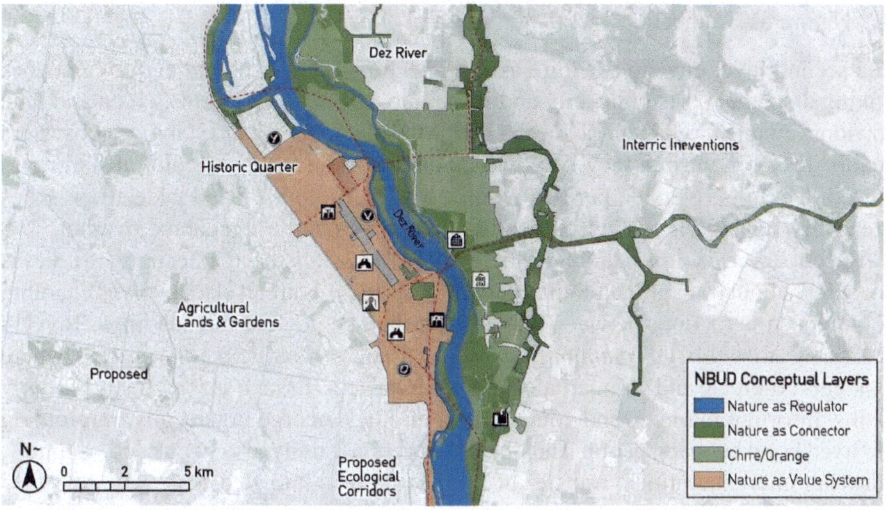

Fig. 3 Analytical map of Dezful: Main natural and architectural infrastructures for Nature-Based Urban Design (NBUD) (Photo made by the authors)

Fig. 4 Conceptual layers of Nature-Based Urban Design (NBUD) in Semi-Arid contexts (Photo made by the authors)

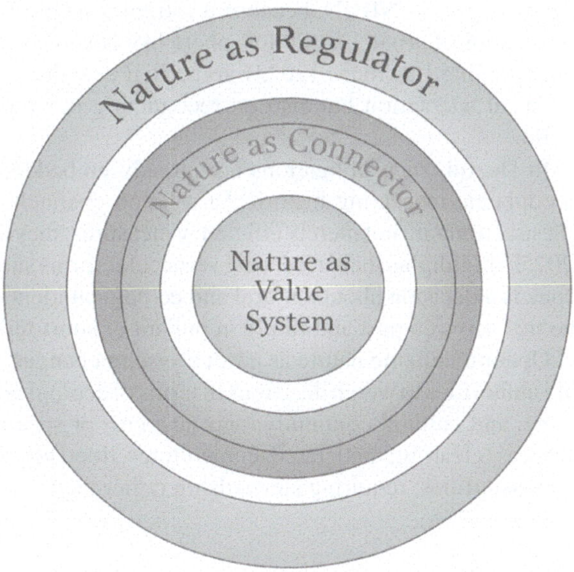

infrastructures—such as riparian buffers along the Dez River, interconnected ecological corridors, and canopy-based shading systems—directly into the functional fabric of the city. Here, vegetation and water systems are not decorative edges but active regulators, reducing flood risks, tempering heat islands, and restoring the hydrological balance.

B. Nature as Connector

The second layer envisions nature as a connector—a physical and cultural network linking disparate parts of the city and its people. In ecological terms, green and blue corridors enable movement of species, improve biodiversity, and stabilize ecosystem health. In social terms, they serve as common grounds where cultural narratives unfold. Dezful's riverside parks, revitalized gardens, and shaded pedestrian routes exemplify spaces where environmental function and cultural meaning overlap.

Recent urban ecology studies (Alberti, 2016) highlight that connectivity in cities is not solely about transport or communications infrastructure; it also involves enabling continuous interactions between humans and their ecological surroundings. Tayyebi (2025a) frames this as *community-driven resilience*: a participatory process where residents co-create and co-maintain nature-rich spaces. In such spaces, festivals, daily walks, informal markets, and educational activities emerge organically, reinforcing the lived bond between people and place. This connectivity also works across time—allowing intergenerational transfer of local ecological knowledge.

C. Nature as Value System

At its deepest layer, nature operates as a value system—an ethical and cultural foundation influencing how urban decisions are made. This is perhaps the most transformative aspect of NBUD, because it reframes nature from a utilitarian resource into a co-author of urban life. Philosophers like Naess and Næss (1990) and geo-cultural theorists like Lawrence (2023) argue that sustainable living depends not only on technical adaptation but also on recognizing our moral obligations to the natural world.

In Dezful, such obligations are already embedded in practices like conserving floodplains, respecting historical irrigation channels, and planting native species. These actions are not merely efficiency measures; they articulate a civic ethic. Tayyebi (2025b) highlights that preserving vernacular forms such as Shavadan or reusing local stone is a decision about cultural and ecological identity, signaling a refusal to sever the ties between past and future in pursuit of short-term gains.

Operationalizing nature as a value system changes the planning paradigm. It asks city authorities to weigh decisions in terms of ecological reciprocity, community well-being, and cultural continuity—not just cost or speed. Under this layer, ecosystem services (clean air, moderated temperatures, flood mitigation) are reframed as cultural responsibilities, requiring stewardship rather than exploitation.

5.1 Integrating the Three Layers

These three layers—regulator, connector, value system—do not operate in isolation. They co-evolve, reinforcing each other. Climatic regulation enhances the habitability of connective spaces; strong social bonds increase willingness to steward ecological systems; embedding values ensures long-term resilience beyond political cycles.

NBUD in Dezful thus becomes an integrated design philosophy, merging traditional adaptive strategies with modern ecological infrastructure planning.

The city's transformation lies not in monumental projects, but in re-understanding its biological and cultural ecosystems as interdependent infrastructure. This synthesis positions nature as both hardware (physical systems regulating and connecting) and software (values, ethics, cultural narratives) of urban resilience. By working through these layers in harmony, other semi-arid cities can adapt the model to their own environmental and cultural contexts, advancing broader sustainability goals such as SDG 11, SDG 13, and SDG 15.

6 Discussion: Translating Concept to Urban Design Visions

The conceptual foundations of Nature-Based Urban Design (NBUD) acquire practical meaning only when they inform the making of places—when ecological ideas turn into guiding principles for design and planning. Translating the theory established in earlier sections into urban visions for Dezful requires a synthesis between scientific reasoning and lived cultural intelligence. Urban design here is not a mechanical application of ecological formulas but a dialogue among environmental processes, human values, and inherited traditions.

According to Tayyebi and Mome (2024), the vernacular intelligence embedded in Dezful's built fabric already provides a code of design principles capable of guiding resilience. The Shavadan, Qomesh, and Kat systems illustrate how climatic awareness, material empathy, and spatial modesty grounded urban life. The challenge of contemporary planning lies in re-articulating these ideas into adaptable frameworks rather than replicating their forms. Each element becomes a generative metaphor—a way of thinking about the city as an organism responsive to its environment.

Moving from concept to design requires identifying guiding principles that embody NBUD's three layers discussed earlier.

Climatic Harmony: The principle that every intervention should sustain the city's regulatory balance—through shading structures, permeable surfaces, and water-sensitive streets reflecting the logic of Shavadan cooling and Qomesh ventilation.
Social Connectivity: Spaces must serve as ecological connectors and social commons simultaneously. Green corridors and riparian parks are conceived not as isolated zones but as living rooms for community interaction (Tayyebi, 2025a).
Cultural Reciprocity: Projects should honor nature as a value system, allowing local materials, crafts, and rituals to inform design aesthetics. Here, Kat masonry becomes a symbol of continuity, reminding designers that beauty stems from endurance rather than luxury (Lawrence, 2023; Naess & Næss, 1990).

Realizing these principles demands an ethical reorientation of the designer's role—from controller to co-learner. To visualize how these abstract principles translate into concrete urban interventions, Fig. 5 presents an analytical illustration of the proposed NBUD design visions along the Dez River. The translation process does

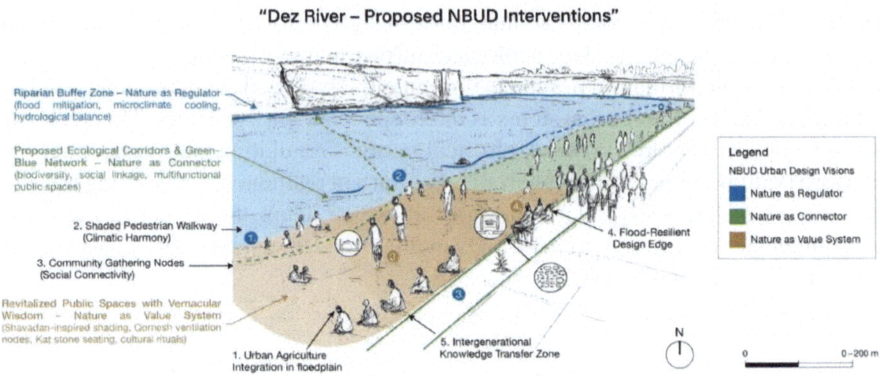

Fig. 5 Analytical illustration of NBUD urban design visions for Dezful: Urban life and climatic wisdom along the Dez River (Photo made by the authors)

not follow a top-down pattern; instead, it follows an interpretive spiral in which human perception and environmental feedback continuously shape design outcomes. Alberti (2016) notes that resilient design emerges when the city "thinks like a planet," recognizing the invisible flows linking nature, culture, and infrastructure. For Dezful, that planetary thinking begins at the local scale—in workshops that engage artisans, students, and residents to narrate what resilience means in their everyday experience.

Human wisdom (khirad-e ensani) and tradition form the ethical core of resilient design. They remind us that adaptation is not only a technical skill but a cultural memory. Incorporating oral narratives, collective rituals, and local craftsmanship into planning transforms resilience from a strategic term into a lived philosophy. As Tayyebi (2025b) argues, preserving the past is a means of shaping the future; design becomes an act of remembrance and anticipation at once.

Ultimately, translating NBUD concepts into design visions signifies an ongoing exchange between the seen (physical infrastructure) and the unseen (human belief and ecological consciousness). When planning is guided by humility, empathy, and continuity, Dezful can evolve as a prototype of climate-resilient living—a city that builds with its memory, not against it.

6.1 Towards Cultural–Ecological Urbanism (CEU)

Building on the three-layer NBUD model and the practical design visions we have just explored, this chapter puts forward a broader theoretical lens that we call Cultural–Ecological Urbanism (CEU). CEU grows naturally out of NBUD, but it takes one important step further: it moves the emphasis from purely spatial and infrastructural solutions to the active co-production of knowledge between culture and ecology. While NBUD shows us how to weave green–blue systems and vernacular elements into the city, CEU asks how values, stories, rituals, and collective memory can themselves become drivers of design. In this view, the city is no longer just a place to manage—it becomes a living archive where ethics, climate, and creativity constantly shape one another.

Tayyebi and Momeni (2024) have already shown that truly resilient cities arise from an unbroken conversation between community intelligence and the demands of the local climate. Tayyebi (2025a) takes this idea further, arguing that participatory processes turn ordinary residents into active stewards of nature. CEU builds directly on these insights. It sees design as both a moral act and an ecological practice at the same time. In Dezful, this means revitalizing the riverfront not only with technical buffers and corridors but also by re-embedding Shavadan-style thermal comfort, Qomesh ventilation points, and Kat stone details into everyday public life. The result is a form of urbanism that is technically sound yet culturally alive—one that honors the past while remaining open to the future.

7 Conclusion and Future Research Directions

The exploration of Dezful's nature-based urban design, grounded in the constructivist–interpretive methodology developed earlier, has illustrated that resilience is not a mere technical attribute but a cultural phenomenon rooted in long-standing human–environment relationships. The city's vernacular systems—Shavadan, Qomesh, and Kat—reveal how ecological harmony was historically achieved through collective wisdom and adaptation to local climate. These lessons have proved that sustainable urban futures depend on integrating traditional knowledge with scientific ecology, making urban design both forward-looking and memory-sensitive (Fig. 6).

In conceptual terms, the NBUD framework developed throughout this study introduces a methodology that merges spatial ecology with socio-cultural narratives. Yet Dezful's case represents only one facet of a broader paradigm shift emerging in

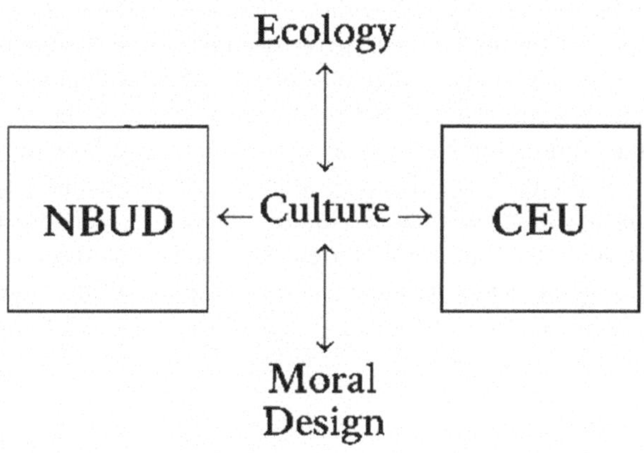

Fig. 6 Conceptual evolution: From Nature-Based Urban Design (NBUD) to Cultural–Ecological Urbanism (CEU) (Photo made by the authors)

semi-arid cities across the Global South. Future research should extend this inquiry by comparing different climatic contexts—such as Yazd in central Iran, Aswan in Egypt, or Fez in Morocco, as well as other geographical areas sharing similar semi-arid climatic characteristics—to identify shared adaptive logics. As Alberti (2016) reminds us, cities that "think like planets" must embrace their ecological limits and cultural diversities simultaneously. Comparative studies would help generalize NBUD principles into globally relevant, climate-aware design toolkits (Table 1).

Seen in this light, the NBUD framework developed here opens the door to a wider perspective—what we have called Cultural–Ecological Urbanism (CEU). In CEU, values, narratives, and rituals are not decorative extras; they become active partners in shaping design logics alongside ecological processes. This evolution positions Dezful not merely as one case study among many, but as a living prototype for culturally grounded climate resilience in semi-arid cities around the world.

In the context of Dezful and its sister climates, theoretical development of CEU demands interdisciplinary collaborations among architecture, geography, anthropology, and climate science. Researchers should explore how cultural practices—festivities, water rituals, and architectural craftsmanship—mirror ecosystem services such as cooling, shading, and water regulation. This cross-mapping of culture and ecology could reveal new design grammars—where every spatial decision embodies cultural memory and environmental ethics.

From a future-oriented perspective, Tayyebi (2025b) underscores that sustainability must include time as an active design material; that is, projects should evolve with local feedback rather than remain static. Long-term monitoring of NBUD interventions, citizen co-creation labs, and digital modeling of vernacular climate responses can strengthen the predictive capacity of urban design in semi-arid regions.

Ultimately, this chapter has addressed a critical gap in contemporary urban resilience literature: while many studies in semi-arid regions have treated climate adaptation either as a purely technical challenge or as a matter of heritage conservation in isolation, few have shown how traditional ecological wisdom and modern scientific approaches can be meaningfully woven together into a single, coherent design framework. By grounding the Nature-Based Urban Design (NBUD) model in Dezful's living vernacular systems—Shavadan, Qomesh, and Kat—and by developing it through a constructivist–interpretive methodology that treats local knowledge as a genuine source of innovation rather than nostalgia, the study has filled that gap. It demonstrates that resilience is not an either/or choice between engineering and culture, but a relational practice in which both must inform one another.

In doing so, the chapter points toward a humanistic approach to urban resilience—one in which the wisdom of tradition and the precision of ecological science converge. Developing Cultural–Ecological Urbanism in similar climatic contexts is therefore not merely a research agenda; it is a moral calling to rediscover the balance between city and nature, heritage and innovation, past and future.

Table 1 Comparative framework linking Nature-Based Urban Design (NBUD) and Cultural–Ecological Urbanism (CEU) (made by the authors)

Stage	Core focus	Conceptual Axis	Outcome
Nature-Based Urban Design (NBUD)	Ecological infrastructure, participatory adaptation	Ecology → Community → Morphology	Spatial–environmental resilience rooted in natural systems
(Conceptual Evolution Column)	Integration layer	Ecology ↔ Culture ↔ Moral Design	Mutual transformation—nature understood through cultural ethics
Cultural–Ecological Urbanism (CEU)	Cultural–moral dimensions of design, co-production of knowledge	Culture → Ethics → Creativity	Resilience as cultural continuity and ecological consciousness

8 Declaration on the Use of Artificial Intelligence Tools

In preparing this book chapter, the author made limited and transparent use of artificial intelligence tools. Specifically, Grok (developed by xAI) was employed to assist with the following tasks only: refining the academic English phrasing and flow of selected paragraphs, suggesting structural improvements in response to reviewer comments, and generating/editing the analytical maps and diagrams (Figs. 1, 2, 3, 4, and 5). All core ideas, theoretical arguments, interpretations of vernacular architecture, and final wording are entirely the author's own. No AI tool was used to generate original research content, analyze data, or develop the central conceptual framework. The author takes full responsibility for the content and integrity of the chapter.

References

Ahern, J. (2011). From fail-safe to safe-to-fail: Sustainability and resilience in the new urban world. *Landscape and Urban Planning, 100*(4), 341–343.

Alberti, M. (2016). *Cities that think like planets: Complexity, resilience, and innovation in hybrid ecosystems*. University of Washington Press.

Babaee, F., & Heidari, S. (2025). Thermal comfort and structural optimization of Shavadans: A multi-criteria approach: Case study–Hot and semi-humid climate, city of Dezful-Iran. *Tunnelling and Underground Space Technology, 162*, Article 106620.

Bahadori, M. N. (2018). Passive cooling systems in Iranian architecture. *In Renewable energy* (pp. Vol1_87–Vol1_101). Routledge.

Bennett, J. (2020). *Vibrant matter: A political ecology of things*. Duke University Press.

Carmona, M. P. (2011). Mostafavi M & Doherty G (2010) Ecological Urbanism. *Urban Design Quarterly, 118*, 41.

Corner, J. (1999). *Recovering landscape: Essays in contemporary landscape theory*. Princeton Architectural Press.

Dadfar, T., Taban, M., & Mehrakizadeh, M. (2014). Shavadan, a passive and low-energy design technique in traditional buildings of Dezful, Iran. *Journal of Academic and Applied Studies*, 2(2), 86–95.

Edwards, B. (Ed.). (2006). *Courtyard housing: Past, present and future*. Taylor & Francis.

Fry, T. (2017). Design for/by "The Global South." *Design Philosophy Papers, 15*(1), 3–37.

Heidegger, M. (1951). *Building Dwelling Thinking*. Harper & Row.

Holling, C. S. (1973, November). *Resilience and stability of ecological systems*. University of British Columbia.

International Union for Conservation of Nature (IUCN). (2021). *The global standard for nature-based solutions*. IUCN Publishing.

Jonas, H. (1984). *The imperative of responsibility: In search of an ethics for the technological age*. University of Chicago press.

Kabisch, N., Korn, H., Stadler, J., & Bonn, A. (2017). Nature-based solutions to climate change adaptation in urban areas—linkages between science, policy and practice. *Nature-based solutions to climate change adaptation in urban areas: Linkages between science, policy and practice* (pp. 1–11). Springer International Publishing.

Kellert, S., & Calabrese, E. (2015). The practice of biophilic design. *London: Terrapin Bright LLC, 3*(21), 2021-09.

Latour, B. (2017). *Facing Gaia: Eight lectures on the new climatic regime*. John Wiley & Sons.

Lawrence, R. J. (2023). Oecumene: Repositioning ourselves in our habitat. *World, 4*(1), 95–109.

Lincoln, Y. S. (1985). *Naturalistic inquiry* (Vol. 75). Sage.
Memarian, G., & Sadoughi, A. (2011). Application of access graphs and home culture: Examining factors relative to climate and privacy in Iranian houses. *Scientific Research and Essays, 6*(30), 6350–6363.
Mostafavi, M., & Doherty, G. (Eds.). (2016). *Ecological urbanism.* Lars Müller.
Moynihan, E. B. (1979). *Paradise as a garden: In Persia and Mughal India* (No Title).
Naess, A., & Næss, A. (1990). *Ecology, community and lifestyle: Outline of an ecosophy.* Cambridge University Press.
Pirnia, M. K., & Memarian, G. H. (2005). *Introduction to Islamic architecture of Iran.* Soroosh Danesh.
Raymond, C. M., Frantzeskaki, N., Kabisch, N., Berry, P., Breil, M., Nita, M. R., & Calfapietra, C. (2017). A framework for assessing and implementing the co-benefits of nature-based solutions in urban areas. *Environmental Science & Policy, 77*, 15–24.
Roaf, S., Crichton, D., & Nicol, F. (2009). *Adapting buildings and cities for climate change: A 21st century survival guide.* Routledge.
Sadoughi, A., Kibert, C., Sadeghi, F. M., & Jafari, S. (2019). Thermal performance analysis of a traditional passive cooling system in Dezful Iran. *Tunnelling and Underground Space Technology, 83*, 291–302.
Sanders, R. A. (1985). *The granite garden: Urban nature and human design.* Basic Books
Schutz, A. (1967). *The phenomenology of the social world.* Northwestern University Press.
Seddon, N., Chausson, A., Berry, P., Girardin, C. A., Smith, A., & Turner, B. (2020). Understanding the value and limits of nature-based solutions to climate change and other global challenges. *Philosophical Transactions of the Royal Society B: Biological Sciences, 375*(1794), 20190120.
Seixas, P. C., & Lobner, N. (2024). The Ecumene: A research program for future knowledge and governance. *Encyclopedia, 4*(2), 799–817.
Tayyebi, A. (2025a). Building climate-resilient cities through community engagement and participatory design. In *Urban climate and urban design* (pp. 233–252). SpringerNature.
Tayyebi, A. (2025b). Preserving the past, shaping the future: Integrating vernacular architecture and sustainable design into Dezful's tourism development. In *Integrating architecture and design into sustainable tourism development* (pp. 309–330). IGI Global Scientific Publishing.
Tayyebi, A., & Momeni, K. (2024). *Healthy building and resilient city in Dezful: A study of passive climatic design.* In Designing healthy cities: Integrating climate-resilient urbanism for sustainable living (p. 133). Springer.
Waldheim, C. (Ed.). (2012). *The landscape urbanism reader.* Chronicle books.

Rethinking Nature in Cities: Perspectives from Southeast Asia

Rachel R. Y. Oh, Yiwen Zeng, Perrine Hamel, Adrian Loo, and Janice S. H. Lee

Abstract Urbanisation in Southeast Asia is unfolding at an unprecedented pace, driving the expansion of infrastructure-intensive *grey cities* at the cost of natural ecosystems. This rapid loss of urban nature raises urgent concerns for environmental sustainability, climate resilience, and public health. Here, we examine the relationships between urban nature, economic productivity, and population health across rapidly developing cities in Southeast Asia. While the concept of *nature in cities* from Global North perspectives emphasises integrating green spaces, ecosystem services, and biodiversity into well-planned, post-industrial and often high-income settings, such models are not easily transferable to many Southeast Asian cities, which face distinct constraints. This include the need to provide basic infrastructure and services to rapidly growing populations, often with weaker governance and lower financial resources. As such, their urbanisation trajectories face direct challenges and trade-offs between economic, environmental and human health dimensions. We explore: what extent does investing in urban nature in Southeast Asia represent a trade-off, or synergy, with economic growth and public health outcomes? Drawing on multiple examples from cities in the region, we explore associations between urban nature extent and ecological integrity, and economic productivity and population health outcomes. We highlight how relationships between people and nature extend beyond cities' administrative boundaries, and that well-integrated urban nature is a key determinant of human survival, health and well-being. We argue that underinvestment in

R. R. Y. Oh (✉)
Department of Geography, National University of Singapore, Kent Ridge, Singapore
e-mail: r.oh@nus.edu.sg

Y. Zeng · P. Hamel · J. S. H. Lee
Asian School of the Environment, Nanyang Technological University, Nanyang, Singapore

Y. Zeng
School of Social Sciences, Nanyang Technological University, Nanyang, Singapore

A. Loo
School of Biological Sciences, National University of Singapore, Kent Ridge, Singapore

P. Hamel · J. S. H. Lee
Earth Observatory of Singapore, Nanyang Technological University, Nanyang, Singapore

A. Cheshmehzangi et al. (eds.), *Nature-based Urbanism*, Urban Sustainability,
https://doi.org/10.1007/978-981-95-6864-2_4

urban ecosystems can erode long-term urban sustainability and public health priorities. We call for a reframing of urban development strategies in Southeast Asia—one that positions urban nature as a critical component for building future-ready resilient, inclusive, and ecologically sustainable cities in Southeast Asia.

Keywords Global south · Asia · Nature-based solutions · Ecosystem services · Human health and well-being

1 Urbanisation in Southeast Asia

Urbanisation is reshaping the global landscape at an unprecedented speed and scale. Over 75% of today's urban population reside in the Global South, a region expected to absorb nearly 96% of future urban population growth (UN DESA, 2018). By 2050, an estimated 2.5 billion additional people will be living in urban areas, with approximately 90% of this increase concentrated in Asia and Africa. Southeast Asia will become one of the fastest urbanising regions, where the urban population will grow from 51% in 2021 to 66% by 2050 and that 40% of the urban infrastructure needed to accommodate this growth has yet to be built (Hajer et al., 2020).

Currently, more than 40 cities across Southeast Asia are home to 51% of the region's population (Sugiyarto, 2014). These cities exhibit considerable heterogeneity in size, function, and form. They include global financial and logistics hubs such as Singapore and Kuala Lumpur; administrative and economic megacities such as Metro Manila and Greater Jakarta; and rapidly growing secondary cities such as Surabaya (Katherina & Indraprahasta, 2019). The dynamism of these cities is shaped by a complex interplay of environmental conditions, governance structures, institutional legacies, and economic transitions. Unlike many cities in the Global North, where urbanisation is relatively stable and governed by robust regulatory systems (Mahtta et al., 2019), Southeast Asian cities often contend with postcolonial development, weak governance, and extractive systems that have resulted in unique urbanisation trajectories for each country.

On one end of the urbanisation spectrum, Singapore is fully urbanised while Cambodia remains predominantly rural, with just about a quarter considered as urban infrastructure (UN DESA, 2018). Malaysia represents one of the most urbanised middle-income countries in the region, with approximately 78% of its population residing in urban areas (Lechner et al., 2020). Indonesia has the largest population at 271 million (Badan Pusat Statistik, 2021). The urban agglomeration around its capital city, Jakarta, comprises the Daerah Khusus Ibukota (Capital Special Region) and the Jabodetabekjur (metropolitan area), which has 11 and 30 million residents, respectively (Badan Pusat Statistik, 2021). Similar urban concentrations are evident in Bangkok and Manila, both of which have surpassed the ten million population threshold, while other urban centres such as Greater Kuala Lumpur and Ho Chi Minh City are rapidly approaching megacity status. These urban cores additionally promote the growth of nearby urban areas, resulting in a broader regional trend where development is concentrated around dominant city-regions (Suzuki, 2019).

Economic growth has been a central driver of urbanisation across Southeast Asia. The region has experienced some of the highest rates of rural-to-urban migration globally (Hugo, 2019). While this has expanded the urban labour markets, produced a more substantial middle class, and catalysed measurable improvements in poverty alleviation, healthcare access, educational attainment, and sanitation infrastructure, it also poses acute challenges for urban governance, environmental and human health.

Box: What defines a city, and why it matters

Typically, cities can be defined by their population size, infrastructure or services, or administrative boundaries (Eurostat, 2021; UN DESA, 2018). The definition of a city has profound implications for governance, urban planning and infrastructure development, such as land use, density, zoning, and transport networks.

Definitions that account for functional urban areas are especially valuable to foster integrated and sustainable spatial planning in ways that capture the socio-economic interdependencies between urban cores and surrounding areas. Similarly, definitions based on ecological or density criteria can help identify areas (and populations) most vulnerable to environmental impacts and inform the prioritisation of adaptation strategies.

Legal and administrative definitions determine the jurisdictions within which authorities operate, including regulatory power and planning mandates. They influence how policies are implemented across different scales, and how governments coordinate responses to urban challenges. Relatedly, infrastructure development is also directly linked to urban definitions. The allocation of public spending on transportation, water systems, energy grids, and social services often depends on whether an area qualifies as urban. Definitions based on density or connectivity can facilitate infrastructure that is both efficient and equitable, by aligning investment with population needs.

2 Challenges in Southeast Asian Cities

The rapid urban transformation unfolding across Southeast Asia represents one of the most dynamic yet challenging frontiers in global development. While cities in the region are engines of economic growth and innovation, the rate and scale of urban expansion have outpaced the adaptive capacity of institutions, infrastructure, and ecosystems. Conventional definitions of rural to urban transitions are now challenged as new types of urban forms appear, such as megacities like Jakarta and Manila, which are sprawling and polycentric, with fluid boundaries. This has created a highly uneven urban landscape marked by environmental degradation and loss, infrastructural strain, widening socio-economic inequalities, heightened vulnerability to climate change and public health crises, which we elaborate on below.

Environmental issues. One of the most pressing consequences of this rapid urbanisation is environmental degradation and loss. Urban expansion across Southeast Asia has often proceeded through the conversion of ecologically significant areas (e.g. tropical forests, natural rivers) into residential, industrial, and commercial zones. These land-use changes have resulted in the loss of habitat and biodiversity (Sodhi et al., 2010; Zhao et al., 2006) and undermine the provision of critical ecosystem services such as water regulation, air purification, and temperature moderation (McDonald et al., 2011). Cities across Southeast Asia now rank among the most polluted globally (Sakti et al., 2023). Industrial emissions, vehicular traffic, and poorly managed waste systems contaminate air, water, and soil quality. These changes are not merely environmental in nature; they also constitute a significant and growing public health crisis.

Biodiversity loss is of particular concern in Southeast Asia. As one of the world's most biodiverse regions, Southeast Asia supports over 50,000 recorded species—a number that continues to rise as new discoveries are made (Raven et al., 2020). Yet, urban development poses a particularly acute anthropogenic threat to this rich biological heritage (Tan et al., 2022). Species are increasingly displaced or imperilled by habitat loss and degradation, noise pollution, and intensified human–wildlife conflict, raising concerns about long-term ecological resilience. In Singapore, over 95% of original forest cover has been cleared since the early 1800s for agriculture, housing, and infrastructure (Corlett, 1991, 1992). Habitat loss and fragmentation remain the primary drivers of species decline and extinction. Since 1819, Singapore has lost approximately 37% of its native species, including local extirpations of charismatic species such as the Malayan tiger and wild water buffalo (Chisholm et al., 2023). The introduction of invasive species adds further pressure, with the potential to outcompete native species for resources, and introduce novel pathogens (Yeo & Chia, 2010). One example is the freshwater crayfish, *Cherax quadricarinatus*, introduced to Singapore through commercial aquaculture and aquarium trade, and which competes with the native, tropical lowland freshwater crab, *Parathelphusa maculata* (Zeng et al., 2019).

Urban infrastructure and services. The region's urban infrastructure has come under acute strain. Despite impressive economic growth, the rate of infrastructure development has lagged population growth (Hamel & Tan, 2022). In many cities, essential services such as clean water, sanitation, electricity, and public transportation remain inadequate, unreliable, or entirely absent. The consequences of this shortfall are most visible in flood-prone districts, where poorly planned expansion and the disruption of natural drainage systems have exacerbated both flood risk and water scarcity (Kiguchi et al., 2021). For example, flooding characteristics in Bangkok have changed from previously high-depth, long-duration events to more frequent flashfloods that are shallower in depths and shorter durations (Kiguchi et al., 2021; Worawiwat et al., 2021).

Such infrastructural deficits are particularly pronounced in informal settlements, which now house an estimated 80 million people across the region (United Nations Human Settlements Programme (UN-HABITAT), 2012). In these areas, residents contend with insecure tenure, overcrowded living conditions, and severely limited access to clean water, sanitation, healthcare, and education. As a result, urban growth has become a key driver of social inequality in cities. Formal housing markets remain inaccessible to many, while speculative development that is often financed by international capital and oriented towards luxury real estate, continues to proliferate. For example, Jakarta has seen a proliferation of "new cities" driven by the private sector, typically targeting middle to high-income residents (Firman & Fahmi, 2017; Herlambang et al., 2019). This market-driven urban expansion has displaced long-term residents, intensified housing shortages, and exacerbated socio-economic inequalities.

Climate hazards. Cities in Southeast Asia are increasingly situated at the frontlines of climate change because of (i) geographic location, often in low-lying coastal or riverine areas, or in tropical climates; and (ii) high population densities and extensive informal settlements. This makes them highly vulnerable to climate-related events such as sea-level rise, extreme heat, and intensified storm events. Jakarta, for example, has experienced land subsidence rates of up to 20 cm per year, driven by groundwater extraction, rapid urbanisation, and geological instability (Takagi et al., 2021). Similar processes including oil and gas extraction and sediment compaction from urbanisation are affecting other coastal cities across the region (Tay et al., 2022). As climate change accelerates, sea-level rise, land subsidence, and extreme weather events are likely to jointly reshape urban geographies, disrupt infrastructure and displace communities to create a cascade of socio-economic consequences (Rosenzweig et al., 2018).

Human health and well-being. Consequently, these modern, urban lifestyles have produced significant health and well-being pressures. Shaped by sedentary routines, chronic exposure to air and noise pollution, and the widespread consumption of ultra-processed foods, urban lifestyles are contributing to a rising burden of non-communicable diseases such as cardiovascular conditions, respiratory disorders, diabetes, and mental health challenges (Adhikari et al., 2019; Goryakin et al., 2017). In Indonesia, dietary transitions have accelerated in urban areas and areas with high tree cover loss (Nurhasan et al., 2024). Communities consume greater quantities of wheat, chicken, fish, sugar-sweetened beverages, and ultra-processed foods, with a corresponding decline in fresh, leafy greens, and legumes. These changes have resulted in reduced overall dietary diversity and increased environmental footprints, as much of the food are not sourced locally (Nurhasan et al., 2024). More crucially, ultra-processed, animal-based diets are associated with higher risk of cancer and cardiometabolic diseases (Cordova et al., 2023), and over time, may substantially increase national healthcare burden and costs.

Beyond physical health, an estimated one in seven individuals in Southeast Asia is affected by mental health conditions, with depression, anxiety, and stress disorders being the most prevalent (WHO, 2023). This amounts to approximately 260–289 million people. The region accounts for 27% of global depression cases and

23% of global anxiety disorders. While the primary drivers are socio-economic, such as poverty and job insecurity, these challenges are exacerbated by large treatment gaps (up to 90% in some countries), and chronic underinvestment in mental health infrastructure (Szücs et al., 2025; WHO, 2023). Increasingly, environmental crises, including extreme climate events, displacement, and ecological degradation, are amplifying these pressures, creating a compounding burden on already fragile systems (Reyes et al., 2023). These health outcomes are also sometimes unevenly distributed, with vulnerable populations facing the most severe consequences due to limited access to healthcare and poor living conditions. A recent global study of 2300 cities across the Global South found that patterns of urbanisation and land degradation have markedly increased heat stress, further entrenching environmental injustices (Du et al., 2025).

Taken together, these interlinked biophysical and human health challenges demand a fundamental remake of urban development in Southeast Asia. The path forward must be grounded in integrated, equity-focused planning approaches that prioritise resilience, inclusivity, and ecological integrity. Without such a paradigm shift, the urban future of Southeast Asia will remain precarious, with consequences that extend far beyond the region itself.

3 Systems or Telecoupling Approaches to Local–Global Interdependencies

The bidirectional flows of resources, services, and influences between urban and rural areas are essential for their development and sustainability. Cities depend heavily on rural areas for food, water, energy, raw materials, and labour, while rural communities depend on cities to provide markets, infrastructure, and capital. Cities also act as service hubs for rural populations, particularly in education, healthcare, and digital connectivity.

However, these urban–rural interdependencies are frequently asymmetrical, and the relationships are not always equitable. Despite occupying just 3% of the Earth's surface, cities account for approximately 70% of global resource use and greenhouse gas emissions (Pacione, 2009). Cities often externalise their environmental costs to rural areas, through waste dumping, industrial runoff, or land appropriation (Hardoy et al., 2013). For example, the expansion of solar farms, hydropower projects, or biofuel plantations into rural regions, often justified by the urban shift to renewable energy, has displaced rural communities and threatens biodiversity (Kramarz et al., 2021). Urban consumption patterns, characterised by a high demand for processed products and energy, additionally contribute to global supply chain pressures that intensify environmental degradation elsewhere (Hu et al., 2016). For example, China's urban dietary transition has increased the demand for water-intensive crops in regions such as Xinjiang, impacting the sustainable use of water as a natural resource (Liu et al., 2023).

The urban transition in Southeast Asia encompasses a complex, often paradoxical interplay between economic growth, environmental transformation and human health and well-being. Efforts to advance one domain frequently affect others, and sometimes in unintended ways as these interrelationships exist across spatial and temporal scales. As such, adopting a system or telecoupling approach could help to untangle some of these relationships.

Systems approach. It advocates for the analysis of (any) system as a whole, while attending to the components of that system, and how they interact with each other (Waltner-Toews & Kay, 2005). Systems can also be open or close; but most systems are open as they require external inputs, are constantly influenced by external inputs, and generate outputs that shape broader environments around them. Systems can be linear or complex and adaptive. A simple system comprises linear interactions between components that lead to a predictable outcome. In contrast, a complex and adaptive system is one where complexity emerges from a small set of critical processes that create and maintain self-organising properties, such that even if individual components are disrupted, the system can maintain its overall function (Holling, 2001).

In urban contexts, a systems approach allows us to assess, for example, how a single intervention such as investments in street trees can influence environment quality, infrastructure demands, socio-economic conditions, and climate and health vulnerabilities. Using this approach, Alvarado et al. (2024) considered the multiple pathways linking street trees to mental health outcomes. Through using causal loop diagrams, they captured feedbacks, unintended consequences and non-linear interactions within city boundaries (Figure 1a), illustrating how systems thinking can move beyond siloed interventions to inform integrated, cross-sectoral urban designs.

The value of a systems approach extends beyond urban greening to more complex socio-environmental challenges such as the interplay between climate change and mental health in the Philippines. Ranked among the most climate-vulnerable countries in the Global Climate Risk Index 2021 (Eckstein et al., 2018), the Philippines experiences frequent extreme weather events such as cyclones, alongside slower-onset hazards such as accelerated sea-level rise (USAID, 2018). These exposures are directly associated with adverse mental health outcomes, including post-traumatic stress disorder, depression, and anxiety disorders (Crandon et al., 2022). Such exposure indirectly amplifies mental health risks by disrupting livelihoods, creating food and water insecurities, and eroding cultural and place-based identities. These chronic stressors are particularly acute for communities dependent on climate-sensitive resources and who have limited access to mental healthcare (Tito et al., 2024).

Together, these examples illustrate the complex and interconnected nature between biophysical environments and human health and well-being, and how application of a systems approach could develop strategies to safeguard both human–nature systems.

Telecoupling framework. The conventional notion of cities as closed, self-contained system is increasingly inadequate to capture the global reach of coupled processes. Definitions of "urban" refer less to physical boundaries, and more by the global flows and interactions that shape processes (Friis & Nielsen, 2019). As

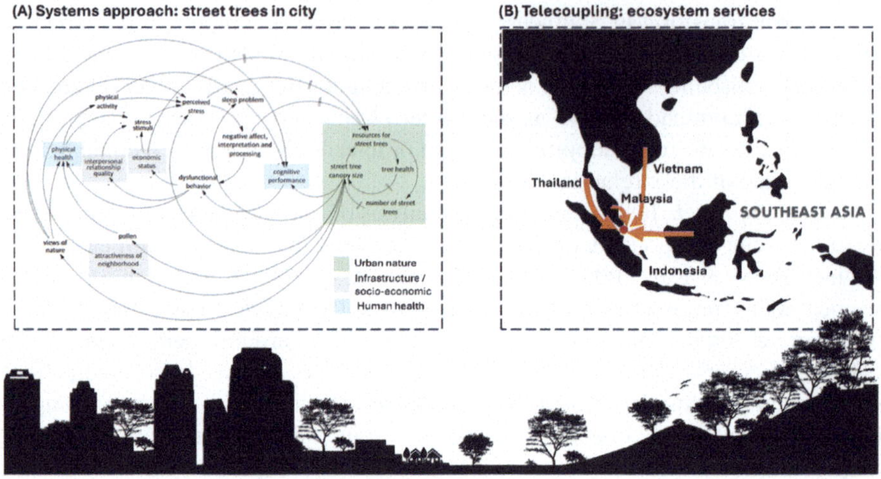

Fig. 1 (a) A *system-based approach* depicting the pathways through which street trees, as a component of urban nature, can influence mental health through interconnected infrastructure and socio-economic pathways. Causal loop diagrams are used to visualise these linkages, with grey nodes representing infrastructure and socio-economic factors, and blue nodes representing (human) physical and mental health outcomes. Arrows indicate the direction of relationships; unless otherwise specified, relationships are positively correlated. Adapted from Alvarado et al. (2024). (b) Application of the *telecoupling framework* to illustrate transboundary flows of ecosystem services associated with Singapore's food imports. Singapore imports 90% of its food products from 187 countries (Singapore Food Agency, 2024), effectively drawing on the production ecosystem services generated beyond its national boundaries. In 2022, key regional contributors included Malaysia (1.94%), Thailand (3.12%), Indonesia (3.29%) and Vietnam (2.02%) (WITS, 2022), reflecting the spatially distributed nature of resource dependencies. Orange arrows represent both direction and relative magnitude of imports

such, the telecoupling framework (Liu et al., 2019) presents a powerful conceptual and analytical approach to examine these long-distance linkages between human and natural systems. Here, urban land teleconnections capture "the distal flows and connections of people, economic goods and services, and land-use change processes that drive and respond to urbanisation" (Seto et al., 2012). When integrated with the concept of *urbanity* (i.e. the magnitude and qualities of livelihoods, lifestyles, connectivity, and place that create urban-ness of intertwined human experiences and land configurations; Boone et al., 2014), this framework reframes the city not just as a physical location, but as a dynamic process connected to distant places, economics and societies.

The telecoupling approach can be applied to examine the loss of species' habitats because of producing export-oriented commodities such as timber, soy, and palm oil (Wiebe & Wilcove, 2025). These habitat losses can occur thousands of kilometres from urban centres and their demand, highlighting the distal yet direct influence of urban lifestyles on ecological systems elsewhere. For example, Singapore, a small, densely populated city state, exerts a disproportionately large environmental footprint through its demand for agricultural and forestry products produced elsewhere in Asia. Such consumption has been linked to substantial deforestation and biodiversity loss in producer countries (Hoang & Kanemoto, 2021; Wiebe & Wilcove, 2025). Another case in point is the palm oil sector in Indonesia and Malaysia. The dramatic expansion of the palm oil industry, driven by rising global demand for edible oils, processed food ingredients (e.g., in countries such as India and Pakistan), and biofuels (e.g., in Indonesia and, previously, the European Union), has had profound consequences for tropical forest ecosystems. The conversion of diverse rainforests into monoculture oil palm plantations has led to significant biodiversity loss, carbon emissions, and a host of socio-political challenges, including land conflicts and the marginalisation of Indigenous and local communities (Meijaard et al., 2020). Beyond deforestation, pollution and waste generated in cities also have transboundary effects, either through physical flows of contaminants or through trade in waste management. For instance, plastic waste from Europe is frequently outsourced to countries such as Vietnam, where lower operational and labour costs make recycling more economically viable (Thapa et al., 2024).

Framing cities as receiving or sending systems within a telecoupled framework enables the tracing of local–global flows linking human and natural systems. Such a perspective illuminates the roles of various stakeholders, such as governments, multinational corporations, local producers, and consumers at both ends of these commodity flows. It also facilitates the assessment of how shifts in governance, corporate policy, or consumer behaviour might improve the socio-ecological outcomes of urban consumption to be sustainable. However, responding to these challenges will require governance mechanisms that extend beyond territorial jurisdictions to engage directly with transboundary flows of goods, information, and authority. In the oil palm sector, flow-based governance approaches in the form of sustainable palm oil initiatives have arisen. Certification schemes like the Roundtable on Sustainable Palm Oil (RSPO), corporate commitments to No Deforestation, No Peat, No Exploitation (NDPE) policies, direct sourcing models, and increasingly, regulatory instruments such as the EU Deforestation Regulation (EU-DR), represent different but complementary strategies for embedding sustainability into global supply chains. These mechanisms, when analysed through the lens of urban telecouplings, reveal the potential and limitations of distant urban actors to influence land-use decisions and environmental outcomes far beyond their borders.

4 Dilemma of Green Spaces in Urban Development

Despite growing recognition of the multiple benefits from urban nature, Southeast Asian cities continue to underinvest in nature-based solutions. This disconnect could result from economic, institutional, and epistemological reasons.

As Southeast Asian cities grapple to accommodate population growth, much of the policy and fiscal attention remain focused on "grey" infrastructure such as roads, housing, energy, and sanitation. Management of natural resources and urban greenery

Fig. 2 Aerial view of densely urbanised Singapore, with the historic Singapore River in the foreground and the high-density, high-revenue financial district situated on reclaimed land in the background. This area exemplifies intensive urban development and land-use transformation in a compact city-state

is often treated as ancillary (Arfanuzzaman & Bharat, 2019). On financing sustainable cities in Southeast Asia, OCED reported that fiscal constraints and political pressure to deliver revenue-generating projects often lead municipal budgets to deprioritise investments in urban nature (OECD, 2025). Here, Singapore represents an exception. Perhaps because of a limited natural hinterland, the country has pursued a deliberate *City in a Garden*, and now *City in Nature* strategy that encompasses a substantial investment on urban nature (Fig. 2).

Urban planning in many Southeast Asian cities remains constrained by outdated regulations, fragmented governance, and limited enforcement (Lourdes et al., 2021). Legal definitions of urban nature remain narrow or ambiguous, and environmental agendas are siloed from economic and spatial planning. Additionally, issues of land tenure, and the lack of urban biodiversity strategies create further barriers for public and private investment. A deeper paradox lies in the perception that countries with expansive hinterlands can always rely on them for uses such as agroforestry, recreation, or conservation. This assumption leads to the systematic neglect of nature in urban areas, reinforcing patterns of dense development with limited access to nature within city boundaries.

Crucially, investments in nature are costly, and their benefits are difficult to quantify as they tend to deliver outcomes that are long-term, diffuse and not easily monetised within conventional cost-benefit frameworks (Chelli et al., 2025). The delayed and frequently intangible nature (of both actual and perceived) returns reduce their appeal to short- and medium-term investors. This challenge is compounded by the fact that many benefits (from nature) manifest indirectly through coupled human–natural system interactions, making attribution and valuation methodologically complex. For example, while public health gains from cleaner air and water are challenging to express in terms of healthcare cost savings, they can be estimated using valuation methods that draw on the best available data (Buckley et al., 2019). This includes metrics such as quality-adjusted life years, and cost-based proxies such as lost manpower hours or caregiving expenditures (Buckley et al., 2019). However, even such metrics rarely capture the full breadth of nature's contributions to human health and well-being. This valuation gap may help explain why global investment in nature remains underdeveloped or poorly aligned with environmental priorities: between 2018 and 2023, global nature spending declined, and fewer than 40% major cities maintained biodiversity action plans (World Economic Forum (WEF), 2025). In Southeast Asia, the shortfall is more pronounced, with biodiversity funding often accounting for less than 0.1% of national budgets (World Economic Forum (WEF), 2025). This has led to contributions from nature being excluded from planning logic, while simultaneously marginalising the cultural, emotional, and recreational values that green spaces hold for diverse urban communities. This systemic undervaluation sustains a cycle in which what cannot be quantified is disregarded, and what is disregarded is not protected.

5 Nature as Essential Urban Infrastructure

In the twenty-first century, integrating nature into cities must go beyond aesthetics. Nature must be treated as vital in urban systems and align with the *Nature Positive* framework. Advocated by the World Economic Forum and Global Biodiversity Framework (Targets 12, 14; Convention on Biological Diversity (CDB), 2025), they call for actions that simultaneously enhance biodiversity and ecosystem integrity while benefiting people. A compelling approach is through the ecosystem services, or nature's contributions to people approach (Díaz et al., 2018). Stakeholders can spatially map and model how different types of nature, such as wetlands or street-level tree canopies, can provide overlapping or distinct services such as urban heat mitigation, stormwater regulation, biodiversity support, and human health benefits, and their synergies and trade-offs (Gillerot et al., 2025; Pienkowski et al., 2024). At the landscape scale, such models help identify areas of concentrated ecosystem services and risk, towards informing land-use decisions for improved urban planning efficiency. Such models can also reveal environmental injustices. By mapping ecosystem services, one can identify inequities in access to clean air, shade, safe green spaces, and climate resilience (Hamel et al., 2021). It could also identify where nature-based investments might have unintentionally widened inequities if benefits accrue disproportionately to wealthier communities.

In Southeast Asia, approaches to integrate nature as essential urban infrastructure must be adaptable and tailored to the diverse urban forms (e.g., new towns, satellite cities; Shaikh & Hamel, 2023). Planned cities offer strategic opportunities to embed nature at the onset (sensu Moser & Côté-Roy, 2021), effectively treating these developments as "blank slates" where best practices in environmentally sensitive urban planning strategies can be implemented. For example, the compact and transit-oriented development of Maja in Banten province (Indonesia) by Ciputra Group, promotes economic growth while limiting urban sprawl (BAPPENAS, 2020). Similarly, the Singapore-based City Developments Limited has integrated the UN Sustainable Development Goals since 1995 through low-carbon and energy-efficient technologies, and nature-sensitive designs (City Developments Limited, 2024). CDL demonstrates how the private sector can drive climate action through establishing public climate education programmes and implementing climate resilience initiatives such as the regenerative tropical microForest to reduce urban heat across their commercial properties (City Developments Limited, 2024). Bottom-up efforts are equally critical. Across Southeast Asia, a groundswell of community-driven initiatives has emerged. These include community gardens, constructed wetlands, and infiltration systems designed to enhance flood control, food security, and improve sanitation (Wolff et al., 2023). In the Philippines, local communities play key roles in mangrove forest rehabilitation and protection, advancing both biodiversity and climate mitigation (Gevaña et al., 2019).

Beyond learning from their city-centric experiences, cities must learn from one another through regional and transboundary collaborations such as C40, Resilient Cities Network (R-Cities), and Local Governments for Sustainability (ICLEI). C40 connects nearly 100 city mayors, and membership is earned and retained by demonstrating evidence-based, collaborative and sustained actions that build healthy, equitable, and climate-resilient communities. R-Cities continues the legacy of 100 Resilient Cities. It supports Chief Resilience Officers across more than 100 cities to develop and implement resilience strategies against growing climate volatility and urban vulnerability. Lastly, the ICLEI represents over 2500 member local and regional governments to bridge the gap between policy ambition and practical delivery. It supports capacity-building, technical support, and scalable knowledge exchange to advance sustainability, climate mitigation and adaptation, and nature-based solutions.

Together, these diverse approaches highlight that "nature in cities" must be understood across multiple types, scales and contexts, and in ways that prioritise both urban development with environmental sustainability and human health (Rauf et al., 2024).

6 Conclusion: Pathways to Resilient and Sustainable Urban Futures

To realise a future where nature is integral to Southeast Asian cities, we must fundamentally rethink how we value, plan, and equitably allocate nature-based infrastructure and resources. The diversity of SEA cities from rapidly urbanising centres to dense megacities renders many Global North models inadequate or inapplicable.

Importantly, nature and natural processes must form the foundation and core of city-making. This requires collaboration across sectors and scales, from local communities to national, regional and global city networks. Urban development must account for distal ecological footprints and transboundary natural resource flows. Encouraging examples from SEA include Singapore's integration of ecological corridors and Manila's community-led forest rehabilitation and protection initiatives. Moving forward, evidence-based yet context-sensitive planning approaches will be vital to enable SEA cities to build urban environments where nature is a pivotal, shared and structuring asset for human resilience, health, and equity.

References

Adhikari, B., Pokharel, S., & Mishra, S. R. (2019). Shrinking urban greenspace and the rise in non-communicable diseases in South Asia: An urgent need for an advocacy. *Frontiers in Sustainable Cities, 1*, 5.

Alvarado, M., Garrett, J., Fullam, J., Lovell, R., Guell, C., Taylor, T., et al. (2024). Using causal loop diagrams to develop evaluative research propositions: Opportunities and challenges in applications to nature-based solutions. *System Dynamics Review, 40*(4), e1756.

Arfanuzzaman, M., & Bharat, D. (2019). Sustainable urbanization in Southeast Asia and beyond: Challenges of population growth, land use change, and environmental health. *Growth and Change, 50*, 725–744.

Badan Pusat Statistik (BPS). (2021). *Badan Pusat Statistik* [WWW Document]. https://www.bps.go.id/

BAPPENAS. (2020). *Rencana Pembangunan Jangka Menegah Nasional 2020–2024.*

Boone, C. G., Redman, C. L., Blanco, H., Haase, D., Koch, J., Lwasa, S., et al. (2014). *Reconceptualizing land for sustainable urbanity* (pp. 313–330). Rethinking urban land use in a global era. MIT Press.

Buckley, R., Brough, P., Hague, L., Chauvenet, A., Fleming, C., Roche, E., et al. (2019). Economic value of protected areas via visitor mental health. *Nature Communications, 10*(1), 5005.

Chelli, A., Brander, L., & Geneletti, D. (2025). Cost-benefit analysis of urban nature-based solutions: A systematic review of approaches and scales with a focus on benefit valuation. *Ecosystem Services, 71*, 101684.

Chisholm, R. A., Kristensen, N. P., Rheindt, F. E., Chong, K. Y., Ascher, J. S., Lim, K. K., et al. (2023). Two centuries of biodiversity discovery and loss in Singapore. *Proceedings of the National Academy of Sciences, 120*(51), e2309034120.

City Developments Limited. (2024). *Integrated sustainability report 2024.* Retrieved May 15, 2025, from https://cdlsustainability.com/pdf/CDL_ISR_2024.pdf

Convention on Biological Diversity (CDB). (2025). *Kunming-Montreal Global Biodiversity Framework Targets 1-23.* Retrieved July 15, 2025, from https://www.cbd.int/gbf/targets

Cordova, R., Viallon, V., Fontvieille, E., Peruchet-Noray, L., Jansana, A., Wagner, K. H., … & Freisling, H. (2023). Consumption of ultra-processed foods and risk of multimorbidity of cancer and cardiometabolic diseases: A multinational cohort study. *The Lancet Regional Health–Europe*, 35.

Corlett, R. T. (1991). Vegetation. In L. Chia, R. Ausafur, & D. Tay (Eds.), *The biophysical environment of Singapore* (pp. 134–154). Singapore University Press.

Corlett, R. T. (1992). The ecological transformation of Singapore, 1819-1990. *Journal of Biogeography*, 411–420.

Crandon, T. J., Dey, C., Scott, J. G., Thomas, H. J., Ali, S., & Charlson, F. J. (2022). The clinical implications of climate change for mental health. *Nature Human Behaviour, 6*(11), 1474–1481.

Dalin, C., Konar, M., Hanasaki, N., Rinaldo, A., & Rodriguez-Iturbe, I. (2012). Evolution of the global virtual water trade network. *Proceedings of the National Academy of Sciences, 109*(16), 5989–5994.

Díaz, S., Pascual, U., Stenseke, M., Martín-López, B., Watson, R. T., Molnár, Z., et al. (2018). Assessing nature's contributions to people. *Science, 359*(6373), 270–272.

Du, H., Zhan, W., Zhou, B., Ju, Y., Liu, Z., Middel, A., et al. (2025). Exacerbated heat stress induced by urban browning in the global south. *Nature Cities, 2*(2), 157–169.

Eckstein, D., Hutfils, M. L., & Winges, M. (2018). *Global climate risk index 2021 who suffers Most from extreme weather events? Wether-related loss events in 2019 and to 2000-2019*. Germanwatch. https://bvearmb.do/handle/123456789/1306

Eurostat. (2021). *Applying the degree of urbanisation: A methodological manual to define cities, towns and rural areas for international comparisons*. European Union.

Firman, T., & Fahmi, F. Z. (2017). The privatization of metropolitan Jakarta's (Jabodetabek) urban fringes: The early stages of "post-suburbanization" in Indonesia. *Journal of the American Planning Association, 83*(1), 68–79.

Friis, C., & Nielsen, J. Ø. (Eds.). (2019). *Telecoupling: Exploring land-use change in a globalised world*. Springer.

Gevaña, D. T., Pulhin, J. M., & Tapia, M. A. (2019). Fostering climate change mitigation through a community-based approach: Carbon stock potential of community-managed mangroves in The Philippines. In *Coastal management* (pp. 271–282). Academic Press.

Gilbert, M., Bickford, D., Clark, L., Johnson, A., Joyner, P. H., Ogg Keatts, L., et al. (2012). Amphibian pathogens in southeast Asian frog trade. *EcoHealth, 9*(4), 386–398.

Gillerot, L., Landuyt, D., Bourdin, A., Rozario, K., Shaw, T., Steinparzer, M., et al. (2025). Forest biodiversity and structure modulate human health benefits and risks. *Nature Sustainability*, 1–13.

Goryakin, Y., Rocco, L., & Suhrcke, M. (2017). The contribution of urbanization to non-communicable diseases: Evidence from 173 countries from 1980 to 2008. *Economics & Human Biology, 26*, 151–163.

Hajer, M., Pelzer, P., van den Hurk, M., Ten Dam, C., & Buitelaar, E. (2020). *Neighbourhoods for the future: A plea for a social and ecological urbanism*. Trancityxvaliz with Urban Futures Studio, University Utrecht.

Hamel, P., Guerry, A. D., Polasky, S., Han, B., Douglass, J. A., Hamann, M., et al. (2021). Mapping the benefits of nature in cities with the InVEST software. *Npj Urban Sustainability, 1*(1), 25.

Hamel, P., & Tan, L. (2022). Blue–green infrastructure for flood and water quality management in Southeast Asia: Evidence and knowledge gaps. *Environmental Management, 69*(4), 699–718.

Hardoy, J. E., Mitlin, D., & Satterthwaite, D. (2013). *Environmental problems in an urbanizing world: Finding solutions in cities in Africa, Asia and Latin America*. Routledge.

Herlambang, S., Leitner, H., Tjung, L. J., Sheppard, E., & Anguelov, D. (2019). Jakarta's great land transformation: Hybrid neoliberalisation and informality. *Urban Studies, 56*(4), 627–648.

Hoang, N. T., & Kanemoto, K. (2021). Mapping the deforestation footprint of nations reveals growing threat to tropical forests. *Nature Ecology & Evolution, 5*(6), 845–853.

Holland, D. S., & Ginter, J. J. (2001). Common property institutions in the Alaskan groundfish fisheries. *Marine Policy, 25*(1), 33–42.

Holling, C. (2001). Understanding the complexity of economic, ecological, and social systems. *Ecosystems, 4*, 390–405. https://doi.org/10.1007/s10021-001-0101-5

Hu, Y., Lin, J., Cui, S., & Khanna, N. Z. (2016). Measuring urban carbon footprint from carbon flows in the global supply chain. *Environmental Science & Technology, 50*(12), 6154–6163.

Hugo, G. (2019). Patterns and trends of urbanization and urban growth in Asia. In *Internal migration, urbanization and poverty in Asia: Dynamics and interrelationships* (pp. 13–45). Springer Singapore.

Katherina, L. K., & Indraprahasta, G. S. (2019). Urbanization pattern in Indonesia's secondary cities: Greater Surabaya and its path toward a megacity. In IOP conference series: Earth and environmental science 338, 1, p. 012018. IOP Publishing.

Kiguchi, M., Takata, K., Hanasaki, N., Archevarahuprok, B., Champathong, A., Ikoma, E., et al. (2021). A review of climate-change impact and adaptation studies for the water sector in Thailand. *Environmental Research Letters, 16*(2), 023004.

Kramarz, T., Park, S., & Johnson, C. (2021). Governing the dark side of renewable energy: A typology of global displacements. *Energy Research & Social Science, 74*, 101902.

Lechner, A. M., Ang, M. L. E., Ooi, J. Y., Azhar, B., Kanai, J. M., Hamel, P., & Sagala, S. (2021). *Urban biodiversity and nature-based solutions in Southeast Asia: Perspectives from Indonesia and Malaysia* (No. 20). ISEAS-Yusof Ishak Institute.

Lechner, A. M., Gomes, R. L., Rodrigues, L., Ashfold, M. J., Selvam, S. B., Wong, E. P., et al. (2020). Challenges and considerations of applying nature-based solutions in low-and middle-income countries in southeast and East Asia. *Blue-Green Systems, 2*(1), 331–351.

Liu, J., Herzberger, A., Kapsar, K., Carlson, A. K., & Connor, T. (2019). What is telecoupling? In *Telecoupling: Exploring land-use change in a globalised world* (pp. 19–48). Springer International Publishing.

Liu, L., Hu, X., Zhan, Y., Sun, Z., & Zhang, Q. (2023). China's dietary changes would increase agricultural blue and green water footprint. *Science of the Total Environment, 903*, 165763.

Lourdes, K. T., Gibbins, C. N., Hamel, P., Sanusi, R., Azhar, B., & Lechner, A. M. (2021). A review of urban ecosystem services research in Southeast Asia. *Land, 10*(1), 40.

Lourdes, K. T., Hamel, P., Gibbins, C. N., Sanusi, R., Azhar, B., & Lechner, A. M. (2022). Planning for green infrastructure using multiple urban ecosystem service models and multicriteria analysis. *Landscape and Urban Planning, 226*, 104500.

Mahtta, R., Mahendra, A., & Seto, K. C. (2019). Building up or spreading out? Typologies of urban growth across 478 cities of 1 million+. *Environmental Research Letters, 14*(12), 124077.

McDonald, R. I., Green, P., Balk, D., Fekete, B. M., Revenga, C., Todd, M., & Montgomery, M. (2011). Urban growth, climate change, and freshwater availability. *Proceedings of the National Academy of Sciences, 108*(15), 6312–6317.

Meijaard, E., Brooks, T. M., Carlson, K. M., Slade, E. M., Garcia-Ulloa, J., Gaveau, D. L., et al. (2020). The environmental impacts of palm oil in context. *Nature Plants, 6*(12), 1418–1426.

Moser, S., & Côté-Roy, L. (2021). New cities: Power, profit, and prestige. *Geography Compass, 15*(1), e12549.

Nurhasan, M., Ariesta, D. L., Utami, M. M. H., Fahim, M., Aprillyana, N., Maulana, A. M., & Ickowitz, A. (2024). Dietary transitions in Indonesia: The case of urban, rural, and forested areas. *Food Security, 16*(6), 1313–1331.

OECD. (2025). *Financing sustainable cities in Southeast Asia: Diversifying instruments and leveraging private investment*. OECD Urban Studies, OECD Publishing.

Pacione, M. (2009). *Urban geography: A global perspective*. Routledge.

Pienkowski, T., Keane, A., Booth, H., Kinyanda, E., Fisher, J. C., Lawrance, E., et al. (2024). Nature's contributions to social determinants of mental health and the role of conservation. *One Earth, 7*(7), 1213–1227.

Rauf, H. A., Wolff, E., Natakun, B., Marome, W., & Hamel, P. (2024). Aligning nature-based solutions and housing policy: A study on stakeholders' perceptions of nature within informal settlement upgrading projects. *Cities, 153*, 105264.

Raven, P. H., Gereau, R. E., Phillipson, P. B., Chatelain, C., Jenkins, C. N., & Ulloa Ulloa, C. (2020). The distribution of biodiversity richness in the tropics. *Science Advances, 6*(37), eabc6228.

Reyes, M. E. S., Carmen, B. P. B., Luminarias, M. E. P., Mangulabnan, S. A. N. B., & Ogunbode, C. A. (2023). An investigation into the relationship between climate change anxiety and mental health among gen Z Filipinos. *Current Psychology, 42*(9), 7448–7456.

Rosenzweig, C., Solecki, W., Romero-Lankao, P., Mehrotra, S., Dhakal, S., Bowman, T., & Ibrahim, S. A. (2018). *Climate change and cities: Second assessment report of the urban climate change research network: Summary for City leaders*. Cambridge University Press.

Sakti, A. D., Anggraini, T. S., Ihsan, K. T. N., Misra, P., Trang, N. T. Q., Pradhan, B., et al. (2023). Multi-air pollution risk assessment in Southeast Asia region using integrated remote sensing and socio-economic data products. *Science of the Total Environment, 854*, 158825.

Seto, K. C., Reenberg, A., Boone, C. G., Fragkias, M., Haase, D., Langanke, T., et al. (2012). Urban land teleconnections and sustainability. *Proceedings of the National Academy of Sciences, 109*(20), 7687–7692.

Shaikh, S. F. E. A., & Hamel, P. (2023). Identifying nature-positive futures in new cities: An application of the urban nature futures framework. *Sustainability Science*, 1–12.

Singapore Food Agency. (2024). *Singapore food statistics 2024*. Retrieved August 9, 2025, from https://www.sfa.gov.sg/docs/default-source/publication/sg-food-statistics/singapore-food-statistics-2024.pdf

Sodhi, N. S., Koh, L. P., Clements, R., Wanger, T. C., Hill, J. K., Hamer, K. C., et al. (2010). Conserving southeast Asian forest biodiversity in human-modified landscapes. *Biological Conservation, 143*(10), 2375–2384.

Sugiyarto, G. (2014). Internal and international migration in Southeast Asia. In *Routledge handbook of southeast Asian economics* (pp. 270–299). Routledge.

Suzuki, Y. (2019). Megacities in Southeast Asia–the progression of unipolar concentration and expansion into the suburbs. In *Mitsui & co. global strategic studies institute monthly report* (pp. 1–7).

Szücs, A., van der Lubbe, S. C., de la Torre, J. A., Valderas, J. M., Hay, S. I., Bisignano, C., et al. (2025). The epidemiology and burden of ten mental disorders in countries of the Association of Southeast Asian Nations (ASEAN), 1990–2021: Findings from the global burden of disease study 2021. *The Lancet Public Health, 10*(6), e480–e491.

Takagi, H., Esteban, M., Mikami, T., Pratama, M. B., Valenzuela, V. P. B., & Avelino, J. E. (2021). People's perception of land subsidence, floods, and their connection: A note based on recent surveys in a sinking coastal community in Jakarta. *Ocean & Coastal Management, 211*, 105753.

Tan, Y. L., Chen, J. E., Yiew, T. H., & Habibullah, M. S. (2022). Habitat change and biodiversity loss in south and southeast Asian countries. *Environmental Science and Pollution Research, 29*(42), 63260–63276.

Tay, C., Lindsey, E. O., Chin, S. T., McCaughey, J. W., Bekaert, D., Nguyen, M., et al. (2022). Sea-level rise from land subsidence in major coastal cities. *Nature Sustainability, 5*(12), 1049–1057.

Thapa, K., Vermeulen, W. J., De Waal, M. M., Deutz, P., & Nguyễn, H. Q. (2024). Towards a just circular economy transition: The case of European plastic waste trade to Vietnam for recycling. *Circular Economy and Sustainability, 4*(2), 851–876.

Tito, V. R., Kazem, H., Kadia, S. O., & Paquito, B. (2024). A systematic review of mental health and climate change in The Philippines. *Asian Journal of Psychiatry, 101*, 104191.

United Nations Department of Economic and Social Affairs (UN-DESA). (2018). *2018 Revision of the World Urbanization Prospects*. Accessed 14 April 2025.

United Nations Human Settlements Programme (UN-HABITAT). (2012). *Global Urban indicators database*. Accessed 14 July 2025.

USAID. (2018). *Climate change risk profile: Philippines* (No. AID-OAA-I-14-00013).

Waltner-Toews, D., & Kay, J. (2005). The evolution of an ecosystem approach: The diamond schematic and an adaptive methodology for ecosystem sustainability and health. *Ecology and Society, 10*(1).

WHO. (2023). *Mental health conditions in the WHO South-East Asia region*. Retrieved August 8, 2025, from https://www.who.int/publications/i/item/9789290210788

Wiebe, R. A., & Wilcove, D. S. (2025). Global biodiversity loss from outsourced deforestation. *Nature, 639*(8054), 389–394.

WITS. (2022). *World Integrated Trade Solutions: Singapore 2022 food imports by Country*. Retrieved August 9, 2025, from https://wits.worldbank.org/CountryProfile/en/Country/SGP/Year/LTST/TradeFlow/Import/Partner/by-country/Product/16-24_FoodProd

Wolff, E., Rauf, H. A., & Hamel, P. (2023). Nature-based solutions in informal settlements: A systematic review of projects in southeast Asian and Pacific countries. *Environmental Science & Policy, 145*, 275–285.

Worawiwat, A., Chaleeraktrakoon, C., & Sharma, A. (2021). Is increased flooding in Bangkok a result of rising local temperatures? *Journal of Hydrology X, 13*, 100095.

World Economic Forum (WEF). (2025). *Nature Positive: Financing the transition in cities*. Retrieved July 15, 2025, from https://reports.weforum.org/docs/WEF_Nature_Positive_Financing_the_Tranisition_in_Cities.pdf

Yeo, D. C., & Chia, C. S. (2010). Introduced species in Singapore: An overview. *Cosmos, 6*(01), 23–37.

Zeng, Y., Shakir, K. K., & Yeo, D. C. (2019). Competition between a native freshwater crab and an invasive crayfish in tropical Southeast Asia. *Biological Invasions, 21*(8), 2653–2663.

Zhao, S., Peng, C., Jiang, H., Tian, D., Lei, X., & Zhou, X. (2006). Land use change in Asia and the ecological consequences. *Ecological Research, 21*(6), 890–896.

Nature-Driven Placemaking in Rapidly Urbanizing Landscapes: Exploring Green Space Transformation in Ho Chi Minh City, Vietnam

Thanh Hung Dang, Minh Trang Khong, Viet Anh Vu, and Nigel K. Downes

Abstract As Ho Chi Minh City undergoes rapid urbanization, its green spaces are increasingly under pressure, resulting in significant transformations in land use, ecological functionality, and social value. This chapter examines the evolution of urban green spaces in HCMC through a combination of spatial analysis, historical review, and block-level case studies. We trace the shift from integrated natural systems and colonial-era public parks to fragmented, engineered green infrastructure, and canal encroachments. Particular attention is paid to the interaction between land-use planning, infrastructure development, and socio-economic transitions, revealing systemic patterns of green space decline and disconnection. We propose the concept of nature-driven placemaking—an approach that integrates biophilic design, community engagement, and adaptive governance to foster inclusive and multifunctional urban landscapes. Drawing on examples from HCMC and comparative case studies across Asia, this chapter highlights how nature-based solutions (NbS) can enhance urban resilience, particularly in climate-vulnerable megacities. A strategic framework is proposed for revitalizing historic green assets, rethinking public space design, and embedding ecological priorities into future development. The findings offer insights for planners and policymakers seeking sustainable pathways for green space transformation in fast-growing cities across the Global South.

Keywords Nature-based solutions (NbS) · Urban green space transformation · Placemaking, Rapid urbanization · Sustainable urban development · Ho Chi Minh City

---

T. H. Dang (✉) · M. T. Khong (✉) · V. A. Vu
University of Architecture, Ho Chi Minh City, Viet Nam
e-mail: hung.dangthanh@uah.edu.vn

M. T. Khong
e-mail: trang.khongminh@uah.edu.vn

N. K. Downes
College of Environment and Natural Resources, Can Tho city, Viet Nam

A. Cheshmehzangi et al. (eds.), *Nature-based Urbanism*, Urban Sustainability,
https://doi.org/10.1007/978-981-95-6864-2_5

1 Introduction

Ho Chi Minh City (HCMC), Vietnam's economic engine and most populous urban centre, is undergoing a profound spatial and socio-ecological transformation (Downes et al., 2016). As the city's footprint expands rapidly into low-lying peri-urban zones, the once-integrated mosaic of green and blue infrastructure is giving way to fragmented and engineered urban forms. The resulting tensions—between densification and ecological continuity, economic growth and environmental degradation, real estate speculation and public realm provision—have significant implications for urban resilience and liveability. Nowhere is this more evident than in the city's shifting landscape of green space (Quang, 2023).

Recent landscape urbanism perspectives frame HCMC's low-lying terrain, canal systems, and mangrove reserves as critical ecological assets that are rapidly being lost to urban expansion (Shannon, 2024). Urban green spaces in HCMC have historically served multifunctional roles: cooling tropical streetscapes, absorbing floodwaters, supporting biodiversity, and providing spaces for recreation, cultural practices, and everyday life (Downes, Storch, et al., 2024). However, these roles are increasingly undermined by land conversion, canal encroachment, and governance fragmentation. Despite ambitious national and local agendas promoting sustainable urban development, the degradation of green space—both in quality and quantity—suggests a persistent disconnect between planning visions and implementation on the ground.

This chapter argues that rethinking green space transformation in HCMC requires a shift toward nature-driven placemaking: an approach that foregrounds ecological logic, community participation, and spatial justice in the planning and design of urban green infrastructure. By combining spatial analysis, historical context, and block-level case studies, we explore how both colonial legacies and contemporary urban development have reshaped the city's relationship with nature. Drawing from comparative examples and local innovation, we propose a strategic framework for integrating Nature-based Solutions (NbS) into the evolving urban fabric—not merely as technical fixes, but as cultural and governance interventions that reclaim the role of nature in everyday urban life.

2 Historical Evolution of Green Spaces in HCMC

2.1 Traditional Public Parks and Tree-Lined Boulevards (French Colonial Period)

The spatial morphology of HCMC (formerly Saigon) reflects a colonial logic that actively integrated green infrastructure into the urban fabric as part of a broader project of bioclimatic control, social ordering, and territorial display. During the French colonial period, urban planners—most notably Ernest Hébrard and later

Coffyn—introduced tree-lined boulevards and structured public parks as both functional and symbolic elements within a new civic vision. Designed for a projected population of 500,000, the city plan responded to tropical climatic conditions by mandating the systematic planting of trees at 5-metre intervals along major streets. This green armature served not only to ameliorate the effects of intense solar radiation but also to demarcate spatial hierarchies and control urban circulation.

Archival records and period photography reveal the extent of early afforestation: tamarind trees lined the Charner Canal, while major arteries such as Rue Catinat (now Đồng Khởi Street) featured dense canopies of sao, bang, and mango trees. These boulevards formed part of a deliberate colonial aesthetic that framed nature as an instrument of civility—offering shade, order, and a visual contrast to the perceived disorder of vernacular landscapes. Their enduring physical imprint has contributed to what Lefebvre might describe as a "representational space", where greenery operates both materially and symbolically as an urban ideal.

In parallel, a number of green spaces were formalized as proto-public parks. Among the most notable were the Botanical Garden (Thảo Cầm Viên) and L'Esperance Park, which combined horticultural experimentation with recreational function. These spaces were hybrid in character: they served as sites of botanical knowledge production—housing plant nurseries and exotic specimens—while also functioning as enclaves of leisure for the colonial elite. Their spatial configuration and limited accessibility reflected the racialized and hierarchical structure of the colonial city, where green amenities were distributed unevenly across social groups and neighbourhoods.

The legacy of these early interventions endures in contemporary HCMC, not only in physical remnants but also in planning culture. The use of linear green elements, broad shaded sidewalks, and boulevard typologies continues to shape expectations around urban amenity and climatic adaptation. Yet, the historical context complicates the narrative of green infrastructure as inherently inclusive or democratic. It suggests instead a more ambivalent genealogy—where ecological functionality was often secondary to colonial spatial order and where green spaces were less public than they appeared.

2.2 The Role of Water Surfaces and Canals in Shaping Urban Nature

The inherent fertility and resource richness of this territory are fundamentally linked to its estuarine location. The land's formation, a direct result of fluvial processes and the interconnected network of inland canals, established a hydrographic system conducive to human settlement. The protective function of the Can Gio mangrove forest is critical, acting as both an ecological buffer for the inner city and a transitional zone mitigating marine and anthropogenic pressures. The low-lying alluvial plain, formed by the Saigon River's downstream flow into the estuary, fostered a highly diverse natural water system, with sediment deposition creating a fertile foundation for habitation. While the sea presented avenues for trade, it also posed inherent risks. Consequently, the urban morphology of the city became inextricably tied to the Saigon River and its associated canal system. Tributaries of the Saigon River initially served as primary sources for water supply and centralized trade activities. Recognizing the limitations of solely relying on natural hydrography, successive colonial and post-colonial administrations implemented significant hydraulic engineering projects to expand and optimize the regional waterway network (Khong Minh et al., 2023).

The French colonial administration notably initiated the construction of an extensive canal network to enhance connectivity and stimulate economic growth. In Cholon, a distinct urban centre, the Arroyo Chinois canal system developed from the Thi Nghe Creek, exemplifies this intervention (Minh, 2024). These predominantly east-west-oriented canals forged a crucial waterway link between the western Mekong Delta provinces and the Saigon River. This infrastructure was paramount for the efficient movement of goods, playing a central role in the economic integration and development of the broader region.

The historical "Gia Dinh Geography" (Đức, 2020) provides a compelling depiction of the unique socio-spatial organization of Saigon-Cholon, encapsulated by the phrase "urban above, boats below". This distinct lifestyle was deeply embedded within the city's intricate canal network. The waterways functioned as dynamic arteries of transportation and commerce. Residential patterns along the canal banks facilitated direct waterfront trade. These canals were the lifeblood of the urban environment, not only enabling the circulation of goods but also fostering a vibrant and interconnected community. The historical account underscores the profound influence of these waterways on daily life, with the emergence of floating markets and stilt dwellings creating a fluid urban landscape where aquatic and terrestrial realms were seamlessly integrated. However, this historical hydro-urban identity is rapidly eroding, as canal banks are increasingly privatized, enclosed, or filled, undermining both ecological value and cultural memory.

2.3 The Gradual Loss of Green Infrastructure Due to Land Conversion

HCMC has witnessed a concerning trend of green infrastructure loss driven by relentless land conversion. The escalating population exerts significant pressure on land resources, fuelling urban expansion and encroaching upon vital ecological zones. The rapid population increase in HCMC has been a primary catalyst for widespread land conversion. As the number of residents grows, so does the demand for housing, commercial spaces, and transportation networks. This demographic shift directly translates into the expansion of urban area, often at the expense of existing green spaces and natural habitats. The periphery of the city, once characterized by agricultural land, wetlands, and open spaces, is increasingly being absorbed into the urban fabric to accommodate this burgeoning population.

Furthermore, the pattern of urban development adhering to roads and turning its back on waterways exacerbates the loss of green infrastructure along riparian corridors. Historically, as described earlier, the city's relationship with its canals and rivers was integral to its identity and functioning. However, contemporary development often prioritizes road access, leading to construction that neglects or even obstructs natural waterways and their adjacent green buffer zones. Canal banks, once potentially vibrant green corridors, are increasingly hardened with concrete, losing their natural vegetation and ecological value. This linear form of development fragments natural habitats and diminishes the potential for blue-green infrastructure integration within the urban environment. The escalating demand for land driven by population growth and urban expansion in HCMC has resulted in a tangible decline in green infrastructure. The conversion of natural wetlands into built environments and a development pattern that often disregards its historical connection to waterways are key indicators of this trend (Fig. 5.1).

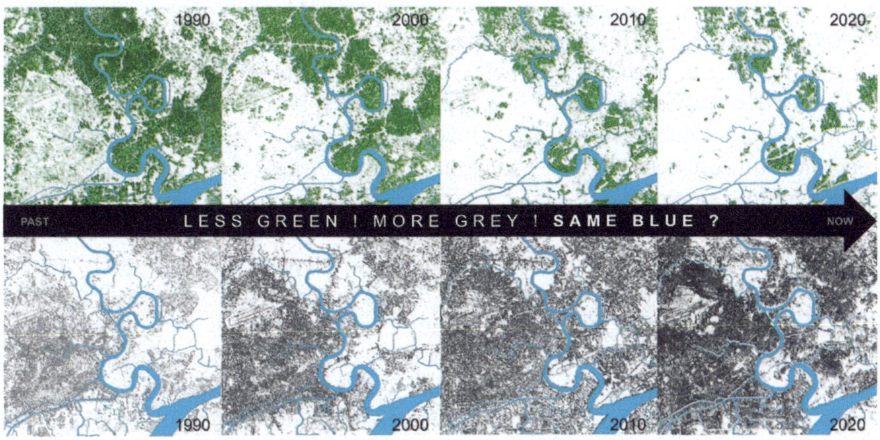

Fig. 5.1 Transition of green, grey, and blue in Ho Chi Minh City from 1990 to 2020 (*Source* by the authors)

3 The Transformation of Urban Nature: A Quantitative and Qualitative Assessment

3.1 Evaluating Changes in Public Parks: Reduction in Size, Accessibility, and Maintenance

The increasing trend of increasing urbanization is a primary driver behind this deterioration. As the city grows, valuable green areas are converted into urban construction land. This direct conversion of natural and semi-natural green spaces into concrete and infrastructure inherently reduces the overall area dedicated to public parks and green spaces within the city. Consequently, existing parks face mounting pressure from the surrounding dense urban fabric, often becoming isolated pockets of green within a predominantly grey landscape.

Furthermore, the planning and distribution of remaining green spaces often fall short in terms of genuine accessibility. While green patches are arranged in the centre of residential areas, with per capita targets, the focus appears to be on numerical targets rather than practical access. The crucial aspect of accessibility is often overlooked. The reality is that opening the door does not lead directly to green space; one has to pass through many urban spaces to reach the park. This spatial disconnect diminishes the usability and perceived value of these green areas for many residents, effectively reducing their accessibility despite their theoretical presence within residential zones. The journey to reach a park can involve navigating congested streets and built-up areas, deterring frequent use and limiting their function as readily available public amenities.

Finally, the maintenance and ecological resilience of these remaining green spaces are increasingly compromised. These green patches depend on operation by irrigation and human care, reducing natural resilience in terms of water resources and biological resources. This reliance on artificial systems for sustenance highlights a departure from more self-sustaining natural ecosystems. The filling of canals, which historically contributed to groundwater recharge and local biodiversity, further exacerbates this issue. Parks that depend solely on human intervention for irrigation are more vulnerable to water scarcity and may lack the natural ecological processes that support biodiversity and long-term health. This dependence also implies a need for consistent and adequate funding and management, which may not always be guaranteed, leading to the degradation of park quality over time.

In conclusion, the evaluation of public parks in HCMC reveals a concerning trend of reduction in size due to land conversion, diminished accessibility despite their intended placement, and a decline in natural resilience due to their dependence on artificial maintenance. Addressing the spatial disconnect and fostering more naturally resilient green infrastructure are crucial steps in reversing this detrimental trend.

3.2 Water Surfaces and Public Spaces: Canal Encroachments, Land Reclamation, and Loss of Biodiversity

The process of urbanization has placed great demands on construction, which means the creation of very large impervious surface areas. At the same time, the domestic activities of a large population in the urban environment have resulted in many natural water components being channelled, piped, and filled. In the 18th and 20th centuries, canals were fixed into the natural landscape, but now due to urbanization this has gradually become a place for garbage collection, pollution, and is losing its urban flow.

The water network is the territorial flow system that helps navigate water supply and drainage, but it is inherent that the canal system is increasingly disappearing. Instead, it is only serving circulation and instant drainage needs on the spot and neglecting congestion on most of the overall stream. With the development of mechanized wastewater and stormwater drainage networks, changes due to a decrease in the permeable surface area for drainage, and changes in water surface and stream morphology, this phenomenon has changed many natural hydrological cycles—this effect can be summed up in the change of stream morphology and vessels becoming broken and volatile.

When technical advancement and economic conditions allow bridges to be built connecting the two riverbanks, the connection between the city and the water surface embarks on a new era of urban development. Likewise, illegal encroachment on rivers and canals is increasingly displacing ecosystems and inevitably has severe consequences together with frequency of floods, not to mention increased degradation and environmental contamination. And, the water's millennial worth of spatial quality is set aside for urgent concerns. Urbanization has changed the urban landscape, altered the urban water system, the balance of water availability and the distribution of water in urban functions.

HCMC has undertaken a series of infrastructural projects aimed at enhancing urban water management, including the completion of flood-control measures and drainage canals, coupled with intensified dredging of sewer and canal networks to optimize drainage capacity. Current and future strategic investments focus on comprehensive dredging, environmental upgrades, and infrastructure development along critical waterways such as the Xuyen Tam Canal (Go Vap and Binh Thanh Districts), with phased implementation. A significant embankment construction is also planned for the Tham Luong—Ben Cat—Nuoc Len Canal (Ho Chi Minh City People's Committee, 2002). Moreover, the city is allocating resources to four key projects involving the renovation and environmental improvement of the Hy Vong, Van Thanh, Ba Lon, and Doi Canals, scheduled for completion within the 2026–2030 timeframe (Khong, 2024).

Contemporary canalization projects in HCMC often incorporate integrated design principles aimed at enhancing both functionality and environmental quality. These include the installation of litter barriers or screens to mitigate solid waste pollution

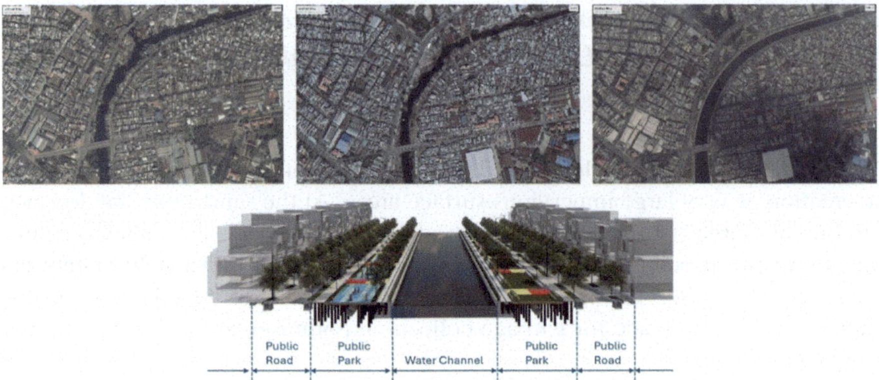

Fig. 5.2 Contemporary canal engineering in Ho Chi Minh City emphasizes canalization strategies aimed at optimizing hydraulic flow and mitigating pollution. These interventions typically include funnel-like geometries that direct water movement and reduce debris accumulation, reflecting a shift toward more controlled and infrastructural approaches to urban water management. 2020 (*Source* by the authors)

within the waterways, complemented by regular maintenance protocols to ensure hydraulic efficiency. Recognizing the critical importance of continuous water flow, the design strategies also integrate features that accommodate natural hydrological fluctuations, thereby enhancing resilience to seasonal variations and mitigating flood risks. The revitalization of canals in HCMC presents a significant opportunity for the strategic integration of blue infrastructure with the urban public realm. The development of canal-adjacent areas into parks, walking trails, and recreational facilities offers valuable green spaces that enhance urban liability. This transformation of formerly neglected or polluted sites into dynamic community nodes underscores the potential of blue-green infrastructure to improve the quality of urban life and necessitates further rigorous examination within professional urbanism discourse (Fig. 5.2).

3.3 Comparing the Quantity and Quality of "Old" Vs. "New" City Nature

Official data from the city illustrate the current inadequacy of park and greenery provision across HCMC (People's Committee of Ho Chi Minh City, 2021). In the densely populated inner core (comprising 13 districts), the existing average of 0.67 m2/person falls significantly short of both the local standard (QCXDVN 01:2020/BXD) (MoC, 2020) recommending at least 2 m2/person within neighbourhoods and a more expansive 7 m2/person outside, as well as the national standard (TCVN 9257: 2012) which suggests a comprehensive 12–15 m2/person. This substantial deficit indicates a severe lack of accessible green spaces for a significant portion of the city's population.

Even in new development areas (spanning 6 districts), where one might expect more contemporary planning principles, the existing 1.72 m2/person still lags behind the recommended minimum of 2 m2/person within neighbourhoods according to QCXDVN 01:2020/BXD (MoC, 2020), and is considerably lower than the more holistic targets outlined in TCVN 9257: 2012 (MoC, 2012). This suggests that even recent urban expansion has not fully prioritized the integration of sufficient green infrastructure. The situation is most critical in the outskirt areas (5 rural districts), where the existing greenery provision is a mere 0.3 m2/person. This negligible figure highlights a profound lack of accessible green spaces for residents in these less central districts, falling drastically below all established benchmarks. Aggregating these figures, the total existing per capita greenery for HCMC stands at 0.55 m2/person, underscoring a city-wide deficiency in providing essential green amenities for its inhabitants (Fig. 5.3).

Comparing both zones with the HCMC General Masterplan to 2025 (Prime Minister of the Government, 2010) further emphasizes the existing deficits and the desired improvements in both quantity and potentially quality. The Masterplan targets of 2.4 m2/person for the inner core and 7.1 m2/person for new development areas represent significant increases from the current provision. Achieving these targets

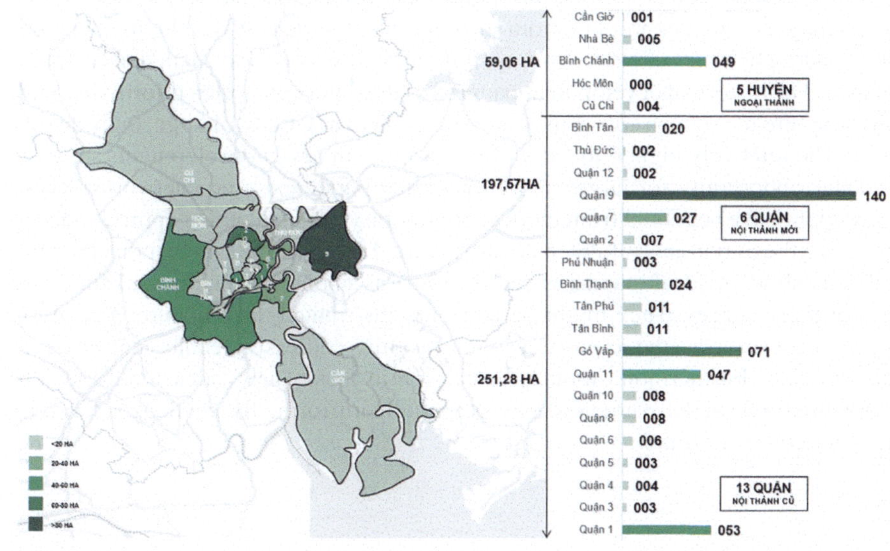

Fig. 5.3 Distribution and per capita availability of public park land across Ho Chi Minh City, based on administrative districts. The figure highlights stark disparities between the urban core, new development zones, and peri-urban districts, revealing a significant shortfall in green space provision relative to national planning standards. *Source*: Program for Developing Public Parks and Trees in the City, 2020–2030 (*Source* by the authors)

Table 5.1 Park and Greenery Planning Comparison

Area	Existing	Standard	In HCMC General Masterplan to 2025 (2010)
Inner core (13 districts)	0.67 m2/ person	QCXDVN 01:2020/BXD:Neighbourhood: ≥ 2 m2/person TCVN 9257:2012:12–15 m2/ person (Public Park Area 7–9 m2/person)	2.4 m2/ person
New development (6 districts)	1.72 m2/ person	QCXDVN 01:2020/BXD:Out of Neighbourhood: ≥ 7 m2/person TCVN 9257:2012:Public Greenery 3–3.6 m2/person Street Greenery 1.7–2.0 m2/person	7.1 m2/ person
Outskirt (5 rural districts)	0.3 m2/ person		12 m2/ person
Total	**0.55 m^2/ person**		**≥ 7.0 m^2/ person**

would necessitate not only increasing the amount of green space but also strategically planning for well-designed, accessible, and ecologically functional parks and green networks—implying a focus on quality.

The extremely low per capita green space in the inner core likely implies smaller, more fragmented, and potentially more intensely used parks and green areas. These spaces may face greater pressure from surrounding high-density development, potentially leading to issues of overcrowding, limited biodiversity, and challenges in maintenance. The historical development patterns of the inner core, often prioritizing built infrastructure over green space, may have resulted in a scarcity of large, high-quality parks. The relatively higher quantity of green space in newer developments suggests a greater opportunity for larger, more integrated parks and green networks. These areas might have benefited from more contemporary urban planning approaches that recognize the importance of green infrastructure. However, the fact that this quantity still falls short of ideal standards raises concerns about the quality and functionality of these spaces. They might be unevenly distributed, lack mature vegetation, or primarily serve aesthetic purposes without fully addressing ecological or recreational needs. Furthermore, rapid urbanization in these "new" areas could lead to future pressures on these green spaces, potentially mirroring the challenges faced by the inner core over time.

4 The Consequences of Green Space Loss

4.1 Climate Resilience

The degradation of green infrastructure in HCMC has significantly eroded the city's climate resilience (Downes & Storch, 2014). Already ranked among the ten most climate-vulnerable cities globally, HCMC faces escalating risks from urban heat, flooding, and environmental degradation, largely driven by unregulated land conversion (Nicholls et al., 2008; Storch & Downes, 2011). The Urban Heat Island (UHI) effect has intensified as tree canopy and permeable surfaces have been replaced by dense construction, contributing to a marked rise in temperature. Average rainy-season temperatures have increased by 2 °C over the past five decades, with projections indicating a further rise of 1.4 °C (ADB, 2010). Thermal stress is further exacerbated by the UHI effect, which is reinforced by the city's limited vegetation and compact urban form (Katzschner et al., 2023). However, flooding represents an even more critical threat (Downes, Viet, et al., 2024; Storch, Downes, Katzschner, & Thinh, 2011). Without substantial adaptation measures, projections for 2050 suggest that 61% of the city's area could be subject to recurrent flooding, with extreme events potentially affecting up to 3.2 million residents, increasing in both depth and duration (ADB, 2012). Social vulnerabilities further compound these environmental risks, including limited access to green space, constrained livelihood options, and exclusion from adaptation planning processes (Tran & Downes, 2023).

4.2 Social and Cultural Implications: Well-being and Identity

The erosion of green space diminishes public access to recreation, reduces environmental comfort, and erodes the city's cultural landscapes. Parks, gardens, and tree-lined streets are vital for physical activity, mental health, and social cohesion—particularly in dense, low-income districts. Beyond function, many green areas hold historical and symbolic meaning. Their removal severs ties to the city's past, weakening collective memory and a sense of place. The loss of such spaces contributes to psychological stress and detachment, especially where urban form becomes increasingly alienating and privatized.

4.3 Economic Trade-Offs: Real Estate Vs. Sustainability

Current development priorities in HCMC emphasize real estate expansion over long-term liability. While plans target 507.91 hectares of new green space by 2030, implementation remains limited, and no dedicated funding or policy mechanism supports sustained investment. Green infrastructure is often treated as residual or aesthetic rather than integral to urban systems. Most investment flows toward high-return commercial and residential projects, sidelining green space in favour of short-term economic gains. This approach risks embedding long-term vulnerability into the urban fabric. Without a shift toward integrated green space planning, HCMC faces worsening environmental degradation, rising inequality, and declining quality of life.

5 Opportunities for Integrating Nature-Based Solutions (NbS)

5.1 Biophilic Design and Urban Greening Initiatives

Vietnamese people, like many other ethnic groups, always love nature, but Vietnamese people especially love a lifestyle surrounded by trees and greenery. When living in townhouses or apartments, having plants on the terrace or balcony is an essential need. In traditional and important events and festivals of each household, community, or organization, the presence of flowers and decorative plants is always a mandatory component. For instance, displaying peach branches, kumquat pots, apricot trees, chrysanthemum vases, etc. on Tet; flower bouquets on birthdays, flower baskets on events. This cultural affinity for greenery offers a promising foundation for mainstreaming biophilic and nature-based design strategies in Vietnamese urban development. This cultural characteristic is a favourable premise for Integrating NbS in urban areas in Vietnam. The necessary discussion now is for a framework of actions to activate and promote the strategies of greening cities.

Currently, biophilic design is emerging as a dominant trend in the world of architecture. Solutions such as integrating greenery into architecture, roof gardens, and green roofs are gradually becoming popular. Initiatives such as Patrick Blanc's vertical garden have quickly spread, replacing the grey colour of the usual concrete walls with the unusual green colour of the plant wall. At the level of each individual or family, vertical greenery solutions like in the world seem to have been popular for a long time in Vietnam. Trends such as urban agriculture, edible landscape, "from garden direct to table" that are emerging in developed countries have actually existed for a long time in Vietnamese cultural life, and are a daily story in many Vietnamese households. These cultural foundations of respect for nature and closeness to greenery of the Vietnamese people need strategic solution frameworks to be widely transformed in urban spaces, to public works and community spaces. Experience from international case studies can be valuable references.

5.2 *Case Studies of Community Parks, Waterfront Developments, and Urban Agriculture*

5.2.1 Water-Sensitive Design: Restoring Hydrological Flows through Urban Landscapes

The Cheonggyecheon restoration project in Seoul exemplifies a shift toward water-sensitive urban design, where a formerly buried stream was uncovered and transformed into an accessible public realm with ecological, climatic, and cultural functions (Robinson & M. H., 2011; Department of Planning, 2025). Through terrain-sensitive design, the project reestablished local hydrological rhythms, improved microclimates, and fostered walkability in one of Asia's most densely built urban cores. Its success offers a model for HCMC's own canal revitalization initiatives, such as those underway along Xuyen Tam and Tham Luong–Ben Cat corridors, which currently emphasize engineered drainage but could be reoriented to support biodiversity and social activation.

Similarly, Singapore's ABC Waters Programme integrates stormwater management with landscape design, creating multifunctional green-blue corridors that manage floods while offering aesthetic and recreational benefits (Public Utilities Board, 2025). The program's integration of bioswales, vegetated basins, and canal-side parks aligns with emerging calls in HCMC for "multi-blue-green belts" as seen in recent plans (Khong, 2024). These cases suggest that water infrastructure in HCMC needs not remain hidden or purely technical but can be designed as landscape assets embedded in daily urban life.

5.2.2 Community co-Production: Activating the Public Realm through Participation

Community-led placemaking has become an important strategy for reclaiming fragmented urban space. In New York City, programmes like the Plaza Program and POPS (Privately Owned Public Spaces) encourage partnerships between municipalities, private developers, and local communities to repurpose leftover or underused spaces for public benefit (New York City Department of Transportation, 2025; New York City Department of City Planning, 2025). These interventions prioritize accessibility, temporary activation, and co-design—principles that are increasingly relevant in HCMC where informal interventions such as Nguyen Hue Flower Street or Nguyen Van Binh Book Street demonstrate the potential of tactical urbanism in the Vietnamese context.

The transformation of parking lots or amorphous leftover spaces into dynamic community nodes shows how even small-scale interventions can significantly contribute to liveability and place identity. In HCMC, where space is contested and urban form is rapidly shifting, empowering residents and neighbourhood groups to co-produce public space can ensure greater cultural resonance and stewardship. As

Scheuer et al. (2024) emphasize, even narrow verges and limited sites can serve critical climate and social functions when strategically designed and scaled.

5.2.3 Governance Incentives: Creating Enabling Conditions for Nature-Based Solutions

Singapore's Skyrise Greenery Incentive Scheme is a leading example of how governance can align public goals with private-sector capacity. By offering financial support for rooftop gardens and vertical greenery, the scheme incentivized developers to integrate biophilic design into real estate products—ultimately contributing to the vision of a "City in a Garden" (Kishnani et al., 2011; National Parks Singapore, 2025). This strategic linkage of urban policy with ecological outcomes illustrates the value of clear, enforceable incentives.

In contrast, HCMC's green planning is often hindered by fragmented governance and a lack of enforceable development incentives. While Vietnamese culture demonstrates a longstanding affinity for greenery—expressed through home gardens, balcony planting, and floral symbolism—these values remain underutilized in formal urban planning. Translating cultural familiarity into institutional frameworks requires zoning bonuses (e.g., FAR incentives), tax relief for developers investing in public green infrastructure, or streamlined approvals for green architectural innovations.

To move from scattered greening to systemic implementation of NbS, HCMC must create an enabling governance environment. This involves not only revising technical standards (e.g., TCVN 9257:2012) but also establishing flexible, transparent incentive structures that empower both public and private actors to invest in nature-driven placemaking.

5.3 Revitalization of Historic Green Spaces and Contemporary Innovations

Returning to the situation of HCMC, we have seen that history has left a system of valuable green spaces. The question is, with the current urban planning and urban management methods, have we had similar valuable green spaces? The tree-lined boulevard, wide and shady sidewalks, small public gardens, and great parks that are tightly integrated into the network of historical urban forms seem to have no place on current planning system.

In addition, the current urban planning and urban management methods in Vietnam are separating constructed greenery from natural ecological zones and agricultural spaces. The indicator of 0.5 m2 of green space per person as mentioned in the previous section is a very worrying figure for the situation of HCMC as mentioned in many popular articles as well as cited in scientific documents. In fact, this number is of the *planned park*, meaning that green spaces are used for the recreational function

Fig. 5.4 The existing green belt in Ho Chi Minh City has not been considered as public greenery (*Source*by the authors)

of the city, not the entire green space. That also means that when we convert and realize the planned "park" areas on the existing green covering seen from satellite (which can be agricultural land, natural ecological tropical, low land), we are likely to reduce the real green cover of the city.

Currently, the trend of developing valuable contemporary landscape architecture is promoting theories such as Sponge City, Water Sensitive Urban Design, Waterfront Revitalization. From the current situation through international experiences, it shows that the landscape of HCMC needs to innovate its approach, from "constructed park" to "touch the earth lightly" with public spaces where natural green space is the main character, a place for the community to "back to wild", with the ecological operating rhythm of the rule "let the nature speak", i.e. allow ecological processes to structure design priorities (Fig. 5.4).

6 Strategic Framework for Nature-Driven Placemaking

To achieve the goal of creating nature-oriented places, it is necessary to approach simultaneously at both strategic and tactical levels, both in terms of top-down management policies and bottom-up social community participation. A comparable challenge is evident in Bengaluru, India, where fragmented institutional responsibilities for green and blue infrastructure among state departments have led to inefficient maintenance, conflicting mandates, and reduced ecological functionality. As Thapa et al. (2025) describe, this fragmentation mirrors the barriers faced in HCMC and underscores the importance of integrated governance frameworks for multifunctional green space delivery.

Strategic Framework—Policy Recommendations for Balancing Urban Growth with Ecological Conservation.

At the strategic level, it is necessary to change the perception, approach and implementation methods of green spaces, encouraging nature-friendly architectural models.

First, with the geographical location and terrain, geology and hydrology of Saigon—HCMC, it is necessary to promote the conservation of lowland and implement a strategy to bring the community "back to wild" with the man-made features in the public landscape being minimally integrated, suitable for the natural ecological appearance. Second, with new green spaces such as public parks and public gardens that are planned and designed, it is necessary to implement the criteria of taking the "green" ("viên" in Vietnamese) as the basis, as the foundation for the "public" ("công" in Vietnamese) to be integrated later, in a multi-purpose and diverse direction. In other words, the natural ecological function of public parks and gardens must come first, then the function of human recreation activities. Third, every new infrastructure needs to integrate the greening solution like bioswale, tree-covered design to create multi-blue-green belts for the city.

Fourth, with existing public greenery, it is necessary to renovate and improve to maximize the green and permeable surface; focusing on trees being on a real ground, where their roots can freely penetrate deep into the ground, to survive for a long time, in accordance with the natural survival laws of plants. Fifth, it is necessary to prioritize the conservation and development of self-growing trees to reduce the cost of tree care and maintenance; increase the width of sidewalks to add more growing space for trees; encourage the replacement of vertical embankment walls and box culverts with bioswale ditches and soft embankments. Scheuer et al. (2024) emphasize that even narrow verges or limited vertical spaces can contribute to microclimate regulation when applied strategically and at scale—an approach highly applicable to HCMC's alley-heavy urban morphology. These strategies are not only based on the system of urban planning in Vietnam, which cannot cover all the necessary principles, but also need to be updated into standards and regulations on parks and public green areas such as TCVN-9257. Standards and regulations on public green areas also need to be accompanied by specific design guidelines, suitable for the Vietnamese context, to increase the feasibility and implementation of Nature-Driven Placemaking strategies (Fig. 5.5).

6.1 Tactical Solutions—Community Participation in Urban Greening and Placemaking

At the tactical level, integrating bottom-up community participation through solutions such as pop-up landscape architecture, public art, … are feasible solutions

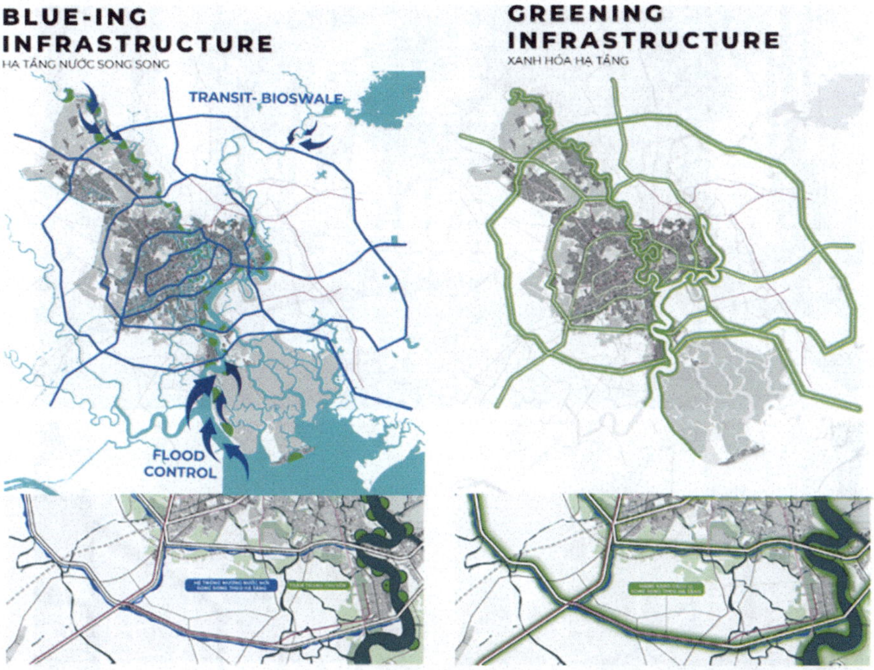

Fig. 5.5 Multi Blue-Green Belt integrated to the new infrastructure of Ho Chi Minh City (*Source*by the authors)

in HCMC as seen in Nguyen Hue Flower Street, Nguyen Van Binh Book Street (Fig. 5.6).

HCMC needs to research to promulgate and implement innovative projects such as the incentive programs for green architecture, biophilic architecture. Investors and developers really need a mechanism to join hands in creating places that focus on nature, contributing to the overall green and clean environment of the city, while also increasing the unique value of their real estate. Initiatives to contribute public art, sponsor the implementation of public landscape projects from the finance capable corporates, with the ability to proactively implement mechanisms that need to be paved by the government.

Some specific solutions that should be studied and implemented in HCMC include:

- Incentive policy of rewarding Floor Area Ration (FAR) or exempting tax for projects that contribute more public space, integrating green solutions into architecture.
- Renovating and embellishing existing green boulevards, parks, and public gardens in the direction of increasing green surface, permeable surface, and reducing hard paving.

Fig. 5.6 The re-invention public realm case of Ho Chi Minh City Book Street—from a parking space to a new public place (*Source*by the authors)

- Restoring waterflows that have been covered or engineered during urban transformation, returning parallel green-blue space to increase public activities.
- Converting amorphous spaces, abandoned spaces such as landfills, mines, former-industrial facilities, or under elevated infrastructure, … into green and public spaces.
- Reconstructing small interstitial spaces, interspersed in the existing urban morphology system or under-used spaces that are being forgotten into green spaces for the community.

6.2 *The Role of Adaptive Governance in Ensuring Long-Term Sustainability*

In both of the above nature-driven placemaking approaches, the role of adaptive urban governance is very important. Changes in strategic thinking from the top down through the promulgation of urban management policies, the organization of integrated strategic planning, taking green ecological factors as the core, will create a solid framework for development.

At the same time, every time there is a green initiative, there must also be a flexible mechanism to accept unprecedented innovations and creations. The harmonious and balanced approach of adaptive governance is the key motivation for any scenarios, to champion the core strategy of green development. It can be said that the adaptive governance approach in placemaking with nature as the driving force is the most suitable way, ensuring sustainable urban development in the long term.

7 Conclusions and Outlook

The transformation of green space in HCMC underscores an urgent need to reconceptualize urban nature in the face of rapid urbanization, climate risk, and deepening spatial inequalities. Once defined by a relatively integrated network of green and blue infrastructure, the city's ecological foundation has become increasingly fragmented—eroded by speculative land use, infrastructural overreach, and disjointed planning. These patterns not only undermine environmental resilience but also threaten social equity, cultural identity, and long-term urban sustainability.

As this chapter has shown, addressing these intertwined challenges demands more than technical upgrades or isolated interventions. What is required is a fundamental reorientation of urban governance, design, and planning—toward a model of nature-driven placemaking. This paradigm centres ecological logics, supports community co-production, and leverages adaptive governance to integrate nature not as a backdrop, but as a structuring principle of urban life. Revitalizing historic green corridors, embedding NbS into spatial planning, and investing in inclusive, multifunctional public spaces must become strategic priorities for HCMC's future development.

HCMC's trajectory reflects broader patterns across Southeast Asia's rapidly growing megacities. In Jakarta, subsiding land and unchecked extraction have necessitated massive infrastructural interventions. In Bangkok, engineered drainage systems have proven insufficient in the face of intensifying flood events. In Manila, efforts to retrofit green infrastructure within dense, informal settlements have illuminated the social and institutional barriers to transformative planning. These shared conditions underscore the need for cross-city learning and regional collaboration on nature-based urbanism in contexts marked by socio-ecological complexity and development pressures.

Looking forward, HCMC must shift its development logic—from residual greening to systemic ecological integration, from engineered control to ecological collaboration, and from top-down plans to shared stewardship. In doing so, the city can chart a path toward a more resilient, inclusive, and life-affirming urban future—where nature is not peripheral but foundational to the urban condition. As Shannon (2024) argues, reconceptualizing urban water systems and peri-urban green assets as the backbone of a metropolitan-scale ecological network is essential for building climate resilience in deltaic megacities. HCMC, with its vast canal systems, mangrove zones, and micro-topographic diversity, holds untapped potential for such an integrated landscape approach. Yet, even the most visionary frameworks will falter without political commitment, institutional coordination, and active community empowerment. Building on Shannon's call for regionally grounded park constellations, HCMC must recalibrate its urban trajectory by embedding terrain-sensitive design and vegetative diversity into long-term planning.

Acknowledgments The authors gratefully acknowledge the support of TA Landscape Architecture Vietnam for their valuable insights and contributions to the research and design discussions that informed this chapter. KMT and VVA were primarily responsible for the conceptual development, methodological framing, analysis, and writing of the chapter. THD and NKD contributed to writing, reviewing, and editing.

References

Asian Development Bank (ADB). (2010). **Ho Chi Minh City: Adaptation to climate change**. Asian Development Bank.

Downes, N. K., & Storch, H. (2014). Current constraints and future directions for risk adapted land-use planning practices in the high-density Asian setting of Ho Chi Minh City. **Planning Practice and Research, 29**(3)*, 220–237.

Downes, N. K., Storch, H., Schmidt, M., Nguyen, T. C. V., Dinh, L. C., Tran, T. N., & Hoa, L. T. (2016). Understanding Ho Chi Minh City's urban structures for urban land-use monitoring and risk-adapted land-use planning. In *sustainable Ho Chi Minh City: Climate policies for emerging mega cities* (pp. 89–116).

Downes, N. K., Storch, H., Viet, P. Q., Diem, N. K., & Dinh, L. C. (2024). Assessing peri-urbanisation and urban transitions between 2010 and 2020 in Ho Chi Minh City using an urban structure type approach. *urban Science, 8*(1), 11.

Downes, N. K., Viet, P. Q., Huynh, T. N., Dang, T. H., & Storch, H. (2024). The urban growth challenges of implementing nature-based solutions for stormwater management in Ho Chi Minh City, Vietnam. In **rethinking Stormwater management through sustainable Urban Design** (pp. 75–88). Singapore.

Đức, T. H. (2020) *Gia Định thành thông chí* [Gia Dinh Geography]. Ho Chi Minh City: Tổng hợp Publishing House.

Ho Chi Minh City People's Committee. (2002). *2064/QĐ-UB: Approval of investment project for water drainage and pollution improvement of Tham Luong–Ben Cat–Nuoc Len canal (Phase 1)*. Ho Chi Minh City.

Katzschner, A., Diem, N. K., Dang, T. H., & Downes, N. K. (2023). The challenge of cooling rapidly growing cities: The case of densification and peri-urbanisation in Ho Chi Minh City and adaptation responses. In **climate change and cooling cities** (pp. 115–127). Singapore.

Khong Minh, T., Nguyen Huynh Dieu, T., & Pham Ngoc, T. (2023). Water integrated in development of infrastructure space for the expansion of urban area in Ho Chi Minh City. In *Proceedings of the 1st SEAUS: Southeast Asian Urbanisms*.
Khong, T. M. (2024). Water flow morphology as continuous development motivation in the urban evolution of Ho Chi Minh City. In *proceedings of the international conference on sustainable carbon neutrality in smart urbanism (SCNSU 2024)*.
Kishnani, N., Kailankaje, V. B., Basu, A., Hui, L. Y., & A. (2011). **green building design training guide for Vietnam: Primer on the green design process**. United Nations Environment Programme & Ministry of Construction.
Minh, T. K. (2024). Water-based development for the expansion of urban area in Ho Chi Minh City. *lecture notes in civil engineering*.
Ministry of Construction (MoC). (2012). **TCVN 9257:2012—Greenery planning for public utilities in urban areas: Design standards**. Ministry of Construction.
Ministry of Construction (MoC). (2020). **QCXDVN 01:2020/BXD–national technical regulation on construction planning**. Ministry of Construction.
National Parks Singapore. (2025, May 4). *Skyrise Greenery Incentive Scheme: About*. https://skyrisegreenery.nparks.gov.sg/incentive-scheme/about/
New York City Department of Transportation (2025, May 4). *NYC Plaza Program*. https://www.nyc.gov/html/dot/html/pedestrians/nyc-plaza-program.shtml
New York City Department of City Planning. (2025, May 4). *Privately owned public spaces*. https://www.nyc.gov/content/planning/pages/our-work/plans/citywide/privately-owned-public-spaces
Nicholls, R., Hanson, S., Herweijer, C., Ranger, N., Hallegatte, S., Corfee-Morlot, J., Chateau, J., & Muir-Wood, R. (2008). *ranking port cities with high exposure and vulnerability to climate extremes: Exposure estimates* (OECD environment working papers, no. 1). OECD publishing. https://doi.org/10.1787/011766488208
People's Committee of Ho Chi Minh City. (2021). **312/QĐ-UBND: Program for developing public parks and trees in the city for the period 2020–2030**. Ho Chi Minh City.
Prime Minister of the Government. (2010). **24/QĐ-TTg: Approval of adjustment to the general planning of Ho Chi Minh City construction to 2025**. Government of Vietnam.
Public Utilities Board. (2025). **ABC Waters Design Guidelines** (5th ed.). Public Utilities Board.
Quang, T. H. (2023). **Hạ tầng đô thị Sài Gòn buổi đầu**. Ho Chi Minh City General Publishing House.
Robinson, A., & M. H. (2011). *Cheonggyecheon Stream Restoration Project*. *Landscape Performance Series*, Landscape Architecture Foundation. doi:https://doi.org/10.31353/cs0140
Scheuer, S., Sumfleth, L., Nguyen, L. D. H., Vo, Y., Hoang, T. B. M., & Jache, J. (2024). A systematic assessment of greening interventions for developing best practices for urban heat mitigation—The case of Huế. *Vietnam. Urban Science, 8*(2), 67. https://doi.org/10.3390/urbansci8020067
Shannon, K. (2024). Metropolitan park constellations of ecological systems: Lessons from Ho Chi Minh City. In *Towards a metropolitan public space network: Lessons, projects and prospects from Lisbon* (pp. 216–221). Routledge. https://doi.org/10.4324/9781003408611-24
Storch, H., Downes, N., Katzschner, L., & Thinh, N. X. (2011). Building resilience to climate change through adaptive land use planning in Ho Chi Minh City, Vietnam. In K. Otto-Zimmermann (Ed.), **Resilient Cities** (pp. 349–363). Springer.
Storch, H., & Downes, N. K. (2011). A scenario-based approach to assess Ho Chi Minh City's urban development strategies against the impact of climate change. *Cities, 28*(6)*, 517–526.
Storch, H., Downes, N. K., & Rujner, H. (2011). A scenario-based approach to assessing the exposure and flood risk of Ho Chi Minh City's urban development strategy in times of climate change. In *In *innovations in sharing environmental observations and information**. Shaker Verlag.

Thapa, P., Lakshmisha, A., Bhaskar, D., Nagendra, H., Plieninger, T., & Thiel, A. (2025). Social-ecological dynamics of green and blue infrastructure in the Indian megacity of Bengaluru. In F. Angeoletto et al. (Eds.), *Ecology of tropical cities* (pp. 277–298). Springer.
Tran, H. K., & Downes, N. K. (2023). Narratives of women's resilience to flood risks in Ho Chi Minh City. *Vietnam. Environment and Urbanization ASIA, 14*(1), 90–103.

Governance, Participation and Multispecies Planning

Prioritizing Nature: Integration of Nature-Based Solutions in Urban Environmental Governance for Enhanced Urban Liveability in India

Basudatta Sarkar, Nabanita Saha, and Padmaja Chakravarty

Abstract Urban environments in India face unprecedented anthropogenic pressures, manifesting in environmental deterioration, climate vulnerability, and diminished quality of life. While the concept of "liveability" has emerged as a critical policy imperative, its environmental dimension—specifically the integration of Nature-based Solutions (NbS)—remains insufficiently examined in scholarly discourse. Building upon the theoretical framework of urban liveability articulated by Panda and Thakur (2019), this chapter presents a systematic analysis of NbS contributions to environmental liveability in the Indian urban context. The research addresses a significant lacuna in current literature by transcending generalized discussions of NbS benefits to establish context-specific, quantifiable indicators applicable to Indian cities and their degree of liveability. Through a comprehensive synthesis of extant literature, this chapter establishes the significance of NbS in enhancing urban liveability, with particular emphasis on measurable indicators and atmospheric quality amelioration, microclimate regulation, and hydrological resource management—critical domains identified in the Energy Research Institute's (TERI) urban environmental assessments. The chapter develops a methodological framework for identifying and evaluating pertinent environmental indicators of urban liveability within the Indian context, examining their potential optimization through strategic NbS implementation. The analysis extends to a critical examination of the current integration status of NbS within urban planning and policy frameworks, delineating implementation challenges and institutional constraints. Central to this discussion is an analysis of Urban Local Bodies (ULBs) as primary institutional actors in the promotion, implementation, and scaling of NbS initiatives. Through comparative analysis of exemplar cases and evidence-based practices, the chapter proposes a systematic framework for ULBs to leverage NbS effectively in pursuit of enhanced urban liveability. This framework encompasses methodological approaches for integrating NbS into urban governance mechanisms, policy formulation processes, and implementation protocols, supplemented by recommendations for enhanced monitoring

B. Sarkar (✉) · N. Saha · P. Chakravarty
Department of Planning and Architecture, National Institute of Technology Rourkela, Rourkela, Odisha, India
e-mail: sarkarb@nitrkl.ac.in

A. Cheshmehzangi et al. (eds.), *Nature-based Urbanism*, Urban Sustainability,
https://doi.org/10.1007/978-981-95-6864-2_6

systems utilizing appropriate indicators and data analytics. The chapter concludes by demonstrating that through strategic institutionalization and mainstreaming of NbS, ULBs can significantly enhance environmental quality metrics in urban spaces, thereby facilitating the development of more resilient, sustainable, and salubrious urban environments for their inhabitants.

Keywords Nature-based Solutions (NbS) · Urban Environmental Governance · Nature and well-being · Urban Liveability Indicators · Environmental Sustainability

1 Introduction

The unprecedented population growth and subsequent pressure on urban liveability is leading to a series of interconnected challenges, including climate change-induced vulnerability, environmental degradation, and poor quality of life. The concept of liveability has become a crucial policy objective over the years, guiding urban planning and development efforts to ensure improved quality of life, sustainability, and resilience in cities. However, the integration of social-economic-environmental-cultural indicators related to Nature-based Solutions (NbS) often lacks the focus although they can very much influence the pillars of liveability—healthcare, culture, education, and infrastructure (Panda & Thakur, 2019). Although scholarly literature explores both liveability and NbS as individual streams, there is a lacuna in the comprehensive integration of liveability and NbS, which hinders policy development methodological frameworks for implementation. Therefore, the strategies to mitigate urban crises like increasing water scarcity, frequent and intense flooding events, deteriorating air quality, and urban heat island effect are often insufficient and compromise public health and well-being (Cohen-Shacham et al., 2016; Kabisch et al., 2016). This chapter addresses a critical gap in the existing literature by moving beyond generalized discussions of NbS benefits. It establishes context-specific, quantifiable indicators of environmental liveability that are directly applicable to Indian cities. It aligns with the theoretical framework of urban liveability and establishes requirement of a paradigm shift from traditional "grey" infrastructure approaches in urban planning and governance in India (Cohen-Shacham et al., 2016; Kabisch et al., 2016; Panda & Thakur, 2019) to the blue-green infrastructure using NbS as a mechanism.

India's urban population is projected to reach 600 million by 2030, intensifying pressures on infrastructure, ecosystems, and public health (MoHUA, 2021). Cities like Delhi, Mumbai, and Chennai grapple with air pollution (PM2.5 levels exceeding WHO limits by 10x), Urban Heat Islands (UHI) amplifying temperatures by 5–8 °C, and recurrent flooding due to impervious surfaces covering 70–80% of metropolitan areas (CSE, 2014; TERI, 2022). While India's Smart Cities

Mission (2015) and National Clean Air Programme (2019) prioritize technological fixes, Nature-based Solutions (NbS) remain underutilized despite their potential to address systemic socio-ecological vulnerabilities (Climate & Development Knowledge Network, 2022; Kabisch et al., 2016). Seddon et al. (2020) have argued that using NbS to strategically utilize natural processes can offer a transformative approach to address socio-environmental challenges and subsequent enhancement of urban liveability. It has been further supported by Climate and Development Knowledge Network (2022) that NbS is capable of mitigating environmental risks, fostering social equity, and promoting economic resilience by integrating ecological principles into urban development. This chapter, therefore, argues for the prioritization of nature within urban environmental governance in India. It will systematically analyse the contributions of NbS to environmental liveability, focusing on measurable indicators, for example amount of green space—as studies show a direct correlation between access to green spaces and improved mental and physical well-being in urban populations (Francis et al., 2012; Ghasemi et al., 2018). It will develop a methodological framework for identifying and evaluating relevant indicators within the Indian context as well as examine the potential for optimizing these indicators through strategic NbS implementation. The overall aim is to propose a systematic approach for ULBs to effectively leverage NbS to enhance urban liveability, encompassing methodological approaches, policy formulation processes, and implementation protocols through an effective environmental governance. Thus, this study addresses two objectives:

1. To critically analyse institutional barriers hindering NbS mainstreaming, emphasizing the role of Urban Local Bodies (ULBs) in bridging policy practice.
2. To establish a context-specific methodological framework for integrating NbS into urban governance, focusing on measurable environmental liveability indicators.

By synthesizing global best practices and India's traditional ecological knowledge, this research contributes to SDG 11 (Sustainable Cities) and advances scholarly discourse on adaptive and agile governance in the Global South.

1.1 Urban Liveability—A Multidimensional Approach

The multidimensional (stability, healthcare, culture, environment, education, and infrastructure) concept of liveability has evolved over decades driven by the vitality of fostering—physical, social, and psychological well-being; equitable access to resources and opportunities; associated quality of life of the urban dwellers and holistic development of the cities (Ghasemi et al., 2018; Kashef, 2016; Lowe et al., 2015; Zanella et al., 2015). The dimensions and the indicators representing them vary across cultural and geographic context but are always interdependent in order to have a balanced approach to urban development.

The economic vitality of cities, often measured through employment rates, GDP per capita, and business climate indicators, remains a cornerstone of liveability frameworks (Li et al., 2024). However, economic prosperity alone cannot ensure liveability unless paired with equitable access to opportunities and mitigation of disparities, as reflected in poverty rates and housing affordability metrics (Lowe et al., 2015; Saeed et al., 2022). Similarly, social and cultural elements forge community cohesion and individual fulfilment, reinforcing the notion that liveability is rooted in human-centric design (Almashhour & Samara, 2022; Badland et al., 2014). Infrastructure and governance further underpin these dimensions, with walkability, public transport accessibility, and citizen participation serving as critical indicators of functional and inclusive urban systems (Higgs et al., 2019; Tan et al., 2016).

The environmental pillar transcends these categories, acting as both a determinant and an outcome of urban liveability. Human centric designs and the parallel economic and infrastructural developments are entwined with the ecosystem and natural elements as disruptions to the natural environment causes human discomfort which leads to reduced human efficiency and comprised liveability conditions (Kollmuss & Agyeman, 2002; Tzoulas et al., 2007). For example, management of UHI effect, presence of green covers, efficient water management, development of recreational areas clubbed with infrastructural facilities and others, are directly influencing physical health, mental well-being, and social equity (World Bank, 2023). Such design elements correlate with reduced stress levels enhanced physical activity, and stronger social ties, underscoring the role of natural environment as a public health intervention (Francis et al., 2012; Ghasemi et al., 2018; Lowe et al., 2015). Cities account for over 70% of global carbon emissions. Therefore, rendering decarbonization strategies—such as green infrastructure, renewable energy adoption, and circular economy are critical to long-term liveability (Giannico et al., 2021; Leach et al., 2017). Thus, integration of environmental sustainability into policy frameworks, for inclusive, safe, and resilient cities (SDG 11) (United Nations, 2015) demand immediate attention and action worldwide.

1.2 Environmental Governance

Environmental governance mechanisms may integrate institutions, policies, and actors mediating with human environment such as, market tools, community initiatives and participation, and trans-national collaborations (Armitage et al., 2020; Lebel et al., 2007; Pillay & Buschke, 2020). However, it must also account for key characteristics of ecological challenges, which scholars describe as encompassing "uncertainties, cause-effect mismatches, and norm plurality" (Ingold et al., 2019) alongside traits such as "an emerging policy domain, fragmentation, deep uncertainties, and a long-term horizon" (Termeer et al., 2017). These interconnected features collectively define environmental issues as classic "wicked" problems—complex and resistant to resolution (Van den Ende et al., 2024). Compounding this complexity, the slow development of societal urgency to address the potentially catastrophic impacts of environmental issues, despite prolonged risks, has led researchers to characterize such crises as "creeping" in nature (Boin et al., 2020, 2021). This delayed recognition underscores the unique challenges environmental governance faces in mobilizing timely, coordinated action. Researchers have argued that effective governance requires adaptive, transformative, and equity centric strategies (Armitage et al., 2020; Folke et al., 2005; Lebel et al., 2007; Ostrom, 2010; Panda & Thakur, 2019; Pillay & Buschke, 2020; Van den Ende et al., 2024; Visseren-Hamakers et al., 2021).

Adaptive governance emphasizes agility and iterative learning to address socio-ecological unpredictability, particularly in climate contexts (Folke et al., 2005) by fostering resilience through decentralized decision-making (Pillay & Buschke, 2020). Building socio-ecological resilience further requires diversifying livelihoods and decentralizing resource management, ensuring communities can adapt to environmental shocks while maintaining cultural and ecological integrity (Folke et al., 2005).

Transformative governance, on the other hand, demands systemic shifts in values, power structures, and economies to tackle root causes of degradation (Armitage et al., 2020; Lebel et al., 2007) by prioritizing social justice (Visseren-Hamakers et al., 2021). Policy recommendations emphasize the need to mainstream equity into governance frameworks, ensuring marginalized voices shape decision-making processes (Visseren-Hamakers et al., 2021). Social learning platforms, such as participatory forums and citizen science initiatives, foster knowledge co-production, bridging gaps between scientific expertise and local lived experiences (Armitage et al., 2020; Lebel et al., 2007), and provides an inclusive environment where the marginalized can participate in the decision-making processes leading to equity-centric governance.

Environmental governance has transitioned from command-and-control regulations (Kaur et al., 2022) to market-based tools like carbon trading (Pillay & Buschke, 2020), though critics warn of commodification and marginalization (Büscher et al., 2012). Technical fixes have given way to adaptive frameworks emphasizing social learning, recognizing that holistic responses are essential. Multi-actor systems now involve NGOs, corporations, and grassroots movements (Bibri et al., 2024). Integrated approaches, such as China's green finance integration (H. Wang & Guo, 2024), reflect the need for multi-scale coordination (Armitage et al., 2020). India's judiciary has expanded environmental rights via litigation (Kaur et al., 2022), while China's "Blue Sky Defense War" reduced urban air pollution by 28% through centralized-local collaboration (Jiang et al., 2021; L. Zhang et al., 2024). Polycentric governance, inspired by Nobel Laureate Ostrom's work (2010), decentralizes authority across scales, while Integrated Environmental Governance (IEG) emphasizes cross-sectoral coordination (Visseren-Hamakers et al., 2021). These frameworks address "wicked problems" and "creeping crises" (Van den Ende et al.,2024), with community-based management bridging global goals and local realities (Armitage et al., 2020; Lebel et al., 2007; Panda & Thakur, 2019).

The governance frameworks outlined above are not merely abstract concepts; they provide the essential institutional and procedural foundation for the effective mainstreaming of Nature-based Solutions (NbS) in urban settings (Kabisch et al.,

2016). The "wicked" and "creeping" nature of urban environmental challenges—such as air pollution, urban heat islands, and water scarcity—demands agility, adaptability, and multi-stakeholder engagement that these governance models promote. For instance, adaptive governance is critical for NbS, as it allows Urban Local Bodies (ULBs) to treat NbS projects as iterative experiments, learning from outcomes and adjusting strategies for interventions like permeable pavements or urban forest corridors based on real-time ecological and social feedback (Pauleit et al., 2019). Similarly, transformative governance challenges the dominance of grey infrastructure paradigms, creating the political and policy space for NbS to be recognized as legitimate, multifunctional public infrastructure that delivers co-benefits for equity, health, and resilience (IPCC, 2022).

Furthermore, the success of NbS in directly enhancing urban liveability is contingent upon this robust governance scaffolding. Polycentric and integrated governance systems break down silos between municipal departments (e.g., parks, water, planning), enabling the cross-sectoral coordination required for NbS implementation (Ostrom, 2010). Nbs forstormwater management, such as a constructed wetland, is not just a drainage project but also a public health intervention (reducing waterborne diseases), a social amenity (enhancing recreational space), and a climate adaptation measure (mitigating flooding). Effective environmental governance ensures these multiple liveability benefits are recognized, valued, and managed collaboratively. Without such governance structures, NbS risk being implemented as isolated, marginal projects, failing to achieve the systemic shift required to meaningfully improve environmental quality, public well-being, and overall urban liveability (IUCN, 2020).

Urban liveability represents a dynamic equilibrium between human needs and environmental limits, demanding interdisciplinary collaboration and participatory governance. While economic, social, and infrastructural pillars provide the support for livable cities, the environmental dimension serves as both a precondition and a sustainer of urban well-being. Even though different initiatives are being implemented, the main challenge remains—using or operationalizing the environment-based indicators within the assessment of urban liveability. The hindrance arises from inconsistent methodological framework, data gap, and many others in context-specific adaptations (e.g., culture—environment), which can influence decision-making on traditional versus contemporary approaches, policy framework and as well as policy priorities (Abdelbaset, 2016; Almashhour & Samara, 2022; Saeed et al., 2022; X. Y. Wang, 2012).

2 Enhancing Urban Liveability with Nature-Based Solutions

Urban liveability reflects the capacity of cities to foster residents' quality of life through equitable access to resources, social opportunities, and healthy environments (Mouratidis & Yiannakou, 2022). While economic and social dimensions are critical, the environmental pillar is foundational, as it directly influences public health, ecological resilience, and long-term sustainability. Nature-based Solutions (NbS) are innovative strategies that harness the power of ecosystems and biodiversity to address a spectrum of pressing societal challenges, including but not limited to climate change mitigation and adaptation. By working with nature, NbS provide a multifunctional approach that simultaneously enhances biodiversity, regulates microclimates, improves human health and well-being, and strengthens social equity and economic resilience. They have emerged as a transformative tool for connecting urban development with environmental sustainability (Seddon et al., 2020). The following table (Table 1) illustrates a comprehensive list identified from existing body literature of various indicators under the first two pillars or parameters of liveability. Table 1 further shows how environmental factors are linked with each of the pillars. A list of new additional indicators which can be added to the composite indices to measure urban liveability has been developed for the first two parameters of urban liveability and shown in Table 1. Similarly, measurable indicators for the rest of the three parameters can be identified focusing on the context. However, the additional indicators are a combination of qualitative and quantitative but measurable indicators. The new indicators explain the context as the values will significantly differ from the global north or the developed nations making them context specific. In the similar way, measurable indicators for the other parameters can be developed. For further analysis, multicollinearity needs to be checked, and the final indicators can be listed out.

2.1 The Underrated Interdependence of Liveability Parameters, Environment, and NbS

NbS offer a paradigm shift by embedding natural processes into urban infrastructure and climate change resilience. For instance, green roofs and walls reduce UHI effects through evapotranspiration, lowering rooftop temperatures by 3–40 °C and decreasing building energy use by 15–30% (Roodman & Lenssen, 1995; Berardi et al., 2014; H. Zhang et al., 2024). Urban forests and parks act as "cooling corridors," with tree canopy cover reducing ambient temperatures by 2–8 °C, as demonstrated in Chicago, where a 10% increase in canopy cover mitigated heat-related emergency calls by 7% (Nowak et al., 2006; Ziter et al., 2019). Similarly, permeable pavements and constructed wetlands enhance stormwater management, reducing runoff by 50–90% in cities like Lisbon, while simultaneously filtering pollutants and replenishing

Table 1 Comprehensive indicators of urban liveability, their interconnection with environmental factors

Parameters of urban liveability	Measurable indicators under subparameters		Interconnection with environmental factors	Measurable indicators	Potential NbS interventions
	Sub-parameters	Indicators			
1. Stability	Public safety and security	Prevalence of petty crime; prevalence of violent crime; homicide rate; perception of safety; number of police; officers per capita; response time of emergency services; safety from accidents; safety of property; safety for specific group like children and women	1.1 resource scarcity and conflict: Degradation of environmental resources (water, arable land, forests, fisheries) intensifies competition for these dwindling assets. This scarcity can exacerbate existing social tensions, particularly in areas with pre-existing inequalities or weak governance (Homer-Dixon, 1999). Water Stress: Urban areas facing water scarcity due to drought, pollution, or over-extraction are vulnerable to conflicts between different user groups (residential, industrial, agricultural). This can lead to protests, social unrest, and even violent clashes (Gleick, 1993a, 1993b) Land Degradation: Soil erosion, desertification, and deforestation reduce agricultural productivity, leading to food insecurity and displacement of rural populations. This can put pressure on urban areas as people migrate in search of livelihoods, potentially straining resources and increasing social tensions (FAO, 2011).	1. Equitable access to water 2. Equitable access to land 3. Equitable access to financial assistance 4. Rate of conflict for water scarcity 5. Rate of conflict for land 6. Rate of conflict for financial resource allocation 7. Percentage of drought-affected population 8. Rate of deforestation 9. Rate of farmable soil erosion 10. Change in agricultural productivity due to soil degradation 11. Change in agricultural productivity due to unpredictable rainfall pattern	1. Urban Greening & Community Gardens: Reduce social tensions by creating shared spaces and providing alternative livelihoods, addressing resource scarcity conflicts 2. Sustainable urban drainage systems (SuDS): Bioswales, rain gardens, and permeable pavements mitigate water scarcity by enhancing groundwater recharge and reducing stormwater runoff 3. Mangrove & Wetland Restoration: Act as natural buffers against storm surges and coastal flooding, protecting infrastructure and economic assets (e.g., tourism)

(continued)

Table 1 (continued)

Parameters of urban liveability	Measurable indicators under subparameters		Interconnection with environmental factors	Measurable indicators	Potential NbS interventions
	Sub-parameters	Indicators			
			Fisheries Collapse: Overfishing and pollution can decimate fish stocks, impacting the livelihoods of coastal communities and potentially leading to conflicts over fishing rights (Pauly et al., 2002) Example: The "Water Wars" in Cochabamba, Bolivia (2000), where protests erupted over the privatization of water resources and increased water prices, illustrate the potential for resource scarcity to trigger social instability. 1.2 climate change as a threat multiplier: Climate change acts as a "threat multiplier," exacerbating existing vulnerabilities and creating new ones. Extreme weather events (floods, droughts, heatwaves, cyclones) displaced populations, damage infrastructure, disrupt economies, and increase competition for resources (IPCC, 2022)	12. Change in agricultural productivity due to temperature rise 13. Change in the rate of food consumption 14. Conflict for fair price food consumption 15. Change in BMI 16. Percentage of displacement for development 17. Percentage of migration for livelihood change 18. Frequency of urban flooding 19. Frequency of drought 20. Frequency of heatwave 21. Frequency of cyclonic activities 22. Percentage of displacement for flooding 23. Infrastructure damage	4. Agroforestry & Soil Conservation: On urban peripheries to improve soil fertility, secure food production, and reduce displacement pressures 5. Equitable Green space distribution: Targeted development of parks and green corridors in marginalized neighbourhoods to foster social cohesion and reduce environmental injustice

(continued)

Table 1 (continued)

Parameters of urban liveability	Measurable indicators under subparameters		Interconnection with environmental factors	Measurable indicators	Potential NbS interventions
	Sub-parameters	Indicators			
			Sea-Level Rise: Coastal cities face the threat of inundation and displacement due to rising sea levels, potentially leading to mass migrations and social disruption (Hauer et al., 2016) Extreme Heat: Urban heat islands, intensified by climate change, can lead to increased mortality and morbidity, particularly among vulnerable populations (the elderly, the poor, those with pre-existing health conditions). This can strain healthcare systems and create social unrest (Stone, 2012) Agricultural Impacts: Changes in rainfall patterns and increased temperatures can reduce crop yields, leading to food insecurity and price volatility, potentially triggering social unrest (Food and Agriculture Organization of the United Nations, 2021; (Schlenker & Roberts, 2009)	24. Rate of change or disruption of existing economy 25. Rate of mass migration due to sea level rise 26. UHI effect 27. Rate of exposure to environmental hazard 28. Degree of exposure to environmental hazards based on socio-economic status 29. Rate of exposure to pollution—For different socio-economic strata 30. Access to parks and green open spaces—For different socio-economic strata 31. Number of houses destroyed due to natural disaster	

(continued)

Table 1 (continued)

Parameters of urban liveability	Measurable indicators under subparameters		Interconnection with environmental factors	Measurable indicators	Potential NbS interventions
	Sub-parameters	Indicators			
			Example: Hurricane Katrina (2005) in New Orleans, USA, demonstrated how a climate-related disaster can expose and exacerbate social inequalities, leading to widespread displacement, looting, and social breakdown 1.3 environmental justice and social cohesion: Unequal exposure to environmental hazards (pollution, lack of green space, proximity to industrial sites) often correlates with socioeconomic disparities. This environmental injustice can fuel resentment, distrust in government, and social fragmentation (Bullard, 2000) Pollution Hotspots: Low-income communities and communities of colour are often disproportionately located near polluting industries and waste disposal sites, leading to higher rates of respiratory illnesses, cancer, and other health problems (Mohai et al., 2009)	32. Livelihood loss due to natural disaster 33. Conflict due to access to resources after natural disaster 34. Disruption in water services due to natural disaster or extreme climate events 35. Disruption in electricity supply due to natural disaster or extreme climate events 36. Equitable food distribution after natural disaster or extreme climate events 37. Equitable shelter provision after natural disaster or extreme climate events	

(continued)

Table 1 (continued)

Parameters of urban liveability	Measurable indicators under subparameters		Interconnection with environmental factors	Measurable indicators	Potential NbS interventions
	Sub-parameters	Indicators			
			Lack of Green Space: Unequal access to parks and green spaces deprives marginalized communities of the health and social benefits of nature, contributing to social inequities (Wolch et al., 2014) Example: The environmental justice movement in the United States emerged in response to the disproportionate burden of pollution borne by minority communities, highlighting the link between environmental degradation and social inequality 1.4. Economic stability: Environmental issues have huge impact on economic Well-being Natural disasters: Destroy property and infrastructure (Hallegatte et al., 2013)	38. Percentage of loss in fisheries industries 39. Percentage of loss of tourist due to natural disaster or extreme climate events 40. Percentage of loss of tourists due to social conflicts	

(continued)

Table 1 (continued)

Parameters of urban liveability	Measurable indicators under subparameters		Interconnection with environmental factors	Measurable indicators	Potential NbS interventions
	Sub-parameters	Indicators			
			Agriculture and fisheries: Depends on a healthy environment (Costanza et al., 1997) Tourism: Many cities rely on tourism that is based on natural beauty (Buckley, 2011) Example: The tourism of Maldives is significantly impacted by climate change		
	Social cohesion and civic engagement	Level of social trust; voter turnout in local elections; number of active community organizations per capita; volunteerism rate; instances of civil unrest/conflict; social community connectedness; opportunity to contribute to important issues			
	Governance, law and order	Corruption perception index; effectiveness of the legal system; threat of military conflict; threat of terror			

(continued)

Table 1 (continued)

Parameters of urban liveability	Measurable indicators under subparameters		Interconnection with environmental factors	Measurable indicators	Potential NbS interventions
	Sub-parameters	Indicators			
2.Healthcare	Access to healthcare	Number of hospital beds per 1000 population; number of doctors per 1000 population; number of general hospitals; access to various services for old adults; number of practitioners for a given population; average distance/ travel time to the nearest hospital/clinic; availability of specialized medical services; availability of over-the-counter drugs	2.1 air pollution's toll: Airborne pollutants (particulate matter, ozone, nitrogen dioxide, sulphur dioxide) from vehicles, industries, and power plants irritate the respiratory system, exacerbate cardiovascular disease, and increase the risk of cancer (WHO, 2014) Respiratory Illnesses: Chronic exposure to air pollution is a major cause of asthma, bronchitis, and other respiratory diseases, particularly in children and the elderly (Anderson et al., 2012) Cardiovascular Disease: Air pollution contributes to heart attacks, strokes, and other cardiovascular problems (Brook et al., 2010)	1. Of vehicle with updated PUC 2. Rate/frequency of monitoring airborne pollutants from industries 3. Number of power plants using non-renewable resources 4. Number of power plants using renewable resources 5. Rate of monitoring airborne pollutants from power plants 6. Amount of water getting polluted per capita/cubic meter 7. Percentage increase in respiratory diseases—Elderly, children, and infants 8. Percentage increase in respiratory diseases—Adults 9. Percentage increase in cardiovascular disease	1. Phytoremediation strategies: Strategic urban afforestation and green walls (especially near roads and industries) to capture particulate matter and absorb pollutants 2. Constructed Wetlands & Riparian Buffers: To naturally treat wastewater and filter agricultural/ industrial runoff before it enters water bodies, reducing contamination 3. Public Green Spaces & Urban Forests: To encourage physical activity, provide shade to mitigate heat stress, and offer spaces for mental restoration. Equitable distribution is key

(continued)

Table 1 (continued)

Parameters of urban liveability	Measurable indicators under subparameters		Interconnection with environmental factors	Measurable indicators	Potential NbS interventions
	Sub-parameters	Indicators			
			Cancer: Long-term exposure to certain air pollutants, such as benzene and formaldehyde, increases the risk of lung cancer and other cancers (IARC, 2013) Example: Studies have consistently shown a strong correlation between air pollution levels and hospital admissions for respiratory and cardiovascular illnesses in cities around the world 2.2 water contamination and disease: Untreated sewage, industrial effluent, and agricultural runoff can contaminate water sources with pathogens (bacteria, viruses, parasites) and toxic chemicals, leading to waterborne diseases (Prüss-Ustün et al., 2014) Diarrheal Diseases: Contaminated water is a major cause of diarrheal diseases, which are a leading cause of death in young children, particularly in developing countries (WHO, 2017)	10. Percentage increase in cancer detection 11. Percentage increase in hospitalization for respiratory diseases 12. Rate/frequency of monitoring water pollution 13. Rate/frequency of identification of contamination sources 14. Number of water treatment plants 15. Percentage change in diagnosis of—Cholera, typhoid 16. Rate/frequency of examining water sample for heavy metal pollution 17. Rate/frequency of examining water sample for pesticides	4. Green Roofs & Walls: To reduce the urban Heat Island (UHI) effect, lower ambient temperatures, and improve air quality for vulnerable populations 5. Water-sensitive Urban Design (WSUD): Rainwater harvesting gardens and restored urban lakes to improve water security and quality

(continued)

Table 1 (continued)

Parameters of urban liveability	Measurable indicators under subparameters		Interconnection with environmental factors	Measurable indicators	Potential NbS interventions
	Sub-parameters	Indicators			
			Cholera and typhoid: These severe waterborne diseases can cause epidemics in areas with poor sanitation and inadequate water treatment Chemical Contamination: Exposure to heavy metals (lead, arsenic), pesticides, and industrial chemicals in drinking water can have long-term health consequences, including neurological damage, kidney disease, and cancer (NRC, 2012) Example: The Flint, Michigan water crisis (2014–present), where lead contamination of the city's drinking water led to widespread health problems, highlights the devastating consequences of water pollution 2.3 Green spaces and Mental & Physical Health:	18. Rate of change in kidney and liver diseases diagnosis/ hospitalization 19. Number of accessible green space per neighbourhood 20. Safety of the green spaces—'eyes on street' 21. Safety of green spaces—Open visible and easily accessible spaces 22. Safety of the green space—Lighting design 23. Additional facilities such as jogging track, yoga space, open gym, children play area in the green space 24. Accessibility features of the green space—Universal design	

(continued)

Table 1 (continued)

Parameters of urban liveability	Measurable indicators under subparameters		Interconnection with environmental factors	Measurable indicators	Potential NbS interventions
	Sub-parameters	Indicators			
			Access to parks, green spaces, and natural areas provides opportunities for physical activity, reduces stress, improves air quality, and fosters social interaction, all of which contribute to improved mental and physical well-being (Bedimo-Rung et al., 2005; Hartig et al., 2014) Stress Reduction: Studies have shown that spending time in nature lowers cortisol levels (a stress hormone) and promotes relaxation (Wells, 2000; Lee et al., 2011) Physical Activity: Parks and greenways encourage walking, cycling, and other forms of exercise, reducing the risk of obesity, diabetes, and heart disease (Sallis et al., 2016) Social Cohesion: Green spaces provide places for people to gather, socialize, and build community bonds, which are important for mental health (Maas et al., 2006)	25. Resting spaces 26. Interactive activity spaces—Intra and intergenerational exchange 27. Mobile health checkup vans near parks and green spaces 28. Percentage of inaccessible green spaces 29. Pedestrian pathways—Walkability 30. Number of clean toilets around the green spaces 31. Number of safe drinking water fountains around the green spaces 32. Percentage of patients hospitalized due to heat wave/heatstroke	

(continued)

Table 1 (continued)

Parameters of urban liveability	Measurable indicators under subparameters		Interconnection with environmental factors	Measurable indicators	Potential NbS interventions
	Sub-parameters	Indicators			
			Example: Numerous studies have demonstrated the positive correlation between access to green space and reduced rates of depression, anxiety, and other mental health disorders (Wells, 2000; Bedimo-Rung et al., 2005; Dockreill & Shield, 2006; Roodman & Lenssen, 1995) 2.4. Impact of climate change on health (Tomlinson et al., 2011) Climate change is increasing health risks Heatwaves: Cause heatstroke and worsen cardiovascular conditions (Gasparrini et al., 2015) Vector-borne Diseases: Expand the range of diseases like malaria and dengue fever (WHO, 2014) Air Quality: Worsens respiratory problems due to increased ground-level ozone (Sheffield & Landrigan, 2011)	33. Number of deaths due to heatwave 34. Number of deaths due to malaria 35. Number of deaths due to dengue	
	Quality of healthcare	Infant mortality rate; life expectancy at birth; patient satisfaction with healthcare services; quality of private healthcare; quality of public healthcare; provision of aged care facilities; general healthcare indicators			

(continued)

Table 1 (continued)

Parameters of urban liveability	Measurable indicators under subparameters		Interconnection with environmental factors	Measurable indicators	Potential NbS interventions
	Sub-parameters	Indicators			
	Public health and Well-being	Prevalence of chronic diseases; mental health indicators; access to mental health services; vaccination rates; air quality index; water quality index			
	Housing	Housing affordability; availability of affordable; housing options; homelessness rate; housing quality; house size; living area size; dining area size; bedroom size; kitchen size Number of toilets; ventilation of the house; garage or parking space provided; housing density; mixed land use; residential population; housing stock and tenure; housing adaptability; residential and population density; social and non-profit housing provision; housing stock diversity			
	Utilities and services	Access to electricity; reliability of electricity supply; access to clean water and sanitation; internet access and speed; effectiveness of waste management services; number of emergency shelters; fire incidence rate			

(continued)

Table 1 (continued)

Parameters of urban liveability	Measurable indicators under subparameters		Interconnection with environmental factors	Measurable indicators	Potential NbS interventions
	Sub-parameters	Indicators			
	Overall social and physical infrastructure	Neighbourhood facility; distance to shopping; child and youth services; emergency centres; distance to workplace; total household income; monthly saving; cost of living; number and type of job available locally			

Source Authors [major references for measurable indicators for the sub-parameters: (Abdelbaset, 2016; Badland et al., 2014, 2015; Dempsey et al., 2010; Elmahdy et al., 2021; Evans & Kantrowitz, 2002; Kashef, 2016; Li et al., 2024; Martino et al., 2021; Miller & Hutchins, 2017; Tang et al., 2021; Valcárcel-Aguiar et al., 2019; Van Kamp et al., 2003; Zhao, 2010)]

groundwater (Mentens et al., 2006). Air quality improvements are achieved through phytoremediation, where urban trees sequester 711,000 metric tons of PM10 annually in cities like Beijing, reducing hospital admissions for asthma by 13% (Li et al., 2024; Nowak et al., 2006). Green spaces also mitigate noise pollution, with dense tree belts reducing sound levels by 5–10 decibels, as observed in Rotterdam neighbourhoods where residents reported lower stress and higher life satisfaction (de Vries et al., 2003; Margaritis & Kang, 2017). These interventions exemplify NbS's dual role in addressing environmental stressors and enhancing liveability. Biodiversity gains are equally significant: urban parks in Melbourne increased native bird species diversity by 40% through native plant reintroduction, fostering ecological resilience (Beninde et al., 2015; Chambers-Ostler et al., 2023). Such outcomes underscore NbS's capacity to align environmental health with human well-being.

The intricate relationship between urban environments and human well-being underscores the necessity of adopting NbS to leverage the ecosystem services it offers, thereby, fostering resilient, equitable, and thriving cities. The following section details out the possibilities of introducing new measurable indicators to the existing exhaustive and comprehensive list in order to cover all aspects of equity and well-being.

2.1.1 Stability and Disaster Resilience

NbS play a pivotal role in mitigating environmental stressors that threaten urban stability. Coastal wetlands, mangroves, and urban forests act as natural buffers against climate-induced disasters, reducing vulnerabilities highlighted by events like Hurricane Katrina (Barbier et al., 2011). Recent studies emphasize their cost-effectiveness, for instance, New York City's post-Sandy investments in wetland restoration are projected to save $1.7 billion in flood damages by 2050. Furthermore, NbS address resource scarcity by enhancing water security through green infrastructure, such as rain gardens, which reduce stormwater runoff and recharge aquifers (Foster et al., 2011). In regions like Sub-Saharan Africa, agroforestry initiatives have improved soil fertility and crop yields, curbing displacement and conflict over arable land. These examples illustrate NbS's dual role in disaster risk reduction and socio-economic stability. A few measurable indicators are: number of flood events per year; Percentage of population living in high-risk disaster zones; Number of households with access to early warning systems; Rate of population displacement due to disasters; Public expenditure on post-disaster recovery, etc.

2.1.2 Healthcare and Environmental Equity

Urban green spaces, a cornerstone of NbS, significantly improve public health by reducing air pollution—a leading cause of respiratory and cardiovascular diseases (Nowak et al., 2006). Parks and greenways also mitigate urban heat islands, which disproportionately affect marginalized communities (Stone, 2012). Recent research highlights that equitable access to green spaces can reduce health disparities. Increasing tree cover has a significant impact at the neighbourhood stress level. Moreover, NbS like rooftop gardens and urban agriculture enhance food security, providing fresh produce in "food deserts" and reducing diet-related illnesses. Such initiatives align with the WHO's emphasis on "health in all policies," positioning NbS as critical to inclusive healthcare strategies. A few measurable indicators are: Annual hospitalization rates for respiratory or heat-related illnesses; Percentage of urban population exposed to air pollution above WHO limits; Prevalence of vector-borne diseases (malaria, dengue) per 100,000 residents; Number of accessible healthcare facilities per 10,000 residents; Average distance to the nearest green space/park, etc.

2.1.3 Education and Cognitive Benefits

Educational outcomes are bolstered by NbS through improved learning environments and experiential opportunities. Green schools, featuring natural lighting and ventilation, correlate with higher test scores and reduced absenteeism (Wells, 2000; Kats, 2006). Exposure to nature also enhances cognitive development. A 2023 study demonstrated that students in schools with green playgrounds exhibited improved attention spans and problem-solving skills. Programs like UNESCO's "Eco-Schools" integrate NbS into curricula, fostering environmental literacy and stewardship. Such initiatives underscore the bidirectional relationship between education and NbS as informed citizens are more likely to advocate for sustainable policies, creating a feedback loop of environmental responsibility. A few measurable indicators are: School absenteeism rates in relation to environmental factors; Average test scores before and after introduction of green school environments; Number of schools with access to green playgrounds or outdoor classrooms; Participation rates in environmental education programs; Green space per child at schools etc.

2.1.4 Infrastructure and Economic Innovation

Transitioning to green infrastructure reduces cities' ecological footprints while stimulating economic growth. Permeable pavements and green roofs not only manage stormwater but also lower energy costs through natural insulation (Bowler et al., 2010). The circular economy model, integral to NbS, transforms waste into resources—e.g., Copenhagen's Amager Bakke plant converts trash to energy, powering 60,000 homes (State of Green, 2024). Economically, NbS generate jobs in sectors like renewable energy and eco-tourism. The UNFCC (2018) estimates that NbS could create $3.7 trillion in annual economic value by 2030. Such innovations highlight NbS's role in decoupling urban growth from environmental degradation. A few measurable indicators are: Percentage of city surface covered by permeable infrastructure; number of green jobs created per year; amount of renewable energy generated as a share of total energy use; annual reduction in stormwater runoff due to green infrastructure; number of eco-innovation projects implemented, etc.

2.1.5 Cultural Identity and Social Cohesion

NbS preserve cultural heritage and foster community bonds by integrating natural landscapes into urban identity. The High Line in New York and Singapore's Gardens by the Bay exemplify how NbS blend aesthetics with functionality, attracting tourism while reinforcing civic pride (Chiesura, 2004). Indigenous practices, such as Māori "urupa" (sacred burial grounds) in New Zealand, demonstrate how NbS can honour cultural traditions while enhancing biodiversity. These spaces also serve as communal hubs, reducing social isolation and strengthening resilience against crises like the COVID-19 pandemic. A few measurable indicators are: Frequency and diversity of cultural festivals in green public spaces; resident participation rate in community gardening or urban greening; number of heritage sites preserved or restored with NbS involvement; percentage of residents reporting a sense of belonging linked to natural spaces; community engagement rate in environmental stewardship programs, etc.

2.1.6 Technology Integration

Advanced tools enable the systematic implementation and assessment of NbS. Geographic Information Systems (GIS) and remote sensing identify priority areas for greening by mapping heat vulnerability and green space accessibility (Pauleit et al., 2019). For example, LiDAR-derived canopy maps in New York guided the MillionTreesNYC initiative, optimizing tree placement to maximize shade and air quality benefits (McPherson et al., 2011). Microclimate modelling tools like ENVI-met predict thermal comfort outcomes, informing the design of green corridors in Stuttgart, Germany, which reduced pedestrian-level temperatures by 4 °C (Bruse & Fleer, 1998; Tomlinson et al., 2011). Multi-criteria decision analysis (MCDA) further balances ecological, economic, and social priorities, as seen in Singapore's "City in a Garden" program, which prioritized high-impact NbS like rooftop gardens and vertical greenery (Tan et al., 2016). Participatory tools, including community mapping, ensure NbS align with local needs, as demonstrated in Medellín's Green Corridors project, which engaged residents in co-designing 30 km of shaded walkways (C40 Cities Climate Leadership Group & Nordic Sustainability, 2019). A few measurable indicators are: Percentage of NbS sites monitored by smart sensors or GIS tools; number of urban environmental data points collected via digital platforms; response time for urban environmental hazards detected by sensor networks; number of citizen science initiatives for urban nature monitoring; integration of digital green infrastructure inventories in urban planning, etc.

2.2 Toward an Ecosystem-Centric Urban Future

Despite their benefits, NbS face implementation barriers, including funding gaps and political short-termism. Methodological inconsistencies in quantifying benefits, such as varying metrics for green space per capita, hinder cross-city comparisons (Saeed et al., 2022). Policy fragmentation and short-term financing further limit implementation, particularly in developing cities where competing priorities overshadow long-term environmental planning (Anderson et al., 2004). A 2023 World Bank report notes that only 5% of global urban funding targets NbS, underscoring the need for innovative financing mechanisms like green bonds. Additionally, equitable distribution remains a challenge; without inclusive planning, NbS risk exacerbating gentrification, as seen in Barcelona's "superblocks" initiative. Addressing these issues requires participatory governance, where marginalized communities co-design NbS projects, ensuring accessibility and relevance. However, emerging frameworks like the IUCN Global Standard for NbS provide guidelines for context-sensitive design, emphasizing community engagement and ecological integrity (Cohen-Shacham et al., 2016).

The synthesis of stability, healthcare, education, infrastructure, and culture through NbS presents a transformative vision for urban development. Cities must prioritize NbS not as ancillary projects but as central to their planning paradigms. The key steps in achieving it include:

A. **Policy Integration**: Mainstream NbS into urban policies, aligning with SDGs 11 (Sustainable Cities) and 13 (Climate Action). For example, Medellín's "Green Corridors" program, which reduced temperatures by 2 °C, exemplifies policy-driven NbS success.
B. **Equitable Governance**: Adopt participatory frameworks to ensure NbS benefit all demographics. Cape Town's post-drought water equity initiatives, involving community-led rainwater harvesting, highlight the efficacy of inclusive approaches.
C. **Financial Mobilization**: Leverage public-private partnerships and green financing. The European Union's "Natural Capital Financing Facility" offers a model, blending grants and loans for NbS projects.
D. **Technological Synergy**: Integrate NbS with smart technologies, such as AI-driven monitoring of urban forests, to optimize outcomes (Mahmoud et al., 2024).

NbS represent a critical pathway to enhancing urban liveability by addressing environmental challenges through ecological innovation. Advanced planning tools and participatory approaches ensure these solutions are equitable and adaptive to local contexts. Future research must prioritize standardized metrics for NbS impact assessment and explore synergies with smart city technologies, such as sensor networks for real-time environmental monitoring. As cities confront climate change and population growth, NbS offer a blueprint for resilient, livable urban futures, where nature and infrastructure coexist to support human and ecological well-being. The imperative is clear: NbS are not merely ecological interventions but socio-economic catalysts. By reimagining cities as symbiotic ecosystems, humanity can forge a future where urban resilience and environmental harmony coexist. As the IPCC (2022) warns, incremental changes are insufficient; a radical reorientation toward NbS is essential to avert climate catastrophe and nurture thriving urban societies.

3 Nature-Based Solutions in India

In the face of increasing climate vulnerabilities, rapid urbanization, and socio-economic inequalities, India has turned to Nature-based Solutions (NbS) as a holistic approach to address interconnected challenges across environmental, social, and governance dimensions. By combining traditional ecological knowledge with modern sustainability goals, NbS in India support resilient livelihoods, environmental protection, and inclusive growth across both rural and urban landscapes. The following sub-sections elaborates on the convergence of the traditional and innovative practices in India, which result in efficient NbS as well as the multi-dimensionality of NbS.

3.1 Bridging Tradition and Innovation

India's historical affinity for harmonizing human settlements with nature provides a robust foundation for NbS implementation. Traditional practices such as *johads* (earthen check dams) in Rajasthan and *bawaris* (stepwells) in Jodhpur exemplify community-led water management systems that sustained arid regions for centuries through rainwater harvesting and groundwater recharge (Centre for Science and Environment, 2014; Suutari & Marten, 2005). These systems, rooted in localized ecological knowledge, are now being revitalized to address contemporary urban challenges. For instance, the restoration of *bawaris* in Jodhpur and decentralized rainwater harvesting initiatives in Goa University demonstrate how traditional techniques, augmented by modern engineering, can alleviate water scarcity while enhancing urban resilience (Centre for Science and Environment, 2014; Indian Water Partnership, n.d.). Similarly, the rejuvenation of Hauz Khas Lake in New Delhi through bioremediation highlights the potential of NbS to revive degraded ecosystems, improving water quality and biodiversity while providing recreational spaces for residents (Indian Water Partnership, n.d.). The "urban groves" initiative in Kochi and the Wild Valley Biodiversity Park in Surat illustrate how community-driven afforestation projects mitigate heat islands and reduce flood risks by enhancing permeable surfaces and restoring native flora (India Forum for Nature Based Solutions, n.d.). These efforts are complemented by policy frameworks such as the Atal Mission for Rejuvenation and Urban Transformation (AMRUT), which prioritize green cover expansion and lake restoration (Centre for Science and Environment,

2014). Furthermore, urban agriculture initiatives, documented in the *Sowing Sustainable Cities* compendium, reveal how rooftop gardens and community farms in cities like Indore and Mumbai bolster food security, reduce waste through composting, and create livelihood opportunities, embodying the multi-functionality of NbS (Patil et al., 2023).

The success of NbS in India hinges on several cross-cutting factors. Community participation emerges as a critical driver, as evidenced by the Alwar water conservation project and Trichy's Community-Based Sanitation and Slum Development (CBSSD) initiative, where decentralized wastewater treatment systems (DEWATS) were co-designed with residents (Indian Water Partnership, n.d.; Suutari & Marten, 2005). Such participatory approaches ensure cultural relevance and long-term sustainability. Additionally, the integration of traditional knowledge with scientific innovation—exemplified by the Jamshedpur Utilities and Services Company (JUSCO)'s holistic water management model—demonstrates how hybrid systems can optimize resource efficiency while preserving ecological integrity (Indian Water Partnership, n.d.). Policy support remains pivotal, though challenges persist in institutional coordination and financing. Programs like the Green Lumbini Initiative and Jabarkhet Nature Reserve underscore the importance of public-private partnerships in scaling NbS, leveraging NGO expertise and corporate funding to achieve ecological and social outcomes (Climate & Development Knowledge Network, 2022).

Regionally, South Asia's shared vulnerabilities to climate change underscore the relevance of NbS beyond India. The IUCN's regional assessment highlights cross-border lessons, such as Bangladesh's urban forests and Nepal's sacred groves, which enhance climate resilience while preserving cultural heritage (Climate & Development Knowledge Network, 2022). These examples reinforce the necessity of mainstreaming NbS into transnational urban planning frameworks, transcending pilot projects to achieve systemic change.

Thus, by leveraging traditional wisdom, fostering community agency, and integrating ecological principles into policy, Indian cities can mitigate climate risks while enhancing liveability. The transition from isolated initiatives to scaled implementation demands robust institutional frameworks, innovative financing mechanisms, and interdisciplinary collaboration. As South Asian cities confront escalating environmental crises, NbS stand as a testament to the transformative potential of harmonizing human progress with planetary boundaries.

3.2 Multi-dimensional Approach to Sustainability and Resilience

NbS application in India demonstrates their capacity to not only mitigate environmental risks but also enhance social cohesion, public health, governance systems, and cultural vitality. The deployment of NbS across rural, peri-urban, and urban contexts showcases their flexibility and relevance in building resilient and inclusive communities.

In the domain of public safety and security, NbS initiatives like the *Dhara Vikas* program in Sikkim stand out. By focusing on the revival of natural springs, this project ensures equitable water access across income groups and strengthens water security, especially in hilly regions vulnerable to water scarcity (Sharma et al., 2024). India's broader commitment to *Climate Resilient Agriculture*, through climate-resistant villages and sustainable farming practices, further integrates NbS to secure livelihoods and maintain food security in climate-stressed agrarian economy (Rao et al., 2016). Additionally, mangrove ecosystems**,** as highlighted by the Gujarat State Disaster Management Authority, have been recognized as a crucial NbS for disaster risk reduction, offering natural defense against storm surges and coastal erosion (Shah & Ramesh, 2022). Similar benefits are echoed in other coastal states like Maharashtra, Goa, Odisha, and West Bengal, where mangrove restoration efforts protect not only urban infrastructure but also the livelihoods of coastal communities (Badola & Hussain, 2005; Jyotsna et al., 2024; Singha et al., 2024; Vaz, 2014). Food security has also been advanced in urban areas through roof top agriculture at homes in cities like Bhubaneswar, Bengaluru, Delhi and Mumbai. It provides a means of income through organic food production, increases the green cover in the city, builds community connections and inspire a sustainable lifestyle among the citizens (Patil et al., 2023; Zaffar & Thakur, 2024).

From a social cohesion and civic engagement perspective, NbS have proven instrumental in mobilizing communities towards collective environmental stewardship. For instance, the *Community-Based REDD Strategy* in Northeast India fostered grassroots involvement in forest governance, enhancing both ecological sanctity and social capital through continuous community dialogue and action (Poffenberger, 2015). In Maharashtra's Vidarbha region, people's participation in watershed management has contributed to more sustainable water practices and stronger communal bonds (Bagdi & Kurothe, 2014). Furthermore, green spaces in urban contexts—such as in Bardhaman—support psychological well-being and a sense of place attachment, where cultural values, social identity, and individual psychology are intimately linked with the quality of urban greenery and biodiversity (Sen & Guchhait, 2021).

NbS also intersect significantly with governance, law, and order by promoting sustainable economic alternatives and reducing environmental degradation. The development of ecotourism in regions like Rajasthan, Murshidabad (West Bengal) and other places in India offers pathways for rural employment and environmental protection (Chandel et al., 2023; S. K. Paul & Roy, 2017; Samal & Dash, 2023). Agroforestry practices in rain-fed areas of Southern India, including those promoted in Telangana and Karnataka, reduce the dependency on forests for fuel and fodder, thereby safeguarding biodiversity and livelihood generation and improving community resilience (Kumar & Singh, 2020; Telwala, 2023). These green livelihood models, such as those in Maharashtra, demonstrate how NbS can reinforce decentralized environmental governance while supporting socio-economic development (Purohit et al., 2024).

In terms of public health and well-being, the ecological benefits of urban vegetation are profound. In Bangalore, a study found that suspended particulate matter (SPM) levels were significantly lower in street segments lined with trees, contributing to better air quality (Vailshery et al., 2013). Similarly, the use of vegetation to mitigate traffic noise in Varanasi has led to quieter, more livable urban environments (Pathak et al., 2008). These examples emphasize the role of NbS in enhancing the quality of urban life, particularly in densely populated and polluted cities.

NbS also enrich cultural vibrancy and leisure, playing a vital role in shaping urban landscapes that cater to the emotional and recreational needs of communities. In Delhi, research has shown that urban green spaces are increasingly used for recreation, nature observation, and cultural gatherings, reflecting the growing importance of accessible and well-planned green infrastructure (S. Paul & Nagendra, 2017). The transformation of urban lakes into recreational commons highlights how NbS can serve as interfaces between people and ecological systems in rapidly urbanizing cities (D'Souza & Nagendra, 2011). Moreover, social interaction and community building are significantly influenced by the design and accessibility of green spaces. In Hyderabad, disparities in access to parks—due to safety concerns, entry fees, and social norms—reveal underlying socioeconomic and gender-based inequities in urban green infrastructure (Basu & Nagendra, 2021). Larger parks with rich vegetation facilitate greater biodiversity awareness and recreational use, while smaller or poorly maintained parks fail to engage residents effectively. Studies from Nagpur and Delhi further reinforce the idea that green spaces can enhance mental health, aesthetic satisfaction, and neighbourhood relationships, but only if they are designed inclusively and reflect the preferences of local communities (Anand & Bhattacharya, 2024; Lahoti et al., 2023).

The Indian experience with NbS offers a compelling case for their transformational potential across environmental, social, economic, and governance spheres. Whether through the revival of springs, mangrove restoration, climate-resilient farming, community-led forest governance, or urban greening, NbS have proven effective in enhancing resilience, equity, and sustainability. However, their success hinges on inclusive planning, strong community engagement, and supportive policy frameworks that recognize the interconnectedness of people and nature.

4 Convergence of Urban Liveability Through Nature-Based Solutions and Environmental Governance: A Systemic Framework

The preceding sections have demonstrated that Indian cities face acute and escalating challenges—including severe air and water pollution, rising urban temperatures and extreme heatwaves, increased frequency of urban flooding, loss of green cover, and declining urban biodiversity—making the integration of NbS into urban planning and policy frameworks both urgent and essential for improving public health, urban resilience, and long-term environmental sustainability. A core argument of this chapter is that enhancing urban liveability in the Indian context requires a fundamental shift from conventional, infrastructure-heavy approaches of urban planning and governance to a more holistic, ecologically-informed paradigm. The evidence presented demonstrates that NbS, when strategically implemented, can significantly improve key environmental indicators, such as air quality, temperature regulation, and water resource management (Nowak et al., 2006; Ziter et al., 2019). These improvements, in turn, have direct and positive impacts on public health, social equity, and economic resilience, aligning with the broader goals of sustainable development (Gómez-Baggethun et al., 2013). However, institutional fragmentation, lack of interagency coordination, and limited financial resources often hinder the mainstreaming of NbS in urban planning. The traditional emphasis on "grey" infrastructure, coupled with a lack of awareness and technical capacity regarding NbS, further exacerbates this issue. This underscores the need for a multi-pronged approach that encompasses policy reform, capacity building, and innovative financing mechanisms.

The role of Urban Local Bodies (ULBs) is paramount in this transition. While environmental governance in India involves multiple agencies at various levels (state pollution boards, water resources departments, etc.), ULBs serve as the crucial convergence point. As the primary institutional actors responsible for urban planning, service delivery, and land-use management, ULBs control the key levers—zoning, building bylaws, municipal finances, and public space design—that directly determine the urban environment's quality. Therefore, they are uniquely positioned to operationalize and mainstream NbS by integrating them into core municipal functions. This involves a fundamental shift in mindset: from viewing nature as a separate entity to be regulated, to recognizing it as multifunctional public infrastructure that delivers core services like stormwater management, microclimate regulation, and public health improvement. For this convergence to happen, NbS must be framed not as an isolated environmental project but as a superior solution for achieving the ULBs' existing mandates of service delivery and livability. This necessitates strengthening the technical and administrative capacities of ULBs, providing them with the necessary tools and resources to effectively plan, implement, and monitor NbS projects. Examples of such tools include Geographic Information Systems (GIS) for spatial planning, microclimate modelling for assessing the impact of green infrastructure, and multi-criteria decision analysis (MCDA) for balancing competing priorities (Bruse & Fleer, 1998; Pauleit et al., 2019).

Furthermore, fostering community participation is crucial for the success and long-term sustainability of NbS initiatives. Engaging local communities in the design, implementation, and maintenance of NbS ensures that projects are context-specific, culturally relevant, and responsive to local needs. This participatory approach also fosters a sense of ownership and stewardship, enhancing the long-term viability of NbS interventions. The Indian context, with its rich history of traditional ecological knowledge, offers unique opportunities for integrating NbS into urban planning. Local community participation can revitalize and leverage these traditional practices, while augmenting them with modern scientific insights and technological innovations, thus, providing contextually relevant solutions.

4.1 Methodological Framework for NbS Integration

The integration of NbS into urban governance demands a systematic approach that bridges ecological principles with institutional action. Grounded in adaptive governance theory (Folke et al., 2005) and polycentric institutionalism (Ostrom, 2010), this framework comprises four iterative phases, each tailored to address India's unique socio-ecological challenges while leveraging global best practices.

4.1.1 Phase 1: Indicator Identification and Spatial Prioritization

The first phase focuses on mapping environmental stressors and identifying priority areas for NbS interventions. Spatial analysis tools such as Geographic Information Systems (GIS) play a pivotal role here. For instance, ArcGIS and QGIS platforms integrate satellite-derived data—such as Landsat thermal bands for urban heat island (UHI) mapping and Sentinel-5P for PM2.5 dispersions—to pinpoint vulnerable zones. In Surat, GIS-based hotspot analysis guided the placement of permeable pavements and bioswales in flood-prone areas, reducing runoff by 40% (India Forum for Nature Based Solutions, n.d.). Complementing this, ecosystem service valuation tools like InVEST and ARIES quantify the socio-economic benefits of NbS, such as carbon sequestration and flood mitigation. Delhi's Ridge Forest restoration, valued at ₹2.1 billion annually in avoided healthcare costs, exemplifies how such valuations can inform budgetary allocations (TERI, 2021). These methodologies align with Pauleit et al. (2019), who advocate GIS as a decision-support tool for green infrastructure planning, and Gómez-Baggethun et al. (2013), who emphasize the role of ecosystem service valuation in legitimizing policy decisions.

4.1.2 Phase 2: Stakeholder Engagement and Co-design

The second phase emphasizes participatory governance to ensure NbS align with local needs and traditional knowledge. Participatory workshops, such as the World Café and Delphi Method, facilitate dialogue among residents, NGOs, and technical experts. In Chennai, fisherfolk collaborated with urban planners to co-design mangrove buffers along the coastline, which reduced post-2015 flood erosion by 60% (Patil et al., 2023). Citizen science platforms like iTree and OpenStreetMap further democratize data collection. Mumbai's "Urban Greens" initiative, which mapped 12,000 trees through community inputs, identified low-canopy wards for targeted afforestation (WRI India, 2022). Such approaches resonate with Armitage et al. (2020), who highlight participatory governance as critical for NbS legitimacy, and, who attribute the success of Medellín's Green Corridors to community-driven design.

4.1.3 Phase 3: Policy Integration and Institutional Mainstreaming

The third phase involves embedding NbS into statutory frameworks and funding mechanisms. Green zoning, which overlays ecological priorities with land-use plans, has been instrumental in cities like Pune. Their 2022 Development Plan reserved 15% of municipal land for urban forests, guided by TERI's heat vulnerability index. Eco-budgeting—allocating funds based on NbS cost-benefit analyses—has also gained traction. Kochi's "Green Budget" redirected 20% of drainage funds to rain gardens, yielding cost savings from reduced flood damage (Climate & Development Knowledge Network, 2022). These strategies align with Kabisch et al. (2016), who stress the need for NbS integration into urban bylaws, and Wang and Guo (2024), who demonstrate China's success in linking green finance to municipal bonds.

4.1.4 Phase 4: Monitoring, Evaluation, and Adaptive Management

The final phase ensures NbS efficacy through real-time monitoring and iterative refinement. IoT sensor networks, such as Airveda's low-cost air quality monitors and soil moisture probes, provide granular data for adaptive management. In Bengaluru, IoT-enabled lakes reduced eutrophication by 35% through dynamic algal bloom management (Indian Water Partnership, n.d.). Satellite-based dashboards like Google Earth Engine and Sentinel Hub offer macro-level insights; Ahmedabad's UHI dashboard, updated weekly, informs park placement and cool roof incentives (Centre for Science and Environment, 2014). These tools are validated by Bibri (2024), who advocates IoT-driven governance for transparency.

5 Conclusions: Institutionalizing Nature-Based Solutions for Equitable and Resilient Urban Futures

This chapter has demonstrated the critical importance of prioritizing nature in urban environmental governance for enhancing urban liveability in India. The integration of NbS offers a viable and transformative pathway to address the pressing environmental challenges facing Indian cities, while simultaneously fostering social equity and economic resilience. The evidence presented underscores the significant potential of NbS to improve key environmental indicators, directly contributing to a healthier and more sustainable urban environment. It also emphasises the urgency of systemic reforms to institutionalize NbS as a cornerstone of urban planning.

However, realizing this potential requires a concerted effort to overcome existing barriers to NbS implementation. This includes strengthening the institutional capacity of Urban Local Bodies (ULBs), fostering community participation, integrating traditional ecological knowledge with modern scientific approaches, and promoting inter-agency coordination. Furthermore, standardized metrics and robust monitoring systems are needed to evaluate the effectiveness of NbS interventions and inform adaptive management strategies. Future research should focus on refining these metrics, exploring the synergies between NbS and smart city technologies, and developing innovative financing mechanisms to scale up NbS implementation.

Mainstreaming NbS into statutory policies is a critical first step. To scale up and implement the best practices, state-level policy alignment workshops could harmonize NbS with existing missions like the Smart Cities Programme and AMRUT, ensuring green infrastructure quotas become non-negotiable components of urban master plans. However, it should be supplemented with training programs led by expert institutions to equip municipal staff with GIS tools and participatory design techniques, thus enabling data-driven NbS planning in cities. Yet, persistent inter-departmental fragmentation highlights the need for NbS-specific task forces within ULBs to streamline coordination. Standardized toolkits, detailing cost-benefit analyses and community engagement protocols, could further empower ULB staff to navigate implementation challenges.

Financing remains a persistent barrier, but innovative mechanisms like green bonds offer scalable solutions to fund large-scale NbS. Hybrid models combining public funds, CSR investments, and international grants—as seen in Bengaluru's Bellandur Lake rejuvenation—could democratize funding while ensuring accountability (Indian Water Partnership, n.d.).

Equity must remain central to NbS design. Tools like the Urban Health Equity Assessment and Response Tool (Urban HEART) have enabled cities like Indore to prioritize bioswales and green corridors in low-income flood zones, reducing waterlogging by 25% (WHO, 2010). Similarly, Bhubaneswar's women-led urban farms, which improved food security for 5000 households, underscore the importance of community stewardship in ensuring NbS benefits reach marginalized groups (Patil et al., 2023). Participatory budgeting, allocating 30% of NbS funds to community-proposed projects, could institutionalize this approach nationwide.

The above strategies must be accompanied by robust monitoring systems to validate NbS efficacy and guide adaptive management. Ahmedabad's UHI dashboard, which leverages Google Earth Engine for real-time heat island tracking, provides a replicable model for data-driven governance (Gorelick et al., 2017). Adopting standardized metrics, such as the IUCN Global Standard for NbS, would further harmonize impact assessments, enabling cross-city comparisons of carbon sequestration rates and biodiversity gains (IUCN, 2020).

Ultimately, the successful integration of NbS into the Indian urban fabric relies on a synergistic relationship between environmental governance, urban planning, and NbS implementation, with urban liveability as the central outcome. Environmental governance provides the essential enabling framework—through policies, funding mechanisms, and institutional mandates—that empowers and directs urban planning processes. Urban planning operationalizes this framework by mainstreaming NbS into the core instruments of city development, such as master plans, zoning regulations, building bylaws, and municipal budgets. This institutionalization ensures NbS moves from isolated pilot projects to a default, scalable component of public infrastructure. Finally, the strategic implementation of NbS delivers the tangible co-benefits—improved air and water quality, reduced heat stress, enhanced social cohesion—that directly elevate urban liveability. Therefore, this integration is not a linear process but a feedback loop: effective governance enables planning, planning mandates NbS, and the demonstrated benefits of NbS justify stronger governance and smarter planning. By championing this integrated approach, Indian cities can reframe nature from an external constraint into a vital asset, paving the way for a more resilient, sustainable, and equitable future for all their inhabitants.

References

Abdelbaset, M. S. (2016). Livability of high-rise districts—case study of West Bay in Doha. *2016*(1). https://doi.org/10.5339/qfarc.2016.SSHAPP1266

Almashhour, R., & Samara, F. (2022). Evaluating livability perceptions: Indicators to evaluate livability of a university campus. *Sustainability (Switzerland), 14*(19). https://doi.org/10.3390/su141911872

Armitage, D., Mbatha, P., Muhl, E.-K., Rice, W., & Sowman, M. (2020). Governance principles for community-centered conservation in the post-2020 global biodiversity framework. *Conservation Science and Practice, 2*(2), e160. https://doi.org/10.1111/csp2.160

Anand, & Bhattacharya, P. (2024). Assessing resident's perception towards ecosystem services of urban green spaces in Delhi, India. *International Journal of Sustainable Development and World Ecology, 31*(2), 150–162. https://doi.org/10.1080/13504509.2023.2261013

Anderson, H. R., Atkinson, R. W., Peacock, J. L., Marston, L., & Konstantinou, K. (2004). *Meta-analysis of time-series studies and panel studies of particulate matter (PM) and ozone (O3): Report of a WHO task group*. World Health Organization Europe.

Badland, H., Whitzman, C., Lowe, M., Davern, M., Aye, L., Butterworth, I., Hes, D., & Giles-Corti, B. (2014). Urban liveability: Emerging lessons from Australia for exploring the potential for indicators to measure the social determinants of health. *Social Science and Medicine, 111*, 64–73. https://doi.org/10.1016/j.socscimed.2014.04.003

Badola, R., & Hussain, S. A. (2005). Valuing ecosystem functions: An empirical study on the storm protection function of Bhitarkanika mangrove ecosystem, India. *Environmental Conservation, 32*(1), 85–92. https://doi.org/10.1017/S0376892905001967

Bagdi, G. L., & Kurothe, R. S. (2014). People's participation in watershed management programmes: Evaluation study of Vidarbha region of Maharashtra in India. *International Soil and Water Conservation Research, 2*(3), 57–66. https://doi.org/10.1016/S2095-6339(15)30023-X

Barbier, E. B., Hacker, S. D., Kennedy, C., Koch, E. W., Stier, A. C., & Silliman, B. R. (2011). The value of estuarine and coastal ecosystem services. *Ecological Monographs, 81*(2), 169–193. https://doi.org/10.1890/10-1510.1

Basu, S., & Nagendra, H. (2021 June). Perceptions of park visitors on access to urban parks and benefits of green spaces. *Urban Forestry and Urban Greening, 57*, 126959. https://doi.org/10.1016/j.ufug.2020.126959

Bedimo-Rung, A. L., Mowen, A. J., & Cohen, D. A. (2005). The significance of parks to physical activity and public health: A conceptual model. *American Journal of Preventive Medicine, 28*(2 SUPPL. 2), 159–168. https://doi.org/10.1016/j.amepre.2004.10.024

Beninde, J., Veith, M., & Hochkirch, A. (2015). Biodiversity in cities needs space: A meta-analysis of factors determining intra-urban biodiversity variation. *Ecology Letters, 18*(6), 581–592. https://doi.org/10.1111/ele.12427

Berardi, U., GhaffarianHoseini, A. H., & GhaffarianHoseini, A. (2014). State-of-the-art analysis of the environmental benefits of green roofs. *Applied Energy, 115*, 411–428. https://doi.org/10.1016/j.apenergy.2013.10.047

Bibri, S. E., Krogstie, J., Kaboli, A., & Alahi, A. (2024). Smarter eco-cities and their leading-edge artificial intelligence of things solutions for environmental sustainability: A comprehensive systematic review. *Environmental Science and Ecotechnology, 19*, Article 100330. https://doi.org/10.1016/j.ese.2023.100330

Boin, A., Ekengren, M., & Rhinard, M. (2020). Hiding in plain sight: Conceptualizing the creeping crisis. *Risk, Hazards and Crisis in Public Policy, 11*(2), 116–138. https://doi.org/10.1002/rhc3.12193

Boin, A., McConnell, A., & Hart, P. (2021). *Governing the pandemic: The politics of navigating a mega-crisis*. Palgrave Macmillan. https://doi.org/10.1007/978-3-030-72680-5

Bowler, D. E., Buyung-Ali, L., Knight, T. M., & Pullin, A. S. (2010). Urban greening to cool towns and cities: A systematic review of the empirical evidence. *Landscape and Urban Planning, 97*(3), 147–155. https://doi.org/10.1016/j.landurbplan.2010.05.006

Brook, R. D., Rajagopalan, S., Pope, C. A., Brook, J. R., Bhatnagar, A., Diez-Roux, A. V., Holguin, F., Hong, Y., Luepker, R. V., Mittleman, M. A., Peters, A., Siscovick, D., Smith, S. C., Whitsel, L., & Kaufman, J. D. (2010). Particulate matter air pollution and cardiovascular disease: An update to the scientific statement from the American heart association. *Circulation, 121*(21), 2331–2378. https://doi.org/10.1161/CIR.0b013e3181dbece1

Bruse, M., & Fleer, H. (1998). Simulating surface-plant-air interactions inside urban environments with a three dimensional numerical model. *Environmental Modelling and Software, 13*(3–4), 373–384. doi:https://doi.org/10.1016/S1364-8152(98)00042-5.

Buckley, R. (2011). Tourism and environment. *Annual Review of Environment and Resources, 36*, 397–416. https://doi.org/10.1146/annurev-environ-041210-132637

Bullard, R. D. (2000). *Dumping in Dixie: Race, class, and environmental quality* (3rd ed.). Routledge.

Büscher, B., Sullivan, S., Neves, K., Igoe, J., & Brockington, D. (2012). Towards a synthesized critique of neoliberal biodiversity conservation. *Capitalism, Nature, Socialism, 23*(2), 4–30. https://doi.org/10.1080/10455752.2012.674149

Centre for Science and Environment. (2014). *Urban rainwater harvesting: Case studies from different agro-climate regions*. Centre for Science and Environment.

Chambers-Ostler, A., Walker, H., & Doick, K. J. (2023). The role of the private tree in bringing diversity and resilience to the urban forest. *SSRN Electronic Journal*. https://api.semanticscholar.org/CorpusID:251360947

Chan, I., Lim, M., Islar, M., Gautam, A. P., Williams, M., Mungatana, E.,... & Zaleski, D. (2021). Transformative governance of biodiversity: insights for sustainable development. *Current Opinion in Environmental Sustainability, 53*, 20–28. https://doi.org/10.1016/j.cosust.2021.06.002

Chandel, R. S., Kanga, S., Singh, S. K., Đurin, B., Oršulić, O. B., Dogančić, D., & Hunt, J. D. (2023). Assessing sustainable ecotourism opportunities in Western Rajasthan, India, through advanced geospatial technologies. *Sustainability (Switzerland), 15*(14), 1–18. https://doi.org/10.3390/su151411473

Chiesura, A. (2004). The role of urban parks for the sustainable city. *Landscape and Urban Planning, 68*(1), 129–138. https://doi.org/10.1016/j.landurbplan.2003.08.003

Climate & Development Knowledge Network. (2022). *Nature-based solutions for urban climate resilience in South Asia: Cases from Bangladesh, India and Nepal*. https://cdkn.org/resource/nature-based-solutions-urban-climate-resilience-south-asia-case-studies-bangladesh-india-and-nepal

Cohen-Shacham, E., Walters, G., Janzen, C., & Maginnis, S. (Eds.). (2016). *Nature-based solutions to address global societal challenges*. IUCN. https://doi.org/10.2305/IUCN.CH.2016.13.en

Costanza, R., D'Arge, R., de Groot, R., Farber, S., Grasso, M., Hannon, B., Limburg, K., Naeem, S., O'Neill, R. V., Paruelo, J., Raskin, R. G., Sutton, P., & van den Belt, M. (1997). The value of the world's ecosystem services and natural capital. LK - https://royalroads.on.worldcat.org/oclc/4592801201. *Nature TA-TT, 387*(6630), 253–260. https://www-nature-com.ezproxy.royalroads.ca/articles/387253a0.pdf

D'Souza, R., & Nagendra, H. (2011). Changes in public commons as a consequence of urbanization: The Agara lake in Bangalore, India. *Environmental Management, 47*(5), 840–850. https://doi.org/10.1007/s00267-011-9658-8

de Vries, S., Verheij, R. A., Groenewegen, P. P., & Spreeuwenberg, P. (2003). Natural environments—Healthy environments? An exploratory analysis of the relationship between greenspace and health. *Environment and Planning A, 35*(10), 1717–1731. https://doi.org/10.1068/a35111

Dempsey, N., Brown, C., Raman, S., Porta, S., Jenks, M., Jones, C., & Bramley, G. (2010). Elements of urban form elements of urban form. https://doi.org/10.1007/978-1-4020-8647-2

Dockreill, J. E., & Shield, B. M. (2006). Acoustical barriers in classrooms: The impact of noise on performance in the classroom. *British Educational Research Journal, 32*(3), 509–525. https://doi.org/10.1080/01411920600635494

Elmahdy, N. M., Rowaida, R., Nasreldin, R., Kamel, R. R., Nasreldin, R., Rowaida, R., & Nasreldin, R. (2021). Contextualizing urban liveability indicators to create liveable neighbourhoods. *International Journal of Engineering Research and Technology, 14*(1), 56–68.

Evans, G. W., & Kantrowitz, E. (2002). Socioeconomic status and health: The potential role of environmental risk exposure. *Annual Review of Public Health, 23*, 303–331. https://doi.org/10.1146/annurev.publhealth.23.112001.112349

FAO. (2011). The state of food insecurity in the world 2011 key messages.

Folke, C., Hahn, T., Olsson, P., & Norberg, J. (2005). Adaptive governance of social-ecological systems. *Annual Review of Environment and Resources, 30*, 441–473. https://doi.org/10.1146/annurev.energy.30.050504.144511

Food and Agriculture Organization of the United Nations. (2021). The state of the world's land and water resources for food and agriculture: Systems at breaking point. In *Synthesis report* 2021. https://doi.org/10.4060/cb7654en.

Foster, J., Lowe, A., & Winkelman, S. (2011). *The value of green infrastructure for urban climate adaptation.* https://savetherain.us/wp-content/uploads/2011/10/Green_Infrastructure_Urban_Climate_Adaptation.pdf

Francis, J., Wood, L. J., Knuiman, M., & Giles-Corti, B. (2012). Quality or quantity? Exploring the relationship between public open space attributes and mental health in Perth, Western Australia. *Social Science and Medicine, 74*(10), 1570–1577. https://doi.org/10.1016/j.socscimed.2012.01.032

Gasparrini, A., Guo, Y., Hashizume, M., Lavigne, E., Zanobetti, A., Schwartz, J., Tobias, A., Tong, S., Rocklöv, J., Forsberg, B., Leone, M., De Sario, M., Bell, M. L., Guo, Y. L. L., Wu, C. F., Kan, H., Yi, S. M., De Sousa Zanotti Stagliorio Coelho, M., Saldiva, P. H. N., ... Armstrong, B. (2015). Mortality risk attributable to high and low ambient temperature: A multicountry observational study. *The Lancet, 386*(9991), 369–375. https://doi.org/10.1016/S0140-6736(14)62114-0

Ghasemi, K., Hamzenejad, M., & Meshkini, A. (2018). The spatial analysis of the livability of 22 districts of Tehran Metropolis using multi-criteria decision making approaches. *Sustainable Cities and Society, 38*(October 2017), 382–404. https://doi.org/10.1016/j.scs.2018.01.018

Giannico, V., Spano, G., Elia, M., D'Este, M., Sanesi, G., & Lafortezza, R. (2021). Green spaces, quality of life, and citizen perception in European cities. *Environmental Research, 196*, Article 110922. https://doi.org/10.1016/j.envres.2021.110922

Gleick, J. (1993a). *Genius: The life and science of Richard Feynman.* Knopf Doubleday Publishing Group. https://books.google.co.in/books?id=IWQ_y90P2uIC

Gleick, P. H. (1993b). Water and conflict: Fresh water resources and international security published by: The MIT Press Stable. http://www.jstor.org/stable/2539033 Water and Conflict Peter H. Gleick. *International Security, 18*(1), 79–112.

Gómez-Baggethun, E., Gren, Å., Barton, D. N., Langemeyer, J., McPhearson, T., O'Farrell, P., Andersson, E., Hamstead, Z., & Kremer, P. (2013). Urban ecosystem services. In T. Elmqvist, M. Fragkias, J. Goodness, B. Güneralp, P. J. Marcotullio, R. I. McDonald, S. Parnell, M. Schewenius, M. Sendstad, K. C. Seto, & C. Wilkinson (Eds.), *Urbanization, biodiversity and ecosystem services: Challenges and opportunities* (pp. 175–251). Springer.

Gorelick, N., Hancher, M., Dixon, M., Ilyushchenko, S., Thau, D., & Moore, R. (2017). Google earth engine: Planetary-scale geospatial analysis for everyone. *Remote Sensing of Environment, 202*, 18–27. https://doi.org/10.1016/j.rse.2017.06.031

Hallegatte, S., Green, C., Nicholls, R. J., & Corfee-Morlot, J. (2013). Future flood losses in major coastal cities. *Nature Climate Change, 3*(9), 802–806. https://doi.org/10.1038/nclimate1979

Hartig, T., Mitchell, R., De Vries, S., & Frumkin, H. (2014). Nature and health. *Annual Review of Public Health, 35*, 207–228. https://doi.org/10.1146/annurev-publhealth-032013-182443

Hauer, M. E., Evans, J. M., & Mishra, D. R. (2016). Millions projected to be at risk from sea-level rise in the continental United States. *Nature Climate Change, 6*(7), 691–695. https://doi.org/10.1038/nclimate2961

Higgs, C., Badland, H., Simons, K., Knibbs, L. D., & Giles-Corti, B. (2019). The urban liveability index: Developing a policy-relevant urban liveability composite measure and evaluating associations with transport mode choice. *International Journal of Health Geographics, 18*(1), 1–25. https://doi.org/10.1186/s12942-019-0178-8

Homer-Dixon, T. F. (1999). *Environment, scarcity, and violence*. Princeton University Press. https://doi.org/10.2307/j.ctt7pgg0

IARC. (2013). *IARC Biennial Report 2012–2013*. https://publications.iarc.who.int/Book-And-Report-Series/Iarc-Biennial-Reports/IARC-Biennial-Report-2012-2013

India Forum for Nature Based Solutions. (n.d.). *Case studies*. https://www.nbs4india.org/case-studies

Indian Water Partnership. (n.d.). *IUWM case studies*. https://cwp-india.org/wp-content/uploads/2020/08/Compendium-of-Case-Studies-on-Integrated-Urban-Water-Management.pdf

Ingold, K., Driessen, P. P. J., Runhaar, H. A. C., & Widmer, A. (2019). On the necessity of connectivity: Linking key characteristics of environmental problems with governance modes. *Journal of Environmental Planning and Management, 62*(11), 1821–1844. https://doi.org/10.1080/09640568.2018.1486700

IPCC. (2022). Climate change 2022: Impacts, adaptation and vulnerability. In *Contribution of working group II to the sixth assessment report of the intergovernmental panel on climate change*. Cambridge University Press.

IUCN. (2020). *Guidance for using the IUCN global standard for nature-based solutions* (1st ed.). International Union for Conservation of Nature. https://doi.org/10.2305/IUCN.CH.2020.09.en

Jyotsna, C., Sinha, B., & Bisaria, J. (2024, April). Analysis of policies and programmes for tackling coastal climate risks in India. *Ocean and Coastal Management, 253*, 107141. https://doi.org/10.1016/j.ocecoaman.2024.107141

Kabisch, N., Frantzeskaki, N., Pauleit, S., Naumann, S., Davis, M., Artmann, M., Haase, D., Knapp, S., Korn, H., Stadler, J., Zaunberger, K., & Bonn, A. (2016). Nature-based solutions to climate change mitigation and adaptation in urban areas: Perspectives on indicators, knowledge gaps, barriers, and opportunities for action. *Ecology and Society, 21*(2). https://doi.org/10.5751/ES-08373-210239

Kashef, M. (2016). Urban livability across disciplinary and professional boundaries. *Frontiers of Architectural Research, 5*(2), 239–253. https://doi.org/10.1016/j.foar.2016.03.003

Kats, G. (2006). *Greening America's Schools: Costs and benefits—A capital E Report*. https://www.usgbc.org/sites/default/files/Greening_Americas_Schools.pdf

Kaur, N., Kaur, M., Padhi, S. S., & Singh, K. K. (2022). Geospatial analysis of the distribution of urban green spaces: A study of four Indian cities. *Cities & Health, 6*(3), 443–459. https://doi.org/10.1080/23748834.2021.1941722

Kollmuss, A., & Agyeman, J. (2002). Mind the gap: Why do people act environmentally and what are the barriers to pro-environmental behavior? *Environmental Education Research, 8*(3), 239–260. https://doi.org/10.1080/13504620220145401

Kumar, M., & Singh, H. (2020). Agroforestry as a nature-based solution for reducing community dependence on forests to safeguard forests in Rainfed areas of India. In S. Dhyani, A. K. Gupta, & M. Karki (Eds.), *Nature-based solutions for resilient ecosystems and societies* (pp. 289–306). Springer. https://doi.org/10.1007/978-981-15-4712-6_17

Lahoti, S. A., Lahoti, A., Dhyani, S., & Saito, O. (2023). Preferences and perception influencing usage of neighborhood public urban Green spaces in fast urbanizing Indian City. *Land, 12*(9), 1–18. https://doi.org/10.3390/land12091664

Leach, J. M., Lee, S. E., Hunt, D. V. L., & Rogers, C. D. F. (2017). Improving city-scale measures of livable sustainability: A study of urban measurement and assessment through application to the city of Birmingham, UK. *Cities, 71*(September 2016), 80–87. https://doi.org/10.1016/j.cities.2017.06.016

Lebel, L., Garden, P., Banaticla, M. R. N., Lasco, R. D., Contreras, A., Mitra, A. P.,... & Sari, A. (2007). Management into the development strategies of urbanizing regions in Asia: Implications of urban function, form, and role. *Journal of Industrial Ecology, 11*(2), 61–81. https://econpapers.repec.org/RePEc:bla:inecol:v:11:y:2007:i:2:p:61-81

Lee, J., Park, B.-J., Tsunetsugu, Y., Ohira, T., Kagawa, T., & Miyazaki, Y. (2011). Effect of forest bathing on physiological and psychological responses in young Japanese male subjects. *Public Health, 125*(2), 93–100. https://doi.org/10.1016/j.puhe.2010.09.005

Li, Y., Beeton, R. J. S., Zhao, X., Fan, Y., Yang, Q., Li, J., & Ding, L. (2024). Advancing urban sustainability transitions: A framework for understanding urban complexity and enhancing integrative transformations. *Humanities and Social Sciences Communications, 11*(1), 1064. https://doi.org/10.1057/s41599-024-03598-x

Li, S., Xiao, Q., Teng, M., Qiu, X., Xu, W., Liu, H., Wu, X., & Wu, C. (2023). A comprehensive morphological classification scheme for local ventilation performance zones in spatially heterogeneous urban areas. *Developments in the Built Environment, 15*, Article 100202. https://doi.org/10.1016/j.dibe.2023.100202

Lowe, M., Whitzman, C., Badland, H., Davern, M., Aye, L., Hes, D., Butterworth, I., & Giles-Corti, B. (2015). Planning healthy, liveable and sustainable cities: How can indicators inform policy? *Urban Policy and Research, 33*(2), 131–144. https://doi.org/10.1080/08111146.2014.1002606

Mahmoud, I., Morello, E., Bisello, A., & Kolokotsa, D. (2024). Embedding technologies for improving nature-based solutions performance and fostering social inclusion in urban greening strategies: Augmented NBS for cities. *Urban Forestry & Urban Greening, 93*, Article 128215. https://doi.org/10.1016/j.ufug.2024.128215

Martino, N., Girling, C., & Lu, Y. (2021). Urban form and livability: Socioeconomic and built environment indicators. *Buildings and Cities, 2*(1), 220–243. https://doi.org/10.5334/bc.82

Maas, J., Verheij, R. A., Groenewegen, P. P., De Vries, S., & Spreeuwenberg, P. (2006). Green space, urbanity, and health: How strong is the relation? *Journal of Epidemiology and Community Health, 60*(7), 587–592. https://doi.org/10.1136/jech.2005.043125

Margaritis, E., & Kang, J. (2017). Relationship between green space-related morphology and noise pollution. *Ecological Indicators, 72*, 921–933. https://doi.org/10.1016/j.ecolind.2016.09.032

McPherson, E. G., Simpson, J. R., Xiao, Q., & Wu, C. (2011). Million trees Los Angeles canopy cover and benefit assessment. *Landscape and Urban Planning, 99*(1), 40–50. https://doi.org/10.1016/j.landurbplan.2010.08.011

Mentens, J., Raes, D., & Hermy, M. (2006). Green roofs as a tool for solving the rainwater runoff problem in the urbanized 21st century? *Landscape and Urban Planning, 77*(3), 217–226. https://doi.org/10.1016/j.landurbplan.2005.02.010

Miller, J. D., & Hutchins, M. (2017). The impacts of urbanisation and climate change on urban flooding and urban water quality: A review of the evidence concerning the United Kingdom. *Journal of Hydrology: Regional Studies, 12*, 345–362. https://doi.org/10.1016/j.ejrh.2017.06.006

Mohai, P., Pellow, D., & Roberts, J. T. (2009). Environmental justice. *Annual Review of Environment and Resources, 34*, 405–430. https://doi.org/10.1146/annurev-environ-082508-094348

Mouratidis, K., & Yiannakou, A. (2022). What makes cities livable? Determinants of neighborhood satisfaction and neighborhood happiness in different contexts. *Land Use Policy, 112*, Article 105855. https://doi.org/10.1016/j.landusepol.2021.105855

Nowak, D. J., Crane, D. E., & Stevens, J. C. (2006). Air pollution removal by urban trees and shrubs in the United States. *Urban Forestry and Urban Greening, 4*(3–4), 115–123. https://doi.org/10.1016/j.ufug.2006.01.007

NRC. (2012). *NRC 2012 models for estimating nutrient requirements of Pigs* (pp. 1–41).

Ostrom, E. (2010). Polycentric systems for coping with collective action and global environmental change. *Global Environmental Change, 20*(4), 550–557. https://doi.org/10.1016/j.gloenvcha.2010.07.004

Panda, & Thakur. (2019). *Making liveable cities: Challenges and way forward for India.* https://www.teriin.org/policy-brief/making-liveable-cities-challenges-and-way-forward-india

Pathak, V., Tripathi, B. D., & Mishra, V. K. (2008). Dynamics of traffic noise in a tropical city Varanasi and its abatement through vegetation. *Environmental Monitoring and Assessment, 146*(1–3), 67–75. https://doi.org/10.1007/s10661-007-0060-1

Patil, S., Rao, N., Koduganti, M., Singh, C., Poonacha, P., Sharma, S., Roy, P., Mahalingam, A., & Singh, N. (2023). *Sowing sustainable cities: Lessons for urban agriculture practices in India.* https://iihs.co.in/knowledge-gateway/sowing-sustainable-cities-lessons-for-urban-agriculture-practices-in-india/

Paul, S., & Nagendra, H. (2017). Factors influencing perceptions and use of urban nature: Surveys of park visitors in Delhi. *Land, 6*(2), 27. https://doi.org/10.3390/land6020027

Paul, S. K., & Roy, S. (2017). Tourism-based heritage conservation in Murshidabad: An appraisal. *Journal of Heritage Tourism, 12*(2), 139–154. https://doi.org/10.1080/1743873X.2016.1159685

Pauleit, S., Ambrose-Oji, B., Andersson, E., Anton, B., Buijs, A., Haase, D., Elands, B., Hansen, R., Kowarik, I., Kronenberg, J., Mattijssen, T., Stahl Olafsson, A., Rall, E., van der Jagt, A. P. N., & Konijnendijk van den Bosch, C. (2019). Advancing urban green infrastructure in Europe: Outcomes and reflections from the GREEN SURGE project. *Urban Forestry and Urban Greening, 40*, 4–16. https://doi.org/10.1016/j.ufug.2018.10.006

Pauly, D., Christensen, V., Guenette, S., Pitcher, T. J., Sumaila, U. R., Walters, C. J., Watson, R., & Zeller, D. (2002, August). Towards sustainability in world fisheries. *Nature, 418*, 689–695. www.nature.com/nature

Pillay, Y. P., & Buschke, F. T. (2020). Misaligned environmental governance indicators and the mismatch between government actions and positive environmental outcomes. *Environmental Science & Policy, 112*, 374–380. https://doi.org/10.1016/j.envsci.2020.07.010

Poffenberger, M. (2015). Restoring and conserving Khasi forests: A community-based REDD strategy from Northeast India. *Forests, 6*(12), 4477–4494. https://doi.org/10.3390/f6124382

Prüss-Ustün, A., Bartram, J., Clasen, T., Colford, J. M. J., Cumming, O., Curtis, V., …& Cairncross, S. (2014). Burden of disease from inadequate water, sanitation and hygiene in lowand middle-income settings: A retrospective analysis of data from 145 countries. *Tropical Medicine & International Health: TM & IH, 19*(8), 894–905. https://doi.org/10.1111/tmi.12329

Purohit, S., Khan, Y. D. I., Hande, O. M., & Krishanmurti, & D'Souza, M. (2024). Green livelihoods as a nature-based solution for climate change mitigation and adaptation. In W. L. Filho, G. J. Nagy, & D. Y. Ayal (Eds.), *Handbook of nature-based solutions to mitigation and adaptation to climate change* (pp. 1–21). Springer. https://doi.org/10.1007/978-3-030-98067-2_117-1

Rao, C. S., Gopinath, K. A., Prasad, J. V. N. S., Prasannakumar, & Singh, A. K. (2016). Chapter four—Climate resilient villages for sustainable food security in tropical India: Concept, process, technologies, institutions, and impacts. *Advances in Agronomy,* 140, 101–214.

Roodman, D. M., & Lenssen, N. (1995). In J. A. Peterson (Ed.), *A building revolution: How ecology and health concerns are transforming construction—Worldwatch paper 124.* Worldwatch Institute.

Saeed, U., Ahmad, S. R., Mohey-ud-din, G., Butt, H. J., & Ashraf, U. (2022). An integrated approach for developing an urban Livability composite index—A cities' ranking road map to achieve urban sustainability. *Sustainability (Switzerland), 14*(14). https://doi.org/10.3390/su14148755

Sallis, J. F., Cerin, E., Conway, T. L., Adams, M. A., Frank, L. D., Pratt, M., Salvo, D., Schipperijn, J., Smith, G., Cain, K. L., Davey, R., Kerr, J., Lai, P. C., Mitáš, J., Reis, R., Sarmiento, O. L., Schofield, G., Troelsen, J., Van Dyck, D., … Owen, N. (2016). Physical activity in relation to urban environments in 14 cities worldwide: A cross-sectional study. *The Lancet, 387*(10034), 2207–2217. https://doi.org/10.1016/S0140-6736(15)01284-2.

Samal, R., & Dash, M. (2023). Ecotourism, biodiversity conservation and livelihoods: Understanding the convergence and divergence. *International Journal of Geoheritage and Parks, 11*(1), 1–20. https://doi.org/10.1016/j.ijgeop.2022.11.001

Schlenker, W., & Roberts, M. J. (2009). Nonlinear temperature effects indicate severe damages to U.S. crop yields under climate change. *Proceedings of the National Academy of Sciences of the United States of America, 106*(37), 15594–15598. https://doi.org/10.1073/pnas.0906865106

Seddon, N., Chausson, A., Berry, P., Girardin, C. A. J., Smith, A., & Turner, B. (2020). Understanding the value and limits of nature-based solutions to climate change and other global challenges. *Philosophical Transactions of the Royal Society B: Biological Sciences, 375*(1794). https://doi.org/10.1098/rstb.2019.0120.

Sen, S., & Guchhait, S. K. (2021). Urban green space in India: Perception of cultural ecosystem services and psychology of situatedness and connectedness. *Ecological Indicators, 123*, 107338. https://doi.org/10.1016/j.ecolind.2021.107338

Shah, H., & Ramesh, R. (2022). Development-aligned mangrove conservation strategy for enhanced blue economy: A successful model from Gujarat, India. *Estuarine, Coastal and Shelf Science, 274*(June), 107929. https://doi.org/10.1016/j.ecss.2022.107929

Sharma, G., TelWala, Y., & Chettri, P. (2024, July). Integrating nature-based solutions for water security in fragile mountain ecosystems: Lessons from Dhara Vikas in Sikkim, India. *Nature-Based Solutions, 6*, 100169. https://doi.org/10.1016/j.nbsj.2024.100169

Sheffield, P. E., & Landrigan, P. J. (2011). Global climate change and children's health: Threats and strategies for prevention. *Environmental Health Perspectives, 119*(3), 291–298. https://doi.org/10.1289/ehp.1002233

Singha, P., Dasgupta, R., Hashimoto, S., & Mitra, B. K. (2024). An integrated coastal exposure modelling approach to assist mangrove ecosystem based disaster risk reduction (eco-DRR) in Tamil Nadu, India. *Progress in Disaster Science, 24*(October), 100388. https://doi.org/10.1016/j.pdisas.2024.100388

Stone, B. (2012). The city and the coming climate: Climate change in the places we live. *Cambridge University Press*. https://doi.org/10.1017/CBO9781139061353

State of Green. (2024). https://stateofgreen.com/en/

Sunyer, J., Esnaola, M., Alvarez-Pedrerol, M., Forns, J., Rivas, I., López-Vicente, M., Suades-González, E., Foraster, M., Garcia-Esteban, R., Basagaña, X., Viana, M., Cirach, M., Moreno, T., Alastuey, A., Sebastian-Galles, N., Nieu, M., & Querol, X. (2015). Association between traffic-related air pollution in schools and cognitive development in primary school children: A prospective cohort study. *PLOS Medicine, 12*(3), e1001792. https://doi.org/10.1371/journal.pmed.1001792

Suutari, A., & Marten, G. (2005). *Water Warriors: Rainwater harvesting to replenish underground water (Rajasthan, India)*. https://ecotippingpoints.com/our-stories/indepth/india-rajasthan-rainwater-harvest-restoration-groundwater-johad/

Tan, Z., Lau, K. K.-L., & Ng, E. (2016). Urban tree design approaches for mitigating daytime urban heat island effects in a high-density urban environment. *Energy and Buildings, 114*, 265–274. https://doi.org/10.1016/j.enbuild.2015.06.031

Tang, M., Hong, J., Guo, S., Liu, G., & Shen, G. Q. (2021). A bibliometric review of urban energy metabolism: Evolutionary trends and the application of network analytical methods. *Journal of Cleaner Production, 279*, Article 123403. https://doi.org/10.1016/j.jclepro.2020.123403

Telwala, Y. (2023). Unlocking the potential of agroforestry as a nature-based solution for localizing sustainable development goals: A case study from a drought-prone region in rural India. *Nature-Based Solutions, 3*(March 2022), 100045. https://doi.org/10.1016/j.nbsj.2022.100045

Termeer, C., Buuren, A. van, Dewulf, A., Huitema, D., Mees, H., Meijerink, S., & van Rijswick, M. (2017). Governance arrangements for adaptation to climate change. *Climate Science*. https://doi.org/10.1093/acrefore/9780190228620.013.600

The Energy and Resources Institute (TERI). (2022). *Urban environmental stressors in India: Annual Report 2022*. Retrieved from https://www.teriin.org

Tomlinson, C. J., Chapman, L., Thornes, J. E., & Baker, C. J. (2011). Including the urban heat Island in spatial heat health risk assessment strategies: A case study for Birmingham, UK. *International Journal of Health Geographics, 10*, 1–14. https://doi.org/10.1186/1476-072X-10-42

Tzoulas, K., Korpela, K., Venn, S., Yli-Pelkonen, V., Kaźmierczak, A., Niemela, J., & James, P. (2007). Promoting ecosystem and human health in urban areas using Green infrastructure: A literature review. *Landscape and Urban Planning, 81*(3), 167–178. https://doi.org/10.1016/j.landurbplan.2007.02.001

United Nations. (2015). The 17 goals. *Sustainable Development.*. https://sdgs.un.org/goals

Vailshery, L. S., Jaganmohan, M., & Nagendra, H. (2013). Effect of street trees on microclimate and air pollution in a tropical city. *Urban Forestry and Urban Greening, 12*(3), 408–415. https://doi.org/10.1016/j.ufug.2013.03.002

Valcárcel-Aguiar, B., Murias, P., & Rodríguez-González, D. (2019). Sustainable urban liveability: A practical proposal based on a composite Indicator. *Sustainability (Switzerland), 11*(1), 86. https://doi.org/10.3390/su11010086

Van Den Ende, M. A., Hegger, D. L. T., Mees, H. L. P., & Driessen, P. P. J. (2024). The transformative potential of experimentation as an environmental governance approach: The case of the Dutch peatlands. *Environmental Policy and Governance, 34*(5), 524–537. https://doi.org/10.1002/eet.2098

Van Kamp, I., Leidelmeijer, K., Marsman, G., & de Hollander, A. (2003). Urban environmental quality and human wellbeing: Towards a conceptual framework and demarcation of concepts; a literature study. *Landscape and Urban Planning, 65*(1), 5–18. https://doi.org/10.1016/S0169-2046(02)00232-3

Vaz, E. (2014). Managing urban coastal areas through landscape metrics: An assessment of Mumbai's mangrove system. *Ocean and Coastal Management, 98*, 27–37. https://doi.org/10.1016/j.ocecoaman.2014.05.020

Visseren-Hamakers, I. J., Razzaque, J., McElwee, P., Turnhout, E., Kelemen, E., Rusch, G. M., Fernández-Llamazares, Á., Chan, I., Lim, M., Islar, M., Gautam, A. P., Williams, M., Mungatana, E., Karim, M. S., Muradian, R., Gerber, L. R., Lui, G., Liu, J., Spangenberg, J. H., & Zaleski, D. (2021). Transformative governance of biodiversity: insights for sustainable development. *Current Opinion in Environmental Sustainability, 53*, 20–28.

Wang, D., & Xu, P.-Y. (2024). Urban green infrastructure: bridging biodiversity conservation and sustainable urban development through adaptive management approach. *Frontiers in Ecology and Evolution, 12*. https://doi.org/10.3389/fevo.2024.1440477

Wang, X. Y. (2012). Review of the research on evaluation of livable city. *Advanced Materials Research, 476–478*, 383–387. https://doi.org/10.4028/www.scientific.net/AMR.476-478.383

Wells, N. M. (2000). At home with nature: Effects of "greenness" on children's cognitive functioning. *Environment and Behavior, 32*(6), 775–795.

WHO. (n.d.). World health statistics, monitoring health for the SDG. 2016. https://www.who.int/publications/b/50930

WHO. (2010). *Urban HEART urban health equity assessment and response tool*. https://www.who.int/publications/i/item/9789241500142

WHO. (2014). World health statistics 2014. https://www.who.int/publications/i/item/9789240692671

WHO. (2017). World health statistics 2017: Monitoring health for the SDGs, sustainable development goals. https://www.who.int/publications/i/item/9789241565486

Wolch, J. R., Byrne, J., & Newell, J. P. (2014). Urban green space, public health, and environmental justice: The challenge of making cities "just green enough". *Landscape and Urban Planning, 125*, 234–244. https://doi.org/10.1016/j.landurbplan.2014.01.017

World Bank. (2023). *Nature-based solutions for climate resilience in the World Bank portfolio*. https://documents1.worldbank.org/curated/en/099201003172340531/pdf/P1768250d0db6f0c80bb5b08c648e4d0f18.pdf

World Resources Institute (WRI) India. (2022). *Urban Greens: Community-driven tree mapping in Mumbai.* Retrieved from https://wri-india.org

Zaffar, H., & Thakur, J. (2024). *Growing food and greening cities.* Nature-Based Solutions News. https://india.mongabay.com/2024/11/growing-food-and-greening-cities/

Zanella, A., Camanho, A. S., & Dias, T. G. (2015). The assessment of cities' livability integrating human wellbeing and environmental impact. *Annals of Operations Research, 226*(1), 695–726. https://doi.org/10.1007/s10479-014-1666-7

Zhao, P. (2010). Building knowledge city in transformation era: Knowledge-based urban development in Beijing in the context of globalisation and decentralisation. *Asia Pacific Viewpoint, 51*(1), 73–90. https://doi.org/10.1111/j.1467-8373.2010.01415.x

Zheng, Y. (2023). Urban governance urban governance: A review of intellectual structure and topic evolution. *Urban Governance, 3*(3), 169–185. https://doi.org/10.1016/j.ugj.2023.06.001

Zheng, S., He, A., Yu, Y., Jiang, L., Liang, J., & Wang, P. (2021). Research trends and hotspots of health—related quality of life: A bibliometric analysis from 2000 to 2019. *Health and Quality of Life Outcomes*, 1–13. https://doi.org/10.1186/s12955-021-01767-z

Ziter, C. D., Pedersen, E. J., Kucharik, C. J., & Turner, M. G. (2019). Scale-dependent interactions between tree canopy cover and impervious surfaces reduce daytime urban heat during summer. *Proceedings of the National Academy of Sciences of the United States of America, 116*(15), 7575–7580. https://doi.org/10.1073/pnas.1817561116

Participatory Design—'How' with Nature: Championing Unexpected Outcomes and Activating the Unseen

Ammon Beyerle and Ali Rad Yousefnia

Abstract The notions and approaches of design have dramatically shifted since the early 1990s regarding creative processes in the built environment. This shift manifests through participatory methodologies that influence outcomes by replacing the rationalist mindset, where control and consensus are ideal, with pluralist, agonistic, action-oriented approaches that embrace frictions. In urban design and landscape architecture, this shift coincides with the proliferation of placemaking over defining "place": process over object. (Beyerle, 2013). This chapter examines three case studies within Victoria, Australia: Burnley Backyard Community Garden (2012), Ballarat Civic Hall Site (2015), and Sunbury Community Arts and Culture Precinct (2024). Spanning more than a decade and varying in scale, these projects illustrate how nature-driven placemaking fosters economic, social, and cultural benefits while engaging with heritage contexts. A central focus is how participatory design, challenges, constraints, and lived experiences have shaped these initiatives, particularly through adaptive reuse strategies. Through these projects, the chapter underscores how participatory design and nature-driven placemaking shape expected and unexpected outcomes. It also focuses on the "how" rather than the final product of participatory project creation. By redefining priorities in community-led projects, these case studies demonstrate how placemaking strategies can create meaningful and lasting impacts that extend beyond their immediate communities. To achieve its objectives, the chapter elaborates on first-hand materials such as interviews, community consultation, and various modes of documentation. It also narrates the story of each project by highlighting its history, challenges, constraints, and lived experiences that led to nature-driven placemaking.

Keywords Participatory Design · Nature-driven Placemaking · Adaptive Reuse · Heritage · Urban Regeneration

A. Beyerle
Here Studio, Melbourne, Australia

A. Rad Yousefnia (✉)
School of Architecture, Design and Planning, University of Queensland, Brisbane, Australia
e-mail: a.radyousefnia@uq.edu.au

A. Cheshmehzangi et al. (eds.), *Nature-based Urbanism*, Urban Sustainability,
https://doi.org/10.1007/978-981-95-6864-2_7

1 Introduction

At the time of this chapter's publication, nearly five decades have passed since Hal Foster's influential edited publication *The Anti-Aesthetic: Essays on Postmodern Culture*. The publication marked a profound critique of modernism, questioning the ideological, aesthetic, and cultural underpinnings of modernist design principles. For Foster and other contributors, the modernist top-down, rationalist approaches were insufficient in meeting the demands of their time; it was necessary to have a paradigm shift and rethink multiple aspects of society, particularly design. While the modernist/rationalist approaches were in decay by the time of Foster's publication, more pluralist and participatory approaches were on the rise. Foster and colleagues' work took place concurrently with the scholarly works and the fruitful period of significant movements in civic rights, student protests, postcolonial, and environmental matters around the globe in major cities and locations (Arnstein, 1969; Flyvbjerg, 2001; Friedmann, 1969; Garfinkel, 1967; Jacobs, 1964; Mouffe, 2000; Mouffe et al., 2001). As Sanoff (2021) argues, there is a rooted association between community participation, participatory design, and involving people in development processes. In addition to civic movements, these practices are deeply related to Western social work, community radicalism and core principles of democratic societies.

While community engagements and participatory design experienced a prosperous period in the last couple of decades, Mair (2013) depicts a rather different picture for the future of democratic practices, particularly in Western societies. Mair also highlights that the disengagement and decline in civic activities created a void between the rulers and the ruled, exacerbating the growth of populist backlashes and the hollowing out of democratic institutions through these processes (Mair, 2013). Thus, participatory design and related matters could be potential micro-political apparatuses to serve as sites of critique and political change, and foster connections and bonds among the members of the society—a new notion of *institutioning* through design (Huybrechts et al., 2017). In a broader sense, the alignment of participatory design outcomes across UN Sustainable Development Goals (SDG), such as goals 3, 5, 8, 10, 11, 12, 13, 15, 17, opens up opportunities to enrich the built environment by increasing the green infrastructure, at the same time as enhancing biodiversity together with the processes to maintain it via healthier, hands on communities (Cheshmehzangi et al., 2021; Fuller et al., 2007; Shanahan et al., 2015). The collaboration between community processes and nature-based solutions is now shaping attempts to decolonise built environment practice.

Building on previous research that has largely focused on idealistic and theory-driven perspectives of participatory design and placemaking, this chapter aims to critically examine the *how*: a) what participation actually looks like in practice, b) what it means to enact it, and c) how a creative researcher or practitioner might critique or thoughtfully design a process that meaningfully engages with participatory design theory. While some emerging examples and written accounts shed light on the realities and challenges of this work, they remain limited. Arguably, the frameworks for critically engaging with these practices are fragmented. Despite the growing number of placemaking initiatives and well-intentioned participatory policies, there is a notable lack of academic literature exploring what is truly happening and what can be learnt. Thus, there is value in documenting these projects and the dynamics within them to build a more grounded understanding. This chapter seizes the opportunity to apply previously explored frameworks within a different context—*nature*-driven placemaking.

2 Participatory Design—An Overview

This section provides a brief literature review by referencing the precedents and methodologies related to creative work. While there is still significant focus on form, product, object, and individual ego in design, an urgency to find alternative methods in the forms of creation is more present in current practices. It is a broader invitation to reiterate the positionalities of spatial dimensions and their creators within the process, particularly for those who prefer alternatives over conventions. At this point, "the participatory" is a synonym for placemaking (Certeau, 1984). For decades, the importance of place has been well established and discussed in urban design theory (Norberg-Schulz, 1980; Relph, 1976). By the early 2000s, the notions of placemaking had embraced multiple notions such as process and activity, emphasising the less-debated aspects and looking at design processes by first posing "how" and "why". Notably, in object-dominated and spatial disciplines like architecture, this period marked moving away from fixed and static notions into more dynamic and fluid understandings (Kahn, 1996). In his oeuvre, *Le Droit à la ville* [The Right to the City], originally published in French in 1968, Henri Lefebvre refers to the city by emphasising the meaning of the work rather than a beautiful object (Butler, 2014; Lefebvre et al., 1996). This critical shift in the narrative highlights the importance of process rather than the final product and the beautification of objects. Instead, the design process turns into a holistic activity by bringing people, stakeholders, owners, users, authorities, investors, consultants, residents, neighbours, citizens, occupants into the story (Hughes, 1996; Till, 2009). As Sanoff (1990) analyses, participatory design offers three main potentials: 1) Making the design better fit; 2) Involving people to influence the outcome; and 3) Allowing people to feel like they have influenced the outcome. This thinking not only applies to the process, but also to the notion that there is continuity and creativity in use, which is a more recognised consideration of unfinished architecture, or even looser spaces (Franck & Stevens, 2006; Hertzberger, 1991; Hill, 2005; Lerup, 1977; Miessen & Basar, 2006).

2.1 Embracing Other Ways of Doing

The shifts and changes mentioned above influenced and redirected space-related disciplines like architecture, design and planning; hence, brought a paradigm shift in spatial practices. The formal thinking, consultations, planning, objections, and consensus evolved by embracing the user, everyday life differences and diversity. Major disruptions such as second-wave feminist theory in architecture opened new notions and understandings, such as listening matters, hearing different voices, and even valuing the unspoken, nonverbal communications, body language, etc. (Flyvbjerg, 2001; Friedmann, 1969; Richardson & Connely, 2005).

One of the pioneers in the field was Muf Art and Architecture, by valuing outcomes, narratives and more profound local benefits over hierarchical notions of power, resulting in dismantling the structures of power (Ainley & Shonfield, 2001; Hill, 1998). Regarding the site and nature, Kahn (1996) writes about the site constructions and the importance of a reciprocal relationship with a site, rather than treating the site as an object of desire being imposed by top-down rationalist ideas. For some, tensions between architecture and landscape always showed themselves in certain forms, and the landscape has usually been framed as a feminine other (Meyer, 1994). The most recent theories are coming from the third-wave feminists, such as indigenous women and their role in emphasising the importance of nature and its domains. They highlight the political role of protecting and preserving nature for future generations, which positively contributes to intersectionality (Meyer, 2008; Mies et al., 2022). The metaphors for the creation of nature-driven urban spaces, via participation and placemaking, fold in and over themselves in positive and creative manners. Thus, as one of the main objectives, the chapter considers successful existing case studies from a nature-based lens to be discussed in the following section.

2.2 Nurturing the Work by Recognising Nature

It is essential to unpack theoretical underpinnings related to practices that adapt participatory design that prioritise nature in their work. Spinuzzi (2005) defines participatory design as a methodology characterised by the way of understanding *knowledge by doing*. This definition manifests a multilayered narrative around participatory design and highlights its importance in terms of its practical aspect *vis-à-vis* the theoretical one. The application of participatory design within a socio-ecological consciousness brings attention to arguments such as nature-based thinking and the integration of "nature-based solutions" in works that support societies and communities by bringing nature to the centre of the design principles and frameworks. According to Li et al. (2023), nature-based solutions are novel concepts, replicating and mimicking natural ecosystems to sustainably mitigate anthropogenic effects in the environment. As participatory design brings humans to the centre of the design process, these meta-narratives create a holistic and inclusive perspective, instead of situating humans and nature in front of each other.

Doina Petrescu refers to Deleuze and Guattari's concept of positive desire as what underpins Ecobox, a small public community space. The project utilises the leftover sites between buildings in Paris and transforms them into a community garden. This work by aaa (*atelier d'architecture autogerée*) could be seen as one of the most articulated and creative works related to nature-driven approaches. It is a site where diverse desires intersect and interact, revealing reasons to be present by embedding all differences and interests. These reasons include a communal space for gathering, eating and drinking together, gardening classes, and sharing ideas. The space turns out to be the representation and expression of all positive desires (Petrescu, 2013).

While these spaces have high social values, they are also fundamentally political, considering inclusions and exclusions, defining the place of the community (Mouffe et al., 2001). In their seminal written work, Massey (1994) opened up new conceptions of place that were *positively* "global"—layered, with cultures, individual perspectives, familiarities, cross-purpose meanings and overlapping occupations.

Participatory design could also be applied to large-scale infrastructure projects as well, particularly in addressing the lack of greenery in cities or addressing major problems, such as urban water management. In a rapidly urbanising world where more than 50% of the population resides in cities, it is crucial to rethink some of these major discussions by perhaps combining blue, green, and even grey infrastructure to tackle problems emerging from the urban context (Li et al., 2023). For example, in non-Western contexts, the usage and distribution of water and the construction of blue infrastructure were always community-based projects—place-based participation. Roaf and Cheshmehzangi (2024) frame *qanats*, gently sloping underground channels constructed to lead water from the interior of a hill to a village below, as an example of resilient design. At the current time, *qanat* networks are the main mechanisms that feed water reservoirs and cisterns, known as Ab-Anbar in the Iranian context. Soltan Rahmati et al., (2024) identify these urban infrastructures as vital for providing water to vulnerable and underdeveloped parts of desert communities—maintained step by step. In another multi-layered and culturally rich example in India, designers bring back the people and community to the centre of blue urban infrastructure in Rajasthan province. By ensuring people are involved and at the centre of the Birkha Bawari Stepwell project, the outcomes include a guaranteed long-term life practice, which simultaneously finds success by harnessing the power of nature and storing the rainwater in the areas with high water stress forecast (Sobti & Rad Yousefnia, 2024). The provision of infrastructure in nature may be inspired by, made and maintained through participation processes. Likewise, a decolonising process unfolds, turning back to bottom-up community power in concert with traditional ways of working with nature.

While referring to the successful outcomes of applied case studies helps to contextualise the main arguments, it is essential to discuss further key definitions related to this chapter. Deutsche (1998) describes "excavations", recognising the layers of meaning and history of a place that are uncovered. Valuing community and the public over form, object, and aesthetic finds its locus in landscape architecture theory as well. In this field, compelling discourses explore how genuinely situated relationships with the landscape can give rise to new forms of knowledge that are deeply rooted in context and experience. Muecke (1994) writes about the imperceptible, knowledge, languages, and relationships that emerge when meditating on it. Indeed, through nature, a presence is shaped over aeons of vast skies, leaves, soil, and non-human actors, and offers alternative forms of consciousness that are not something to shirk or avoid. Although this view emerges more clearly within relatively post-colonial perspectives, there are certainly values to the ideas for further exploration. Meyer (2008) argues for a language of sustainability with beauty, urging the recognition of the relationships to change human perception and practices. In particular, there is a level of necessity for nature-based approaches to be aesthetically pleasing to make an open invitation—they need to be beautiful and invite everyone to care for them; their aesthetics and everyone's connection to nature matter. It is timely to think about the necessity of the overwhelming need to (re-)design cities. What would it look like to reframe the built environment processes to consider what nature drives? These actions could benefit and create the foundations of new languages, tectonics, and human/non-human relationships, more importantly, via the critical involvement of people in design processes that benefit and nurture cities, humans, and nature together (Nassauer, 1995).

3 Methodology

The chapter draws on multi-modal examples, both of actions and outputs. In doing so, it takes concepts that are relevant to the case study projects into consideration. These modes of practice are communication, governance, event, site occupation, process, and materialisation. Multiple methods have been implemented to reach the goals, such as data gathering, namely journals, photographs, interviews, key diagrams, timelines, sequences, plans, and public presentations in the public realm. Participatory work like this may often be heuristic, developed in situ in the context itself, striving for a certain aim, yet interacting and bringing about other outcomes through a positively responsive relationship to the different site(s) and stakeholders. The critical analysis is embedded in the creative work and must be considered as an open set of outcomes—research through design, by design, and about design. According to the seminal argument by Clare Bishop and Grant Kestler on critical reception, a positionality lies at the centre of these arguments, suggesting that to understand works, one must understand both their autonomous and antinomous values (Bishop, 2004, 2023, 2025; Gillick & Bishop, 2006). The chapter analyses the case studies by considering the challenges, constraints, and lived experiences and how these shaped the outcomes, the initiatives, and adaptive reuse strategies that matter to all parties involved in the design process. These arguments are taken up in recognition that participatory and nature-based works are fundamentally about (re)constructing the relationship with the relational situation and the site, embracing tensions, and frictions. Finally, the chapter returns to the materials to reflect on what happened, achieved and emerged by a nature-driven agent. More succinctly, for the purposes of this book, what was the advent of the nature/nurture elements in the expression and actions of the project, in relation to examples of community engagement, participatory design and placemaking?

4 Projects and Case Studies by Here Studio

All the projects were led by the design practice Here Studio, a registered architecture practice based in Victoria, Australia. At different times, the team included architects, landscape architects, and urban designers. The practice has its foundations in a collective of designers, social workers, and artists called Urban Village Melbourne, which focused on activating disused or underused semi-public and public buildings and sites, with a not-for-profit and public-interest purpose. The first two projects were undertaken in parallel with a PhD by Creative Works, in which the projects served as case studies, the first as a minor case, and the second as a major one (Beyerle, 2018) (Fig. 1).

4.1 Burnley Backyard Community Garden (2012)—Objectors to Advocates

- *A potted history.*

Forty-nine Tudor St, Richmond, is a small quarter-hectare site owned by the City of Yarra Council, located in a predominantly low-density residential area about 200 metres from the Birrarung (Yarra) River, in Melbourne's inner-eastern suburbs. The site previously housed a small brick police station, then a single-storey Neighbourhood Centre and Maternal and Child Health service. It sat vacant for five years, aside from informal gardening and free-ranging chickens maintained by neighbours. In 2011, after significant community opposition to a proposed sale, the Council called for expressions of interest (EOI) from community groups. Here Studio, a local architectural practice founded in 2010, submitted a proposal in partnership with the Richmond Community Learning Centre and others. Together, they developed a feasibility study via a participatory design process, aiming to "make effective use of the indoor/outdoor space … as a community facility for the next 25 years" and "retain this facility for community use." While the design scope was limited, the proposal focused on a comprehensive business model including governance, organisational structure, fundraising, a startup timeline, risk management, and community support. [A competing bid came from Cultivating Communities, a more established community gardening group, supported by Breathe Architecture.]

- *Challenges and Constraints.*

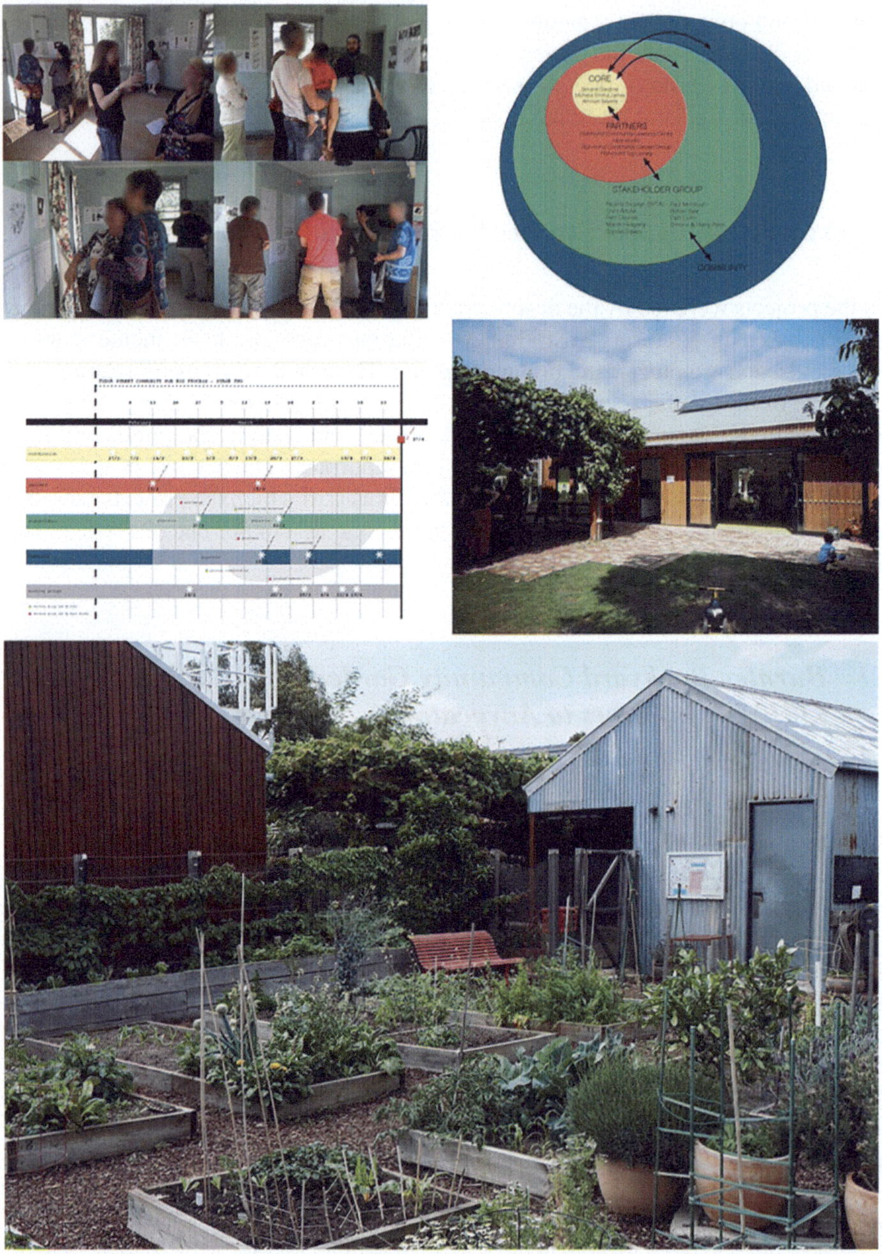

Fig. 1 Burley Backyard community Garden design processes, diagrams, consultations. Images courtesy of Here Studio (post-occupancy photography by Simon Fazio)

Broader community engagement was hampered by mistrust of the Council and burnout from previous campaigns. Residents were reluctant to discuss physical changes to the buildings, and the existing structures were not well-suited to updated community needs. The site had contamination issues, trace metals in the soil and asbestos in the buildings. Demolition was deliberately avoided in public discourse, with partners managing out the risks for the project. Two surveys were distributed to around 900 to 1200 households, and about 50 responses were received, focusing mainly on use and frequency of participation, not design. Aside from one paid manager, all partner organisations were volunteer-run and underfunded. Here Studio, being new, received little funding for its role.

The site fell under a heritage overlay and was embedded within a neighbourhood of detached, single dwellings. Meanwhile, the nearby GTV9 television station site was being redeveloped into high-end housing, including a new community centre just 350 metres away. Funding for the lead organisation was tied to existing programmes at another site, and relocation was under consideration. The EOI process ran for only 3–4 months. A key constraint was the avoidance of discussing demolition, even though the main building, located centrally and slightly north, severely limited outdoor space. The demographic—primarily older people and children—favoured highly accessible outcomes.

- *Lived Experience.*

The architects defined participatory practice through governance, event, and process. They proposed concentric working groups, stakeholder governance, and on-site events. In practice, they worked closely with the project manager, meeting almost weekly. Engagement with the built environment was difficult, shifting the focus to business planning and project operations. Surveys revealed that individuals under the age of 50 prioritised external spaces such as BBQ areas, gardens, craft spaces, and kitchens. They also suggested using these spaces for social functions and venues, such as birthday parties, childcare, and repair workshops, was preferred. On the other hand, individuals over 50 years old favoured internal meeting rooms, craft, and garden spaces, suggesting that they were more likely to be associated with book clubs and environmental groups.

The team held market stalls and two workshops. Shared priorities emerged as a multifunctional indoor-outdoor community space, a "public backyard", and a shared learning space. Architects proceeded with concept designs, coordinating engineering and costing, and producing hand-drawn plans, perspective sketches, and diagrams. These plans propose reusing and extending the old police station on the east, adding new buildings to the west, and deconstructing the main building into a garden with remnants of its form and a shade sail, leaving room for raised garden beds.

In the final month, the team hosted an on-site exhibition and community BBQ, presenting the design, including the proposed demolition, and prioritising open green space. The response was emotional and positive, with a "swell of support". At the Council meeting, residents spoke in favour of the project, which was ultimately awarded the funding. The completed Burnley Backyard features gardens and open rooms, with the learning centre's main activities relocated to the nearby GTV9 site.

Fig. 2 View of the communal area. Image courtesy of Here Studio, photography by Simon Fazio

Photos show designated areas for the events and green spaces with strong indoor-outdoor integration (Figs. 2, 3).

4.2 *Ballarat Civic Hall Site (2015)—Landscape Prioritised*

- *A Potted History.*

In 2014, Ballarat, a regional city 100 km northwest of Melbourne with 100,000 residents, was undergoing change. Once stagnant in its colonial past, it was now exploring more participatory and innovative approaches for its built environment. In 2012–13, tensions peaked when the City of Ballarat proposed demolishing the 1955 Civic Hall to develop government offices on a 1.5-hectare central site. The site included a 1990s library, a skate park and a car park area. The hall had sat derelict for 15 years. Opinions in the community were polarised, from full preservation to complete demolition.

Here Studio, then based in Ballarat, initiated a participatory design process and was later commissioned to develop masterplan options. The approach included an "Open Door Studio" in the Civic Hall, incorporating art and public events. A phased

Fig. 3 Ballarat Civic Hall Site and Community Engagement Stages. Images courtesy of Here Studio (post-occupancy photography by Simon Fazio)

process helped distil hundreds of community ideas down to three key options, each of which was presented to the Council for support.

- *Challenges and Constraints.*

Decades of mistrust between the Council and the public led to entrenched positions and blame among stakeholders. The Civic Hall site had a heritage overlay, poor permeability, contamination, and an unconventional and irregular building layout. Its economic viability was uncertain, especially following the global financial crisis, budget constraints and state-imposed rate caps. Local governments had limited ability to fund infrastructure through value capture, further constraining the project's feasibility.

- *Lived Experience.*

Here Studio structured participation around communication, events, and governance. The design studio operated from the Civic Hall's foyer, complemented by newsletters, artists-in-residence, and frequent events. Outdoor activations were built on a tradition of protest concerts on the site's southern steps. Workshops included a Welcome to Country and community walks.

Designs were presented through exhibitions and summarised to the state government architect. Four design principles were prioritised. ***Public***, including concepts and components such as civic pride, shared resources, shared spaces, community, diversity, dialogue; and hope; ***Sustainability***, including climate change mitigation and adaptation, social cohesion, economic prosperity and emphasis on adaptive reuse; ***Activation and Access*** through revitalisation of the CBD area by implementing commercial activities, and alternating transportation modes, with boosting walkability and cycling, prioritising health and well-being. Finally, ***Innovation*** by leading changes in Ballarat through prototyping, showcasing, and testing new industries; also focusing on library, housing typologies and architectural qualities. Although the process extended to 18 months, Here Studio's involvement ended shortly after the CEO's departure. Within a vacuum of authority, the wider participatory momentum waned. Over a year later, the state government commissioned a major office building, while the Council had begun developing the plaza and landscape areas, retaining the main hall (Fig. 4).

4.3 Sunbury Community Arts and Culture Precinct (2024)—Occupy before Design

- *A Potted History.*

Jacksons Hill, located in the outer-suburban town of Sunbury, 30 km north of Melbourne, is a culturally significant site for traditional owners. Formerly a mental health asylum (1879–1980s), then a university campus until 2018, the 32-hectare site remained vastly underutilised until Hume Council acquired part of the site in 2022.

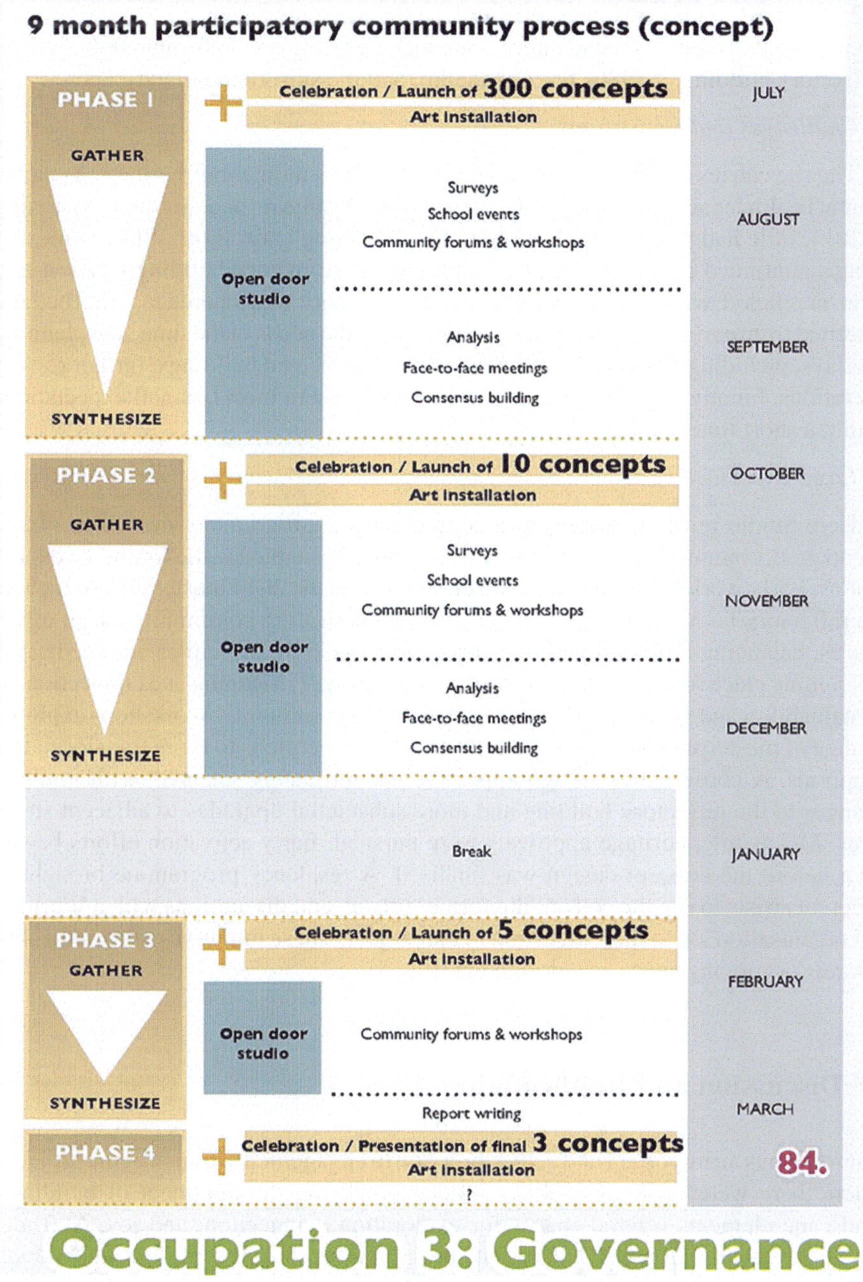

Fig. 4 Nine-month participatory community process and its phases. Diagram courtesy of Here Studio

In 2024, Here Studio was appointed to lead a participatory concept design process for stage one of a new arts and culture precinct, focusing on the former 1894 Female Refractory building, a 1980s brick structure, and the surrounding landscapes.

- *Challenges and Constraints.*

The site carries complex layers of meaning, such as Indigenous heritage, a painful mental health legacy, and years of delayed action. Despite prior community visioning in 2014, little had progressed, leading to long-standing frustration. While some arts groups continued to occupy nearby buildings, the refractory building's prison-like form conflicted with the project's ideals of creativity and openness. The budget, inherited from earlier project phases, did not reflect the needs of the time, and planning overlays, including Aboriginal heritage and state-protected buildings, further restrict alterations. Finally, the project required swift delivery to meet Council expectations within a short timeframe.

- *Lived Experience.*

Here Studio initially developed a comprehensive programme via modes of site occupation, communication, and process, but quickly adapted to the Council's expectations. Early workshops identified the need to revise the 2014 master plan to include the refractory building. The process focused on activating community usage of the spaces, balancing ambition with feasibility. Four guiding principles emerged: 1) A welcoming place for a connected, creative community, 2) Meaning and connection, 3) Sustainability and success, 4) Empowerment of the community. Workshops explored a range of masterplan options, from light-touch interventions to more transformative proposals. A conservative design approach was ultimately selected, with minimal changes to the refractory building and more substantial upgrades to adjacent structures. Meanwhile, heritage approvals were pursued. Early activation efforts began, even before the concept design was finalised. A residency programme brought in different artists to use the cells of the former prison, and discussions with a Wurundjeri organisation led to their inclusion in the project. These initiatives reflected a shift towards occupying space as a design act (Fig. 5).

5 Discussion and Implications

Nature brings an important and active direction to engagement. In each of the projects, where there were some difficulties in conceptualising the treatment of buildings, landscape elements offered spaces for exploration, connection, and joy. At Tudor Street, the discussion centred on activity programming, business models, and indoor-outdoor relationships. At Ballarat Civic Hall, contention over the building stalled progress, yet the community readily embraced outdoor events, where even small acts, like placing potted plants or sitting in the sun, made a visible activation of space by utilising natural elements. At Sunbury, conversations around building form were slow. On the other hand, landscape concepts evolved more freely and frequently.

Fig. 5 Site and context analysis with three design alterations to gather information based on participatory design. Images courtesy of Here Studio

Across all projects, on-site workshops emerged as key moments for understanding and collaboration between people. The natural landscape, alive in contrast to derelict buildings, became a medium for connection.

In nature-driven projects, nature is not only a backdrop but often the defining feature. Participatory design becomes a way to "breathe new life" into both old buildings and the broader site. At Tudor Street, outdoor areas became the focus, with the building playing a supporting role. In Ballarat, public space took precedence, with soft and hard landscapes being the first constructed elements. At Sunbury, early activation came through artist residencies and open grounds, developing elements like entry paths and courtyards in the master plan. Key themes emerge from these interactions: while adaptive reuse often prioritises buildings, it is the landscape that people find less contentious yet more meaningful. Secondly, nature is tangible, relatable, and open to care, elements harder to assert over contested architecture. These open spaces, though rarely positioned as central, are often delivered as the most visible outcomes.

The experience of working in settings like the aforementioned case studies was itself transformative. In Tudor Street, walks from the tram passed birds, gardens, and the sound of domestic animals like chickens. In Ballarat, daily paths from residential areas to the sites framed Mt. Warrenheip, past eucalyptus trees, and familiar streets. At Sunbury, walking through the terraced asylum grounds led to the former prison courtyards, where sky glimpses framed cells that once held women, and after meetings, looking down from the hill towards sacred Wurundjeri lands towards Narrm (Melbourne). Walking in nature anchored the process in place. Here, landscape was not just a setting—it became host, guide, and destination.

While each project aimed for building and site reuse, the momentum often came through nature. Unexpected outcomes emerged—often unplanned, sometimes unimagined. These are not failures but markers of success in participatory placemaking. At Tudor Street, a sunny BBQ during the final exhibition showed how community occupation becomes design in action. In Ballarat, activation of outdoor space persisted despite political tension over the hall. At Sunbury, early artist residencies and site access gave the community something to engage with long before construction. Nature made this possible. These interventions show how landscape occupation fosters culture and a sense of place beyond built forms.

Ultimately, nature introduces agency. When built form reflects unresolved conflict, the living landscape becomes a space of resolution. Participatory practice grounded in place invites embodied, shared regeneration—helping people realign priorities, away from governance, risk and building management towards simply being-in and with-the-world (Heidegger, 2001).

6 Concluding Marks

This chapter shares insights on the space between vision-making and process. It suggests early visions into how nature can act as an agent for change. While in some of the theories and practices, nature is often treated as "other:" and is alienated within the modernist discussions, the case studies discussed in the chapter suggest that nature may be nested within the consciousness of projects. In shaping places, nature is often seen as an external responsibility or collective good, yet it is also an activation mechanism, expressing ways of being. Built environment practitioners may feel the pressure to deliver technical solutions that reintroduce nature into urban contexts or give nature a voice within their work. However, adopting a place-based, participatory approach could serve the purpose and open constructive ways forward, revealing possibilities that extend beyond rational methods, and eventually boost nature-based solutions in all means. Hand in hand, nature can facilitate the reactivation of human-made environments. When integrated meaningfully, it supports sustainability goals while also generating shared and, in some cases, unexpected methods for long-term project continuity. Nature carries an unlocking capacity and a logic of its own, one that humans can participate in as part of nature. Engaging with nature's agency through design is a practice that deepens relationality, especially when it is recognised that rationality alone is insufficient.

References

Ainley, R., & Shonfield, K. (Eds.). (2001). *This is what we do: A muf manual*. Ellipsis.
Arnstein, S. R. (1969). A ladder of citizen participation. *Journal of the American Institute of Planners, 35*(4), 216–224. https://doi.org/10.1080/01944366908977225
Beyerle, A. (2013). Agonistic participation: A political and architectural opportunity. *Journal of Arts & Communities, 5*(2), 147–169. https://doi.org/10.1386/jaac.5.2-3.147_1
Beyerle, A. (2018). *Participation in architecture: Agonism in practice*. The University of Melbourne.
Bishop, C. (2004). Antagonism and relational aesthetics. *October, 110*, 51–79.
Bishop, C. (2023). *Artificial hells: Participatory art and the politics of spectatorship*. Verso books.
Bishop, C. (2025). The social turn: Collaboration and its discontents. In *The performance studies reader* (pp. 250–257). Routledge.
Butler, C. (2014). *Henri Lefebvre: Spatial politics, everyday life and the right to the city* (1. Issued in paperback). Routledge.
Certeau, M. d. (1984). *The practice of everyday life*. University of California Press.
Cheshmehzangi, A., Butters, C., Xie, L., & Dawodu, A. (2021). Green infrastructures for urban sustainability: Issues, implications, and solutions for underdeveloped areas. *Urban Forestry & Urban Greening, 59*, 127028. https://doi.org/10.1016/j.ufug.2021.127028
Deutsche, R. (1998). *Evictions: Art and spatial politics* (2. Aufl (Online-Ausg.)). The MIT Press.
Flyvbjerg, B. (2001). *Making social science matter: Why social inquiry fails and how it can succeed again* (S. Sampson, Trans.). Cambridge University Press. https://doi.org/10.1017/CBO9780511810503
Foster, H. (Ed.). (1983). *The anti-aesthetic: Essays on postmodern culture* (1st ed.). Bay Press.
Franck, K. A., & Stevens, Q. (2006). *Loose space: Possibility and diversity in urban life*. Routledge.

Friedmann, J. (1969). Notes on societal action. *Journal of the American Institute of Planners, 35*(5), 311–318. https://doi.org/10.1080/01944366908977241

Fuller, R. A., Irvine, K. N., Devine-Wright, P., Warren, P. H., & Gaston, K. J. (2007). Psychological benefits of greenspace increase with biodiversity. *Biology Letters, 3*(4), 390–394. https://doi.org/10.1098/rsbl.2007.0149

Garfinkel, H. (1967). *Studies in ethnomethodology*. Prentice-Hall.

Gillick, L., & Bishop, C. (2006). Letters and responses. October, 115, 95–107. JSTOR.

Heidegger, M. (2001). *Poetry language thought* (A. Hofstadter, Trans.). Harper and Row.

Hertzberger, H. (1991). *Lessons for students in architecture*. Uitgeverij 010 publishers.

Hill, J. (Ed.). (1998). Shared ground. In *Occupying Architecture*. Routledge.

Hill, J. (Ed.). (2005). *Occupying architecture: Between the architect and the user* (0th ed.). Routledge. https://doi.org/10.4324/9780203983829

Hughes, F. (Ed.). (1996). *The architect: Reconstructing her practice*. MIT Press.

Huybrechts, L., Benesch, H., & Geib, J. (2017). Institutioning: Participatory design, co-design and the public realm. *CoDesign, 13*(3), 148–159. https://doi.org/10.1080/15710882.2017.1355006

Jacobs, J. (1964). *The death and life of great American cities*. Penguin Books.

Kahn, A. (1996). Overlooking: A look at how we look at site or… site as "discrete object" of desire. In D. McCorquodale, K. Ruedi, & S. Wigglesworth (Eds.), *Desiring practices*. Black Dog Publishing.

Lefebvre, H. (1996). The right to the city. *Writings on Cities, 63181*.

Lefebvre, H., Kofman, E., & Lebas, E. (1996). *Writings on cities*. Blackwell publ.

Lerup, L. (1977). *Building the unfinished: Architecture and human action (1st print)*. Sage Publ.

Li, L., Chan, F., & Cheshmehzangi, A. (2023). Nature-based solutions and sponge city for urban water management. In *Adapting the built environment for climate change* (pp. 371–402). Elsevier. https://doi.org/10.1016/B978-0-323-95336-8.00004-4

Mair, P. (2013). *Ruling the void: The hollowing of Western democracy*. Verso.

Massey, D. B. (1994). *Space, place, and gender*. University of Minnesota Press.

Meyer, E. K. (1994). Landscape architecture as modern other and postmodern ground. *The Culture of Landscape Architecture*, 13–34.

Meyer, E. K. (2008). Sustaining beauty. The performance of appearance: A manifesto in three parts. *Journal of Landscape Architecture (Wageningen, Netherlands), 3*(1), 6–23. https://doi.org/10.1080/18626033.2008.9723392

Mies, M., Shiva, V., & Salleh, A. (2022). *Ecofeminism*. Bloomsbury Academic.

Miessen, M., & Basar, S. (Eds.). (2006). *Did someone say participate? An atlas of spatial practice: A report from the front lines of cultural activism looks at spatial practitioners who actively tresspass into neighbouring or alien fields of knowledge*. MIT Press.

Mouffe, C. (2000). *The democratic paradox*. verso.

Mouffe, C., Deutsche, R., Joseph, B. W., & Keenan, T. (2001). Every form of art has a political dimension. *Grey Room, 2*, 99–125.

Muecke, S. (1994). From honeymoon gap to millions of stars: Aboriginal landscape and the aesthetics of disappearance. In *The culture of landscape architecture* (pp. 69–84). Melbourne, Victoria, Edge Publishing.

Nassauer, J. I. (1995). Messy ecosystems, orderly frames. *Landscape Journal, 14*(2), 161–170.

Norberg-Schulz, C. (1980). *Genius loci: Towards a phenomenology of architecture*. Rizzoli.

Petrescu, D. (2013). Losing control, keeping desire. In *Architecture and participation* (pp. 43–63). Routledge.

Relph, E. C. (1976). *Place and Placelessness*. Pion.

Richardson, T., & Connely, S. (2005). Reinventing public participation: Planning in the age of consensus. In J. Blundell-Jones, D. Petrescu, & J. Till (Eds.), *Architecture and participation* (1st ed., pp. 77–104). Routledge.

Roaf, S., & Cheshmehzangi, A. (2024). Persian vernacular buildings—Lessons on resilient design. In A. Cheshmehzangi & S. Roaf (Eds.), *Persian vernacular architecture* (pp. 1–12). Springer Nature. https://doi.org/10.1007/978-981-96-1116-4_1

Sanoff, H. (1990). *Participatory design: Theory & techniques*. H. Sanoff.
Sanoff, H. (2021). Participatory design: A historical perspective. *Journal of Arts & Architecture Research Studies, 2*(3), 12–21. https://doi.org/10.47436/jaars.2021.67348.1001
Shanahan, D. F., Fuller, R. A., Bush, R., Lin, B. B., & Gaston, K. J. (2015). The health benefits of urban nature: How much do we need? *Bioscience, 65*(5), 476–485.
Sobti, M. P., & Rad Yousefnia, A. (2024). Beyond water: Step Wells and community hydraulics in modern India. In A. Cheshmehzangi, M. Sedrez, & A. Flynn (Eds.), *Rethinking Stormwater management through sustainable Urban Design* (pp. 273–292). Springer Nature). https://doi.org/10.1007/978-981-97-4924-9_14
Soltan Rahmati, H., Torkshavand, H., & Amiribolvardi, M. (2024). Ab-Anbars of Yazd: A study of geographical and historical distributions of cisterns in a Desert City. In A. Cheshmehzangi & S. Roaf (Eds.), *Persian vernacular architecture* (pp. 33–55). Springer Nature. https://doi.org/10.1007/978-981-96-1116-4_3
Spinuzzi, C. (2005). The methodology of participatory design. *Technical Communication (Washington), 52*(2), 163–174.
Till, J. (2009). *Architecture depends*. MIT press.

Multi-Species Urbanism and Coexistence in Small-Scale Nature-Based Solutions: A Comparative Study between Global North and South Cities

Diana M. Benjumea Mejia and Tania Giraldo Ospina

Abstract This paper explores how small-scale nature-based solutions (NbS) can encourage coexistence among multiple species by comparing the cities of Singapore and Manizales, Colombia. The study recognises the differences in agency among species and takes a spatial design approach, assessing five dimensions of multi-species spatial integration in urban environments: spatial representation (agency), data collection, ecological indicators, infrastructure metrics, and citizen participation. A normative and policy analysis was performed to evaluate urban planning strategies that incorporate NbS and support multi-species coexistence and agency in both cities. The research employed principles of multi-species studies to explore further interspecies spatial agency, including walking ethnographies, transect surveys, and ecological mapping. Results reveal a consistent gap between official planning narratives and on-the-ground practices, with multi-species considerations often marginalised in formal planning, even where strong green infrastructure exists. Conversely, in areas with limited formal governance, resident-led initiatives foster diverse multi-species habitats, indicating that informal practices can sometimes fill normative gaps. Factors such as open access and proximity to mixed land uses have been shown to enhance ecological health and social engagement. Strategies that recognise the inherent right of natural biodiversity to coexist in urban and non-urban settings are necessary, as they may offer unique perspectives on designing multi-species urban futures. The article presents a flexible, transferable assessment tool designed for the Global South, offering a practical framework to evaluate green infrastructure from the perspective of multi-species coexistence.

Keywords Multi-species design · Nature-based solutions · Spatial agency · Ecological assessment · Multi-species and coexistence strategies

---

D. M. Benjumea Mejia (✉)
College of Interdisciplinary and Experiential Learning, Singapore University of Social Sciences, Bukit Timah, Singapore
e-mail: dianabenjumea@suss.edu.sg

D. M. Benjumea Mejia · T. Giraldo Ospina
The National University of Colombia, Caldas, Colombia
e-mail: tgiraldoo@unal.edu.co

A. Cheshmehzangi et al. (eds.), *Nature-based Urbanism*, Urban Sustainability,
https://doi.org/10.1007/978-981-95-6864-2_8

1 Introduction

Nature-based solutions (NbS) are recognised as a crucial approach to addressing numerous urban sustainability challenges that cities face. The solutions offer multifunctional benefits, such as enhancing ecological, social, and economic outcomes. Projects developed in both the Global North and South countries emphasise the importance of financial and ecological value, reinforcing a shifting narrative that these benefits are not mutually exclusive (de Groot et al., 2012). However, a critical issue with this perspective is the increasing commodification of nature, which has led to green gentrification in both contexts (Anguelovski et al., 2019). Empirical studies conducted in Colombia, the USA, and Barcelona have demonstrated that large-scale green interventions designed to attract tourism while preserving natural ecosystems can have unintended socio-environmental consequences. A key challenge is the environmental impact associated with mass tourism, which places pressure on fragile ecosystems. Furthermore, these interventions often exacerbate social inequalities, as rising property values and rental costs displace lower-income communities from their neighbourhoods (Anguelovski, 2023). Another significant limitation lies in the governance and management of NbS, which are frequently controlled by private and governmental entities due to the scale of such projects. This top-down approach often minimises meaningful and active public participation, limiting the agency of local communities in the planning and long-term stewardship of these interventions.

In this study, we explore the concepts of participation and agency as interconnected yet distinct in their application to urban spatial studies. Active participation in public spaces, for example, is widely used in planning and urban discourses (Dorst et al., 2019; Qi et al., 2024), with scholars often referring to active citizen participation in public spaces as strategic planning (De Magalhães & Carmona, 2009), which entails a long-term focus encompassing the planning, design, execution, and maintenance of public spaces. This approach, as argued by scholars, embraces democratic governance that is often horizontal and encourages community participation with shared responsibility beyond the state, private sector, and institutions cooperating voluntarily (Dempsey & Burton, 2012; Randrup et al., 2021; Randrup & Jansson, 2020). Agency, on the other hand, refers to the capacity to act and influence change, actively sustaining interventions within the space. Often, agency leads to advocacy and might help to transform social dynamics (Geddes et al., 2024; Kok et al., 2021). The main distinction with participation is that spatial processes may also be led by external stakeholders and unfold intermittently, which might not necessarily lead to advocacy and long-term effects over time.

Participation and agency are not necessarily discussed within NbS, given the prevalent importance of integrating NbS in urban green infrastructure and building design (Dorst et al., 2019). Nevertheless, ensuring citizen participation in NbS projects, particularly in small-scale initiatives in urban spaces, can facilitate meaningful involvement in NbS networks. Communities can play a vital role in supporting and maintaining key ecological functions (Wolff et al., 2022), as well as help to prevent green gentrification and the commodification of nature. Of particular interest for this study is the opportunity to create cohabitation spaces between humans and the biotic diversity.

Several design frameworks have guided the integration of natural elements into buildings and urban green spaces, they, too, are often conceived to maximise the human experience and enhance their well-being (Myers, 2019). Biophilic design is an example; despite being widely accepted, the design strategies often overlook the agency of nature's biotic diversity and focus on the benefits humans can derive, fostering a transactional relationship. The biotic diversity is then reduced to symbolic, decorative, or therapeutic elements rather than being recognised as a network of active agents. In the NbS discourse, an anthropocentric view is celebrated, emphasising the use of nature for the benefit of built environment users rather than promoting a balanced perspective where both humans and biodiversity benefit mutually. Many cases exist where nature is forced to survive in artificial urban spaces or buildings, often undermining the health of the introduced natural ecosystems (Apfelbeck et al., 2020; Myers, 2019). This results in a scarcity of native flora in urban environments, with an overgrowth of synanthropic species and a decline in local species diversity (Russo et al., 2025).

To counter such a dominant view over nature and its biotic diversity, scholars have advocated for more comprehensive strategies that incorporate multi-species design (Apfelbeck et al., 2020; Croeser, 2024; Parris et al., 2018), recognising the contributions, agency, representation, and participation of nature's biological diversity in space as equally important as human participation (Elton, 2021). We propose multi-species agency as an alternative to overcome the opportunistic and reductionist view of natural ecosystems in design endeavours. From this perspective, agency is understood as the capacity of more-than-human entities (e.g., natural ecosystems) to act and influence outcomes within their environment (Cianchi, 2015). This concept highlights the interconnectedness and interdependence of entities, viewing more-than-human beings as active agents with their own forms of agency that influence human experiences (Duerbeck et al., 2015). Building on Atchison's discussion of inter-species agency, both humans and biological diversity exercise agency through established relationships, often without intention. In this context, human agency exists within a network of multiple actants (such as organic biological organisms), which are continuously shaped by ongoing interactions. However, while living agents actively participate in shaping space, their modes of presence and expression, unique to their species, should be recognised and regarded as part of their spatial representation beyond human intervention.

This study highlights the importance of citizen and more-than-human participation in NbS for ensuring equitable spatial coexistence. We expand the idea of multi-species coexistence by recognising the agency of various biotic entities as a form of spatial representation, building on Kohn's (2013) insights. This approach mainly seeks to spark conversations about the dominance of human-centred discourses and the importance of re-evaluating traditional planning methods by including non-human actors (Bragança, 2025). In the end, it fosters the possibility of alternative, relational, and collective worlds via multi-species design.

Two contrasting case studies are presented, each with distinct strengths. Manizales, Colombia, is highlighted for its public spaces that function on multiple levels and align with the specific visions and needs of local communities. Following a community-driven approach, the community organisation plays a central role in planning urban green public spaces and their design. Long-term resource management is also crucial in these approaches that adopt a communal governance model, introducing a more balanced and democratic method of managing urban green spaces. These spaces often serve as venues for citizen engagement and environmental education. Nevertheless, despite the importance of these interventions, their contribution to the NbS city infrastructure for multi-species coexistence is often overlooked by the planning authorities; hence, the importance of evaluating their contributions. Conversely, Singapore is recognised as a pioneer in NbS innovations, integrating solutions into building facades and interiors that emulate natural ecosystems in service buildings and macro urban green infrastructure. Given the numerous successful projects implementing NbS with a strong emphasis on ecosystem services, some design guiding principles could inform further multi-species design strategies (Croeser, 2024; Kabisch et al., 2022; Maller, 2021). It is important to mention that despite its success, most of these large-scale projects receive peripheral contributions from citizens due to the high maintenance of these spaces, which often require careful monitoring to avoid uncontrolled growth of plant species that could compromise the built infrastructure (Tan et al., 2021).

This study introduces a relational and process-based understanding of design, where biotic entities are not merely objects of contemplation or management, but co–constituents of urban ecosystems with autonomous capacities for movement, adaptation, and resilience. Five dimensions are explored to examine principles for multi-species spatial integration in urban areas: representation in space (agency), data collection, ecological indicators, infrastructure indicators, and ultimately, citizen participation. This includes analysing urban planning and design norms to understand urban green infrastructure from a governance perspective. We focus on small-scale NbS interventions in urban gardening and farming, as they encourage citizen engagement and allow us to explore the five dimensions for designing coexistence in small-scale NbS.

1.1 Citizen Participation/Engagement in the Management of NbS

NbS infrastructure has been widely implemented as an alternative to incorporate simulated natural environments into urban spaces, which can support new ecosystems and enhance biodiversity. This helps, among other reasons, to address some challenges of urbanisation, such as climate mitigation (Depietri and McPhearson, 2017). NbS are often incorporated into urban greening projects aimed at increasing urban ecosystem services by utilising design interventions that integrate nature within green and blue infrastructures, also referred to as "natural, semi-natural, and man-made green and blue features" (Nesshöver et al., 2017). Using ecological engineering techniques, these designs insert the human environment into natural ones, becoming one of the most acclaimed strategies in recent years to solve the numerous challenges faced in rapidly urbanising cities. Despite the numerous benefits of such interventions and their widespread popularity in both the global North and South, discussions about citizen and community participation in implementing NbS are widely recognised as essential for ensuring that NbS can integrate in local environments and respond to their needs (J Skodra et al., 2020; Wolff et al., 2022).

Another relevant aspect of enabling participation is its contribution to social systems and the associated ecological benefits, such as food production, income generation, and increased agency in green spaces (Parris et al., 2018). However, establishing or measuring the extent to which human agency is successfully enacted or generated in NbS projects is often difficult, as community intervention occurs through a mediated process among stakeholders, influenced by varying degrees of participation. These processes are subject to stakeholders' participatory mechanisms, methodologies, and agendas, which may or may not foster long-term agency and encourage deeper community engagement, especially among those heavily exposed to climate hazards (i.e., lower-income and informal settlements) (Wolff et al., 2022).

Although efforts to integrate participatory processes within NbS are gaining traction, particularly in the Global South countries of Southeast Asia and Africa, of special interest are the instances where communities already demonstrate proactive dynamics, either by directly looking after biodiversity well-being or through citizen science by surveying a wide range of taxa across different environments to populate databases (Parris et al., 2018).

1.2 Multi-Species in NbS

Further along the line of spatial agency and citizen participation in NbS infrastructure is the question of moving beyond human-centric approaches towards urban spaces that are inclusive of the myriad of more-than-human species that also share urban spaces with us (Benjumea Mejía, 2025). These include animals, plants, fungi, and microorganisms. The concept of multi-species design and urbanism seeks to explore these interactions, recognising the need for design strategies that account for the complexities of interspecies relationships and their diverse ecological requirements (Pineda-Pinto et al., 2023). Several disciplines have delved further into the benefits of designing with and for multi-species interactions (Groutars et al., 2024). Fields such as biodesign and human-computer interaction have explored the role of non-human actors in artefact design, and architecture and urban design have recently begun to investigate how multi-species perspectives can be integrated into spatial planning and ecological interventions in urban areas (Grobman et al., 2023; Shingne, 2022; Van Patter, 2023). However, there is limited discussion in the field of NbS.

Currently, approaches to NbS often focus on computational predictive modelling to assess the performance of newly designed natural environments (Matos & Roebeling, 2022; Mosca et al., 2025). These models generate indicators such as CO2 sequestration, temperature regulation, and rainfall control, among other environmental factors. Simulations enable designers to predict probable scenarios that maximise the benefits of NbS for human-centred environments while optimising resource allocation. Additionally, they provide quantitative insights into the ideal ratios of trees, plant species, and water bodies, offering a data-driven depiction of the projected environmental performance of urban green spaces (Tan et al., 2021).

While these approaches to urban design provide a more detailed overview of ecosystem performance, it is also important to critically assess the extent to which such simulations consider multi-species coexistence (Huang et al., 2024). A predominant focus on quantifiable environmental performance risks reducing biological diversity to a set of inserted variables, potentially overlooking the complexity of ecological systems. The result might be green spaces designed and integrated based on expected ecological outcomes and controlled environmental behaviours rather than fostering adaptive and dynamic interactions among species, which could diminish their resilience and adaptive capacity (Calvo & Lawrence, 2023).

Recent scholarship has raised ethical concerns regarding the instrumental framing of nature. Modelling tools, while valuable, often obscure the agency of non-human species, and the simplification of ecological networks can produce unintended disruptions to urban biodiversity (Buckley et al., 2024; Moyano et al., 2024; Standish & Parkhurst, 2024). Furthermore, the persistent quantification of nature and other species reflects a systematic exclusion and epistemological bias in urban governance, where interspecies relationships are framed primarily in terms of human benefit, and trade-offs are rarely evaluated from a multi-species perspective often neglecting the broader consequences for more-than-human life and ecosystems (Tabara & Chabay, 2013; Boros et al., 2024).

To move beyond human-centred NbS, emerging approaches in multi-species urbanism call for the recognition of more-than-human life forms as active agents capable of shaping their own spatial conditions and ecological interactions (Grobman et al., 2023; Shingne, 2022; Van Patter, 2023). This perspective aligns with frameworks that consider urban space as co-produced by human and more-than-human actors (Robinson et al., 2021). From this perspective, plants, trees, animals, microorganisms, soil, and water are not passive components in ecological design, but rather relational participants with the capacity to influence, adapt, and transform their environments.

Conceptually, multi-species agency challenges the anthropocentric foundations of spatial practice. It invites researchers to view both human and more-than-human agents as co-constitutive actors in shaping the quality, function, and experience of urban green spaces (Hernandez-Santin et al., 2022). Human interactions with nature, including sensory perception, social dynamics, and governance structures, determine not only how environments are utilised but also how they evolve. For instance, community-led green spaces tend to foster inclusive, convivial forms of stewardship shaped by shared norms and adaptive rules (Benjumea et al., 2022). Conversely, self-led green spaces allow for individual, self-directed interactions, with social exchanges driven by personal choice. In state-owned urban areas, spatial features and interactions are mediated and controlled; similar restrictions are also found in privately owned green spaces.

However, defining non-human agency remains an epistemological challenge. Unlike humans, more-than-human entities cannot communicate spatial preferences through verbal or symbolic language. Scholars such as Marder (2013) who attempt to rationalise species experience through human frameworks are inherently limited. Instead, agency should be understood as emerging through relational processes and multi-species communication and spatial representation, whereby organisms respond to ecological cues, form mutual dependencies, and co-adapt in shared environments, participating in the configuration of space (Elton, 2021). While this may be more readily observable in less anthropogenically altered systems, such as the Amazon (Kohn, 2013), the urban context poses unique challenges. Species introduced into fragmented, artificial, or non-native ecologies often depend on human maintenance and intervention for their survival (Atchison et al., 2024).

A core question that persists is how to identify and evaluate biotic agency within urban spaces. Although the concept has gained traction within environmental humanities and posthumanist urban theory (Brits, 2018; Elton, 2021; Haraway, 2010; Jhagroe, 2024; Prudence & Baylee, 2018), practical methodologies for capturing this agency remain underdeveloped. Some scholars propose evaluating agency through spatial affordances—the design characteristics that enable species' survival, movement, and reproduction (Robinson et al., 2021). Others emphasise the role of human-nature interactions, such as participation with plants in urban gardens and farming, which facilitate mutual recognition and co-dependency (Power, 2005; Shimoyamada, 2019; Elton, 2021).

In this study, the focus is on one specific typology of small-scale NbS in urban gardening and farming initiatives, as these are likely to be interventions that enable community engagement and have the potential to facilitate proactive multi-species coexistence (Frantzeskaki, 2019). However, the extent to which this occurs depends on the level of human intervention within these spaces. While not all small-scale interventions can be considered optimal examples of multi-species coexistence, they offer valuable insights into the evolving dynamics between human activities and the natural environment. These interactions inevitably involve tensions but also present opportunities for environmental monitoring. Although their ecological impact may not be comparable to that of large-scale NbS that integrate undisturbed natural environments, small-scale initiatives can create more habitat spatial heterogeneity, also known as habitat patchiness in ecology (Villard & Metzger, 2014). They provide sites for nesting, hibernating, basking, and foster connecting ecological networks, which are often disrupted in urban areas due to habitat fragmentation (Fahrig, 2017) and the widespread use of pesticides and chemicals that hinder the proliferation of organic life.

1.3 Exploring Multi-Species Coexistence in Urban Green Spaces

Guiding principles of multi-species spatial integration in urban spaces are often discussed against the backdrop of resolving conflicts between human comfort and wildlife needs (Grobman et al., 2023) while also addressing gaps in ecological literacy among design experts. In multi-species design, one fundamental principle is the recognition of more-than-human stakeholders as active participants in the design process. This places a strain on designers, as they must consider not only human needs but also the needs, behaviours, and ecological roles of various species, including microbes, plants, and animals (Chamaidi & Stavrakis, 2024; Coskun et al., 2022). Furthermore, the tendency to interpret more-than-human practices through the lens of human practices hinders the ability to appreciate their differences, creating constraints in the design process, as human-centric design-thinking methods are unsuitable for more-than-human stakeholders (Coskun et al., 2022). The latest studies, nonetheless, have introduced several propositions that could be incorporated into the design inquiry aimed at fostering a balanced coexistence of all species in urban environments.

Recent scholarship has proposed several approaches to recalibrate design practices in urban green spaces towards equitable multi-species coexistence (Hernandez-Santin et al., 2022). One such method is the Animal-Aided Design (AAD) framework, which introduces "species portraits" containing essential information for planners or communities about the habitat requirements of target species across all stages of their life cycle. These portraits guide the design process from early analysis, where target species are formally acknowledged as stakeholders, through construction, monitored via ecological supervision, and into occupancy evaluations that assess whether habitats successfully support species settlement, reproduction, and ecological function. While AAD marks a significant advancement in integrating fauna into urban design, it remains limited in scope. It largely overlooks flora and microbial life, failing to account for the broader concept of biodiversity as an integral element of urban placemaking (Benjumea Mejía, 2025). Moreover, the framework does not sufficiently address the dynamic agency of both human and more-than-human actors within placemaking processes. Crucial dimensions such as infrastructure suitability, ecological benefits, and spatial quality are not explicitly incorporated, leaving key criteria for fostering resilient, inclusive, and co-beneficial urban ecosystems unaddressed.

The Biodiversity Inclusive Design (BID) is another proposal that seeks to integrate key areas for multi-species interaction by fostering functional ecosystem systems, enabling species persistence within the built environment (Hernandez-Santin et al., 2022). It has been argued that BID redefines the human-nature relationship by incorporating design features that support the survival and thriving of various species as part of a healthy and resilient ecology. According to Hernandez-Santin et al.,

(2022), approximately fourteen frameworks have been identified that regard biodiversity as an active stakeholder in the design process. These frameworks offer important guiding principles for integrating species as clients in urban and architectural planning, helping to reorient spatial configurations towards ecological inclusivity.

However, despite the growing availability of biodiversity-aware frameworks applied to NbS, a significant gap remains in assessing how these interventions perform after implementation, especially in terms of whether they genuinely promote multi-species spatial agency and coexistence. Most existing design methods focus on planning and implementation phases, with little attention to systematic post-occupancy evaluation that investigates whether the intended ecological functions and more-than-human benefits actually emerge in practice. There is limited empirical guidance on how to measure the success of such spaces once they are inhabited, and whether the environments genuinely accommodate species-specific behaviours, interactions, and needs over time. Additionally, existing post-occupancy studies tend to focus on human satisfaction, neglecting the dynamic and interdependent relationships that underpin more-than-human agency and presence within urban ecosystems. This lack of evaluative continuity hampers the ability of designers, planners, and communities to learn from previous projects and improve future ones.

This study presents a framework that builds upon existing NbS and biodiversity-aware design methods by introducing five dimensions that together offer a more integrative and evaluative approach to urban multi-species coexistence. The first dimension examines more than human agency, focusing on how spatial design supports biotic diversity. A performative view of agency helps establish the level of agency, which does not merely denote presence but rather the capacity of different species to shape, use, and influence space, or what we have termed spatial representation. The second dimension involves data collection and assessments of the spatial and ecological aspects of NbS using a variety of methods such as remote sensing, ethnographic studies, and cartography. These approaches aim to capture the experiences of both humans and biodiversity, yielding a richer understanding of urban ecosystems. The third dimension involves ecological indicators, which assess ecological performance through biodiversity evaluations, habitat connectivity, pollination, and ecosystem health metrics to determine whether spaces are functioning as intended for diverse species, not just symbolically, but also ecologically. The fourth dimension examines the materiality, configuration, and appropriateness of NbS infrastructure for multi-species occupancy and long-term ecological viability, moving beyond aesthetic, or human-centric metrics of green design. Lastly, the fifth dimension addresses the mode of citizen and community involvement in fostering and stewarding multi-species habitats, recognising that local knowledge and care practices are critical for maintaining inclusive green infrastructure.

To further understand some of the components of the preparatory process to design for multi-species coexistence and how small-scale NbS can play a role in this, we aim to extract lessons that promote ongoing discussions regarding this process. Therefore, this study seeks to determine whether policy and planning efforts incorporate multi-species perspectives and whether there is institutional recognition of non-human agency in urban contexts. The discussion will focus on small-scale NbS across different spatial categories and their contributions in various dimensions (i.e., representation in space (agency), data collection, ecological indicators, infrastructure indicators, and ultimately, citizen participation). Finally, we will examine the role that small-scale NbS with high community proactive dynamics play in fostering multi-species coexistence.

2 Methodology

The aim of this study is to analyse two case studies that excel in different aspects: Manizales in strategic planning and democratic, community-driven processes related to small-scale NbS, though often lacking formal planning guidance; and Singapore, an international example of NbS with extensive experience in design principles and guidelines for potential cohabitation with natural ecosystems, as well as formal state-led urban greening mechanisms and controlled access. A comparative case study approach is implemented in two phases. The first phase incorporates normative and policy analysis. The methods include reviewing normative documents related to urban planning strategies that address NbS and multi-species coexistence in both countries (i.e., official urban planning strategies, policy documents, environmental regulations, and municipal master plans).

The second phase involves an observational stage designed to gain deeper insights into the implications of these norms and their reflection in small-scale initiatives. Purposive sampling was employed to identify and assess small-scale urban green initiatives across two geographically distinct urban contexts: Manizales, Colombia, and Singapore. The selection was guided by the typological focus on gardens and urban farming initiatives, as they are key nature-based interventions due to their role in enhancing ecological quality and fostering human-nature interactions (Frantzeskaki, 2019). These interventions are particularly relevant in contexts where urban densification has led to fragmentation of habitats, but where community-driven efforts may mitigate biodiversity loss and promote cohabitation with the biotic diversity.

The selection criteria contemplated accessibility (Unrestricted access/Restricted) as open-access areas are more likely to support everyday human-nature and informal; land-use diversity (Commercial vs. Residential proximity) as it can impact the ecological quality and diversity (e.g., the proximity of a commercial zone might introduce higher traffic stressors, while residential adjacency may support long-term ecological and stewardship care); Practical access for the researcher. Due to logistical constraints, only neighbourhoods accessible to the researchers were included. Areas with limited access because of safety, legal, or logistical reasons were excluded at this stage.

A total of six initiatives were selected in both cities. In Manizales, they were distributed predominantly in residential areas that are publicly accessible and delineated with minimal physical barriers, such as bamboo fencing, maintaining visual and functional permeability (Table 1). In Singapore, the sites were in the Central and Western regions. Most of these are situated in residual green areas or adjacent to pedestrian walkways, with only two initiatives near commercial zones. The remainder are embedded within residential neighbourhoods.

This sampling method inherently presents limitations. By concentrating on accessible neighbourhoods and specific intervention typologies, the sample may not represent the full diversity of small-scale nature-based solutions within either city. Additionally, reliance on purposive sampling can introduce selection bias, as site inclusion was partly based on logistical feasibility rather than random or stratified criteria. However, this approach aligns with the exploratory aims of the study, which seeks to understand the qualitative aspects of urban green space design and utilisation from a multi-species perspective. Future research phases may consider including a broader range of sites to examine various neighbourhood green space typologies and topographic conditions.

The observations adhered to multi-species research principles, where an immersive spatial experience helps researchers discern relationships and instances of agency throughout urban environments (Elton, 2021). Rather than viewing gardens as isolated green elements, they transform into a community spread across the landscape; even individual plants contribute collectively, creating an organic, loose, and "disorganised assemblage" (Marder, 2012).

A combination of methods was employed to facilitate data gathering. A walking ethnography assisted in the annotation phase, focusing on spatial arrangements, sounds, smells, textures, and the presence of more-than-human entities (i.e., insects, birds, and plant diversity). Reflective notes were compiled after each observation, detailing the material composition of the space, including tools, fences, water containers, and improvised structures, as well as how these elements affect multi-species interactions. In each transect, an ecological mapping was performed and qualified using a set of indicators determined to measure: Spatial representation, data gathering and measurement, infrastructure support, ecological potential, and human-nature interaction. Each dimension is given an indicator from 1 (lowest) to 5 (highest) (Table 2).

2.1 Data Analysis

For the normative analysis, the findings are abstracted to highlight the inclusion of multi-species perspectives. The findings assess the extent to which policies account for non-human agency and ecological interdependencies, including their impact on urban planning and NbS. For the fieldwork observations, rather than attempting to establish a direct quantitative comparison across contrived contexts, we abstract the contributions in each scenario; thus, the methodology remains constant, employing the same methods. However, its application and interpretation are open to local conditions, allowing for situated knowledge production while retaining analytical consistency (Haraway, 2016). The descriptive results of the multi-species ethnographic observations are reported for each country.

Table 1 City and neighbourhoods context small NbS selected

		Accessibility/ Typology	**Land Use**	**Ecological Feature**	**Infrastructural features**
Singapore	A	Unrestricted/ Canal-side linear green	Residential	Edible Plants, Butterflies, Spontaneous Vegetation	Canal Edges, DIY Planters, Edge Fencing, Bicycle Paths And Pedestrian Trails
	B	Unrestricted/ Community garden	Residential	Ornamental, Plotted Plants, Low Faunal Diversity	Edge Fencing,, DIY Planters, Repurposed Containers, Basic Tool Sheds, Seating
	C	Semi-restricted	Residential/proximity to urban green infrastructure	Ornamental, edible plants, low faunal diversity	DIY Planters, Repurposed Containers, Basic Tool Sheds, Grasses
	D	Restricted (shop owners and residents only)	Mixed-use (residential + commercial)	Combined ornamental and spontaneous vegetation	DIY Planters, Repurposed Containers, Pedestrian Trails, Seating
Manizales	E	Unrestricted	Residential	Ornamental, edible and spontaneous vegetation. Butterflies, insects and birds.	DIY planters and bamboo fencing.
	F	Semi-restricted	Residential	Combined edible and spontaneous vegetation. Butterflies and insects.	Small bamboo structures and fences

(continued)

Table 1 (continued)

		Accessibility/ Typology	Land Use	Ecological Feature	Infrastructural features
	G	Restricted (residents only)	Residential	Combined ornamental, edible and spontaneous vegetation. Insects and birds.	Edge fencing bamboo, DIY planters, repurposed containers, basic tool sheds.
	H	Semi-restricted	Residential	Combined ornamental, edible and spontaneous vegetation. Insects and birds	Edge fencing bamboo, DIY planters, repurposed containers, basic tool sheds.
Photos					

Table 2 Evaluative dimensions, indicators, and assessment methods for multi-species urban interventions

Dimension	Indicators	Assessment Criteria
Representation in space more-than-human agency	Growth spontaneous; spatial organisation; conducive to biodiversity growth; inner paths for humans; inner paths for more-than-human	Observation of spatial arrangements; presence of informal and formal paths; growth patterns of vegetation, degree to which natural elements are integrated into the built environment in a supportive, aesthetic, or background role (e.g., vegetal species are present but not dominant; they complement architectural or urban features without driving the spatial experience)
Ecological indicators	Provision of living space with abiotic and biotic conditions; undisturbed groves; supports vascular plants (indigenous and exotic); plant integrity; species richness (birds, invertebrates, mammals); pollinator presence	Biodiversity surveys; field notes on plant and animal presence; photographic documentation. Assessment of the presence and quality of wild, unmanaged vegetation patches. Evaluation the presence of areas with minimal human disturbance that offer both physical (abiotic) and biological (biotic) support for species
Green infrastructure appropriateness	Habitat connectivity (corridors, vegetated pathways); rest zones (humans); nesting and foraging areas	Mapping of physical connections; identification of human use zones and animal habitats
Citizen participation	Appropriate for communal exchanges; level of support to sustain the space; autonomous engagement	Observation of usage patterns; documentation of maintenance activities; presence of signage and informal use. Assesses how the space enables or reflects the existence of community-based care, trust, and mutual assistance. High-scoring spaces facilitate emotional and practical support between users, fostering resilience, belonging and therapeutic benefits.

The urban gardens and farms in Singapore and Manizales, though not statistically representative due to their small-scale and dispersed nature, provide valuable insights into localised NbS. This purposive sample captures various intervention types, spatial contexts, and observable ecological attributes across urban conditions. The study aims not for generalisable data but to create an exploratory framework for assessing multi-species spatial coexistence in diverse urban settings. It prioritises diversity in site conditions, accessibility, and interactions between human and non-human agents, aligning with qualitative ethnographic traditions in urban ecology and landscape research (Rishbeth & Rogaly, 2018).

3 Case Studies and Results

This section presents the case studies and results of the normative analysis in Manizales and Singapore to determine the policies and planning efforts that recognise multi-species approaches, as outlined in the five guiding dimensions presented in this study for examining multi-species coexistence in Nbs. The second part will discuss in detail small-scale NbS according to the two spatial categories and their contribution to multi-species coexistence.

3.1 Manizales

Manizales is an Andean city in Colombia situated at an elevation of 2150 metres above sea level, with overcast skies all year round and surrounded by tropical humid forests. Its urban area covers 34.6 km^2 and has an approximate population of 400,000 inhabitants. It is situated in the centre-west of the country, on the western slope of the Central Cordillera. Its discontinuous urban form results from the mountainous topography, the steep slopes of the hillsides, and the presence of areas designated for environmental protection interest.

Most of the urban area was established in an interfluve, a narrow flat plateau (0° to 7° slope) with steep slopes that can exceed 60°, especially in the western sector (Chardon, 2002). These slopes and their areas of environmental interest hinder and simultaneously control urban sprawl (Chardon & Cifuentes Suarez, 2010). It is characterised by scenic routes featuring renowned sunsets and a diverse mountainous landscape. The city confronts risk factors associated with earthquakes, slope instability, flash floods, and violent, intense torrential rains (Bernal et al., 2017; Chardon, 2002). It stands out nationally for having public space indicators that are higher than those of several cities, along with regulatory advancements in linking environmental variables to the planning of its territory, comprehensive risk management, and climate change adaptation measures.

In 2017, the Ecological Support Structure was incorporated as a structural axis in land-use planning, encompassing environmental variables that constitute the main ecological structure and ecological infrastructure. The latter includes the public space system and prioritises areas for providing ecosystem services in the urban area (Alcaldía de Manizales, 2017a). The Land-Use Plan emphasises recognising urban vegetation as a solution for air quality, acoustic quality, and ecological connectivity in the city. By incorporating the Ecological Support Structure into territorial planning, the city introduced a new approach to urban planning focused on environmental variables. This Structure is an interconnected system of areas that sustain ecological processes and ecosystem services, both present and future, to ensure the city's carrying capacity for its population (Mejía, 2016). Manizales prioritised areas for ecosystem service delivery and integrated them with the urban edge, enhancing

Fig. 1 Manizales location and NbS infrastructure: a) Sancancio Hill. b) Ecoparque Los Yarumos on the city's periphery. c) The Maltería plain and the local airport. d) The Olivares stream basin. e) In the foreground, the School of Architecture and Urbanism located along Avenida Santander, the city's main thoroughfare. In the middle ground, the Herveo Tower, a remnant of the historic cable system once linking the city to the Magdalena River (*Source* Produced by the authors)

ecological connectivity and mitigating urban development impacts. This perspective views the city as a system rather than a fragmented development (Fig. 1).

Citizen involvement in community-driven urban spatial interventions is significant, with only a limited number of initiatives supported by the municipal government. For instance, food production through the planting of vegetable gardens has also established a space for citizen encounters and environmental education. The Guardians of the Hillside is an initiative of the Municipal Mayor's Office and serves as an example of citizen intervention led by women who report damage and perform cleaning and maintenance actions in areas of high instability and environmental protection (Coles & Quintero-Angel, 2018; Wesely, 2021).

The city has implemented interventions based on Nature-based Solutions (NbS) in collaboration with various partners. Nevertheless, many of these initiatives, located in public or publicly accessible private spaces, are not officially recorded. Only a handful of interventions with special permits for urban garden development appear in the records. Although there have been attempts to acknowledge these initiatives within the NbS framework, their contributions are frequently overlooked, particularly when there is insufficient technical expertise to showcase their impact on ecological services. Additionally, the primary barriers to small-scale NbS are the lack of state support, causing many interventions to occur informally, even in areas outside of the city's designated public spaces. Institutional support deficiencies, coupled with limited financial resources and land-use planning that prioritises traditional urban development, hinder the integration of natural spaces into urban planning.

The Land-Use Plan (LUP) in Manizales serves as the primary source for gathering information, as it contains the latest norms regarding urban development and the integration of urban green infrastructure. Unlike in Singapore, urban planning and design strategies do not place a strong emphasis on policies that consider multi-species infrastructure or relevant design principles, and green ratios are lacking. However, the latest version (2017) of the LUP addresses some indicators and principles to determine the connectivity of natural ecosystems: identification of main nodes of natural ecosystems, definition of difficulty levels for connectivity and movement of organisms across nodes, and a definition of the network connecting each node. To establish structural connectivity, an index of complexity and circuit is introduced. The ecosystem function is then measured through this structural connectivity. The complexity index expresses the extent to which all nodes within a system are interconnected by corridors, allowing for inferences about the percentage of network connectivity. Circuitry within the network is described in terms of the presence and extent of each circuit. These indices indicate the degree of connectivity among different land covers and contribute to defining the elements of the Principal Ecological Structure and the associated management guidelines.

However, despite the existence of these indicators, inventories of public spaces and areas of environmental interest are outdated and incomplete, which hampers decision-making processes related to the management of these areas and the projection of vegetation cover. Moreover, these strategies are not designed to address small natural areas within urban environments.

3.2 Singapore

Singapore is a low-lying island city-state located just north of the equator in Southeast Asia, with an average elevation of 15 metres above sea level. Characterised by a consistently warm and humid tropical rainforest climate, the city experiences abundant rainfall and minimal seasonal variation. Once covered by dense lowland dipterocarp forests and mangroves, Singapore has undergone extensive urban transformation while still maintaining a network of protected nature reserves and green corridors integrated into its carefully planned urban layout. The country's notable progress in urban development is widely recognised for its innovative approaches in conserving natural habitats and rewilding grey urban spaces, as well as adopting strategies that enhance the sense of place for city inhabitants. With an approximate population of 5.89 million and a total area of about 728.6 square kilometres (as of 2023, including land reclamation), Singapore is considered one of the most densely populated cities in the world. Despite its spatial limitations, the city-state is internationally acclaimed for pioneering the integration of nature-based solutions (NbS) in urban development, showcasing a variety of types, including vertical greenery systems, green roofs, nature parks, blue-green drainage infrastructure, and rehabilitated mangrove and wetland ecosystems. Singapore is a leading global example of biophilic urbanism, yet it still manages and selectively acknowledges the role of nature. The city showcases how ecological innovation can integrate with technologically advanced planning and real estate development. This prompts urban designers and policymakers to consider moving beyond viewing nature as just scenery or infrastructure, towards seeing it as an active partner in shaping urban futures. It also raises questions about whether multi-species justice can be realised in highly optimised, hyper-urban environments (Fig. 2). These initiatives support the ambition to become a "City in Nature", positioning biodiversity as a strategic element of urban liveability and climate resilience.

Several plans have shaped the urban outlook of Singapore, addressing district character, building heights, landscapes, gateways, activities, night lighting, and the Parks and Waterbodies Plan (Li, 2023). A distinctive programme highlighting the importance of restoring nature lost to urban development is the Landscaping for Urban Spaces and High-Rises (LUSH), introduced in 2009 and later expanded to cover more geographic areas and development types (URA, 2017). This programme includes vertical greenery, green roofs, and rooftop urban farming, which contribute up to 10% of the overall Landscape Replacement requirements and can be classified as part of either the hardscape or softscape components. Planting quality is quantified through Green Plot Ratio and Leaf Area Index metrics, though implementation strategies are often left to the discretion of individual developers.

Fig. 2 Big-scale NbS typologies in Singapore: a) CBD, urban green infrastructure; b) Ang Mo Kio Park, water management; c) Park Royal building, green façade; d) Oasis building, green façade; e) Super Trees, green façade. (*Source* Produced by the authors)

From a normative perspective, Singapore has implemented a variety of guidelines for designing new urban developments that assess the impact magnitude of projects (i.e., BIA), along with detailed design guidelines to measure biodiversity criteria for new designs, including wildlife corridors and native species recolonisation. A key framework developed by NParks is the *Biodiversity-Sensitive Urban Design* (BSUD), which offers a data-driven approach to integrating biodiversity into planning, including GIS-based mapping, sensor tracking of fauna movement, and recolonisation indicators. Likewise, the Biodiversity Impact Assessment Guidelines include habitat surveys, native species prioritisation, and impact matrices (e.g., the Leopold Matrix) to identify trade-offs. Citizen participation is actively fostered through outreach, education, and citizen science initiatives, bridging formal planning with community-based ecological stewardship.

Singapore excels at visibly integrating nature, with a wide variety of guidelines developed by the Urban Redevelopment Authority (URA) for designing urban spaces that include nature, along with detailed instructions on incorporating vertical green spaces. NParks, on the other hand, oversees biodiversity protection, and the careful integration of new urban developments is closely monitored, ensuring the protection of highly valuable natural environments. Despite the technical rigour of these frameworks, there is limited room to interrogate ethical questions surrounding the agency of nonhuman species in design. Nature is widely acknowledged as functional, aesthetic, and even therapeutic (as seen in the design strategies for therapeutic outcomes), but not necessarily as a co-constituent in the production of urban space. The apparent success of Singapore's nature integration conceals deeper tensions. For example, biodiversity is often choreographed to fit aesthetic norms or human recreational preferences, which may conflict with ecological authenticity.

This raises questions about the viability of multi-species design and coexistence in projects that, although carefully overseen by regulatory bodies, may not foster ethical interactions or ensure equal spatial representation and agency within the built environment. It could be argued that citizen participation might help address these issues; however, despite being widely promoted in large-scale public NbS, the spatial politics of participation raise concerns about whose voices and agencies are prioritised. While citizen involvement is apparent in flagship projects, smaller interventions, particularly in privately owned or commercial developments, may lack effective mechanisms for public or multi-species engagement. Additionally, it is important to note that post-participation or interaction beyond the end of development projects could also hinder the possibility of multi-species coexistence, even before the design process begins. Thus, raising awareness of multi-species coexistence in heavily urbanised areas requires efforts that extend beyond citizen science and normative approaches.

Singapore is a leading global example of biophilic urbanism, yet it still manages and selectively acknowledges the role of nature. The city showcases how ecological innovation can integrate with technologically advanced planning and real estate development. This prompts urban designers and policymakers to consider moving beyond viewing nature as just scenery or infrastructure, towards seeing it as an active partner in shaping urban futures. It also raises questions about whether multi-species justice can be realised in highly optimised, hyper-urban environments.

3.3 Spatial Categories of NbS Small-Scale Initiatives

Six initiatives in Manizales are distributed throughout the city (see Fig. 3). All spaces are freely accessible in predominantly residential areas, marked only by bamboo fencing to outline the gardens. Initiatives 1 (NAKSI Community Garden: 500 m^2), 2 (San José Communal Garden: 1100 m^2), and 3 (Sowers of Hope: 170 m^2) arise from community efforts and have received institutional support from the Municipal Government through environmental education campaigns. Initiatives 4 (Palermo: 180 m^2), 5 (Cervantes: 100 m^2), and 6 (Alcázares: 700 m^2) are led by individual citizens. The size difference between Initiatives 1 (500 m^2) and 2 (1100 m^2) does not affect civic engagement; both include volunteer activities involving residents from other parts of the city. This illustrates the capacity of these spaces to foster social interaction and collective learning (Suárez Hormazábal, 2025). Besides environmental benefits, these NbS cultivate a sense of place and transformed abandoned areas, formerly dumping grounds for waste, changing residents' perceptions.

Six initiatives were selected in Singapore, located in Central (spaces 3 and 6) and Western regions (spaces 1, 2, 4, and 5) (see Fig. 3). All are publicly accessible, typically in residual green areas or adjacent to pedestrian pathways. Only spaces four and six are near commercial zones; the others are in residential neighbourhoods. Each initiative shows resident engagement with these small green spaces, which, despite their size, highlight the potential for coexistence between human and non-human species. Notably, spaces one and three have incorporated portions of pedestrian paths into their design, extending gardening along less-frequented walkways (approximately 300 m^2). These interventions creatively adapt urban infrastructure, blending mobility and cultivation. In contrast, space two, located in the Central-Western area and bordering key ecological networks, is compact (approximately 200 m^2) but still supports multi-species interactions near residential developments. Spaces five and six, each under 150 m^2, are located near commercial areas and utilise small-scale green infrastructure, featuring a variety of potted plants that demonstrate how minimal interventions can enhance local biodiversity and community stewardship.

Regarding Spatial Representation (Agency), the spatial configurations of small-scale Nbs initiatives in Manizales demonstrated significant intentionality for multi-species engagement. This was anticipated due to their considerable size. Spatial organisation and pathways for more-than-human species facilitated biodiversity coexistence and its mobility, as well as supporting spontaneous vegetation (5, 5, 4 respectively) (See Table 3). These indicators are particularly relevant for multi-species coexistence, as they consider the organic propagation of natural ecosystems that can enable healthy biodiversity, indicating an emergent ecological agency, aligning with multi-species design principles. Singapore presented scores that indicate substantial contributions to biodiversity growth and inner connectivity for

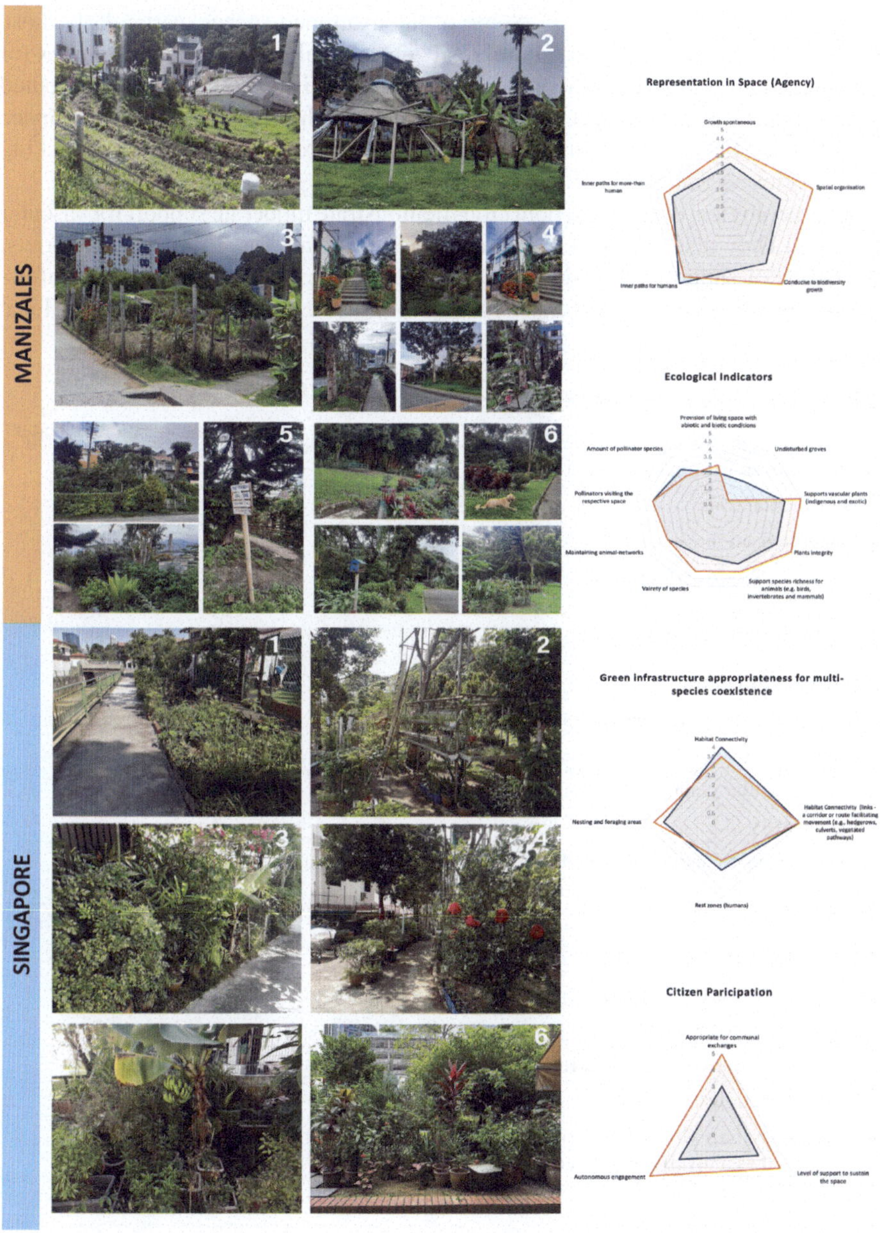

Fig. 3 Small-scale NbS in Manizales and Singapore (*Source* Photos and illustrations by the authors)

humans (3.5 and 5, respectively). Some of the examined spaces, particularly those located in the central area near commercial zones, exhibited fewer opportunities for spontaneous growth and spatial organisation (3 respectively). This may be attributed to the limited size of the spaces, often situated near high-traffic areas (i.e., Space 4 and 6) (See Fig. 3), providing better infrastructure for human access but lacking multi-species accommodation.

In the ecological dimension, both cities demonstrate moderate support for vascular plant life. They also received moderate scores for plant integrity, species variety, and pollinator support, alongside a strong potential for nesting and foraging infrastructure. In Manizales, biodiversity in neighbourhood spaces often relies on local

Table 3 Spatial measurement: Singapore and Manizales small NbS

Item	Singapore	Manizales
Representation in space (agency)		
Growth spontaneous	3	4
Spatial organisation	3	5
Conducive to biodiversity growth	3.5	5
Inner paths for humans	5	4.5
Inner paths for more-than human	3.5	4
Ecological indicators		
Provision of living space with abiotic and biotic conditions	2.5	3
Undisturbed groves	2.5	1
Supports vascular plants (indigenous and exotic)	4	5
Plants integrity	4	5
Support species richness for animals (e.g. birds, invertebrates and mammals)	3.5	4
Variety of species	3	4
Maintaining animal networks	3.5	3.5
Pollinators visiting the respective space	4	4
Amount of pollinator species	3.5	3
Green infrastructure appropriateness for multi-species coexistence		
Habitat connectivity	4	3.5
Habitat connectivity (links—a corridor or route facilitating movement (e.g., hedgerows, culverts, vegetated pathways)	4	4
Rest zones (humans)	2.5	2
Nesting and foraging areas	3	3.5
Citizen participation		
Appropriate for communal exchanges	3	5
Level of support to sustain the space	2.5	4
Autonomous engagement	3	5

1 = Deficient; 5 = Excellent. Summary of values for all six spaces in each city.

knowledge and community planning rather than formal strategies. However, both cities encounter challenges: nearly all undisturbed groves, which represent optimal ecological integrity, are absent (1 in Manizales and 2.5 in Singapore), highlighting systemic issues in incorporating "urban wilderness" into green infrastructure.

The green infrastructure appropriateness for multi-species coexistence dimension showed potential in the initiatives located near residential areas, with moderate to high scores in habitat connectivity for nodes and links, and nesting and foraging areas. Resting zones for humans, on the other side, were particularly low in Manizales, with only one space (i.e., 6) receiving the maximum score. Furthermore, water, essential for multi-species interaction and ecological functionality, was absent in all observed gardens in Manizales. This reflects a missed chance to integrate hydrological features into the nature-based solutions (NbS) narrative. The lack of rainwater harvesting systems and habitat-level water features highlights the infrastructural constraints on these grassroots ecological practices.

Community participation is a strength in Manizales. Gardens are pedagogical, cultural, and territorial tools maintained by civic actors opposing exclusionary urban renewal. One initiative gained attention for its "defensive green design", featuring elements like "vegetal barricades" from banana plants and a "green shell" that serves as shelter and classroom (space 3) (Suárez Hormazábal, 2025). These tactics signify collective resistance and reflect multi-species care in socio-political vulnerability. Conversely, Singapore received lower scores but shows potential for autonomous engagement and appropriation. Communal engagements may be influenced by location, size, and spatial characteristics (Benjumea et al., 2022). The regulated nature of spaces could explain this, yet several dedicated communal green initiatives, both community-led and government-promoted (i.e., Communities in Bloom), are not included in this study, but are part of efforts to integrate nature in urban areas. Additionally, engagement in green spaces in Singapore occurs through Citizen Science, enabling technology to track and report biodiversity and species behaviour in urban settings, indicating an organised effort to protect and monitor biodiversity adaptation.

Notably, in Singapore, the lowest ecological performance was recorded in sites located adjacent to commercial areas (Sites 4 and 5). These spaces exhibited diminished values across several indicators, particularly those related to biodiversity support and the maintenance of animal networks. This trend may be attributed to high levels of human and vehicular traffic, which can fragment habitats and disrupt more-than-human interactions, thereby undermining the ecological resilience of small-scale NbS. These findings might suggest that urban land-use typologies significantly influence the effectiveness of multi-species spatial interventions. Alternative urban acupuncture interventions also appear in the landscape, showcasing forms of nature coexistence in the urban space. Some of these small-scale interventions are found on the fringes of urban green infrastructure. With a lower land extension, however, it signals human-nature interaction in urban residual spaces.

These findings demonstrate how small-scale, resident-led ecological initiatives can foster more inclusive and ecologically responsive urban environments that recognise the role of non-human actors and the relational nature of local ecological efforts. This approach contrasts with mainstream technology-driven city models, which often prioritise data-driven efficiency and predictive control, thereby largely sidelining the complexities of socio-ecological systems and reducing nature to quantifiable assets such as carbon sinks, climate buffers, visual features, or therapeutic elements. While state-led green infrastructure is environmentally motivated, the risk of imposing standardised, technocratic solutions that neglect the situated knowledge, relational dynamics, and multi-species agency inherent in everyday green spaces remains a challenge. Conversely, this study presents urban gardens and micro-farms as emergent, adaptive infrastructures of care and coexistence. These spaces are not merely functional; they are enacted through ongoing negotiations among human and non-human agents, emphasising the performative and relational dimensions of urban nature. Particularly in Manizales, gardens become forums for biocultural stewardship, resistance to environmental marginalisation, and non-institutional ecological governance. In Singapore, spontaneous interactions among multiple species can occur in leftover spaces, even within a regulated urban environment, and may sometimes conflict with landscape management practices.

4 Discussion of Normative and Spatial Explorations

When comparing Singapore and Manizales across different spatial dimensions for multi-species coexistence, namely, Representation in Space (Agency), Data Collection and Assessments, Ecological Indicators, Green Infrastructure appropriateness for multi-species coexistence, and Citizen Participation, distinct differences arise. Singapore illustrates biophilic design in its architecture and urban planning, yet the role of more-than-human entities is seldom considered in formal ideation processes. Conversely, spontaneous initiatives within the urban environment reveal an alternative reality, where nature is not merely represented as a backdrop or amenity. Manizales, while it lacks formal directives on multi-species design, it boasts grassroots and community-driven efforts that prioritise informal natural elements. Singapore institutionalises data collection and measurement through the Singapore Index on Cities' Biodiversity, which utilises satellite imagery, ecological surveys, and ongoing monitoring. In Manizales, however, data gathering tends to be ad hoc, propelled by NGOs or academics, leading to fragmented community monitoring. On the topic of ecological indicators, Singapore uses structured metrics, such as counts of native species and habitat connectivity, for national planning and international reporting. Manizales does not have a thorough ecological performance framework but includes a range of land-use-related ecological indicators in its Land-Use Plan, accounting for factors such as conservation area proportions, vegetation cover, scenic beauty of its landscapes, sustainable production areas, and ecosystem transformation status. Nevertheless, these indicators focus more on land classification than ecosystem processes, offering limited integration of landscape-level metrics like network connectivity (e.g., gamma and circuitry indices).

From an infrastructure standpoint, Singapore promotes green and blue infrastructure through regulatory incentives, although these are often designed with a focus on human aesthetics and comfort. Conversely, Manizales features context-specific interventions developed in response to local needs, lacking substantial guidelines on incorporating biodiversity into the urban infrastructure through small-scale NbS. Lastly, community involvement in Singapore is primarily educational and consultative, supported by citizen science initiatives aimed at fostering environmental stewardship. In Manizales, however, there is more robust participation, with citizens actively engaging in the creation and maintenance of green spaces, sometimes supported by the government through community programs.

4.1 Small-Scale NbS

In Manizales, small-scale NbS can be seen as "pressure points" in the city, based on local insight, that can create ripple effects. For instance, instead of master planning an entire neighbourhood, citizens can choose to improve a community-run vegetable plot that holds social, cultural, and nutritional significance or an urban garden that enhances the ecological value of the neighbourhood while attracting diverse wildlife and community integration. These interventions, however, require further support to elevate their ecological contributions. It is expected that these forms of interventions excel in the dimensions of spatial representation (agency) and citizen participation; however, they lack in the other three dimensions in which Singapore excels, namely, data gathering and measurement, infrastructure appropriateness, and ecological potential. It was also noted that these interventions are not uniformly distributed throughout the city, and their ecological contributions might not be influenced by socioeconomic context; on the contrary, these interventions contributed to improving multi-species relationships by combining anthropocentric and non-anthropocentric values towards nature, as also observed by Frantzeskaki (2019).

The disparity between normative expectations and observed practices highlights a fragmented implementation of multi-species design. In Singapore, although green infrastructure is abundant and supported by robust governance, its ecological richness is hindered by inadequate representation of more-than-human agency within formal planning processes. Conversely, in Manizales, the absence of normative frameworks does not inhibit the emergence of rich multi-species interactions developing organically through citizen-led interventions. Importantly, unrestricted access and mixed land use in Manizales promote both ecological value and social utility, despite the lack of formal ecological metrics.

Manizales, while lacking formal structure, fosters fertile ground for emerging multi-species networks through participatory making. However, it would benefit from enhanced governance in projects that incorporate NbS. Establishing clear norms and strategies backed by data-rich planning could bolster the participatory innovation already evident in urban planning processes. In contrast, Singapore, with its defined normative and governance frameworks and nationwide NbS strategies, can promote design-thinking approaches that address multi-species coexistence. The vision for urban infrastructure should extend beyond policies and norms to accommodate both human and more-than-human species, allowing for shared urban spaces that include natural biodiversity. Given that design and citizen engagement largely rely on the developer's methodologies, further discussions on redistributing spatial agency during the design phase, as well as before and after occupancy, could be integrated into educational and outreach initiatives, fostering a greater interest and affinity for greenery.

This implies a reorganisation of the relationship between society and nature, acknowledging the need for a new definition of what is considered "natural"—one that transcends the traditional notion of nature as the absence of human influence. Such a shift has implications for governance, particularly regarding sustainability strategies and collective action within the political agenda. In this context, the debate is opened on the equivalence of rights between human inhabitants and nature, considering both as subjects or objects of rights, and recognising that the exercise of rights entails corresponding duties (Arias Maldonado, 2018) without broader biodiversity corridors and inclusive, multi-species-sensitive policies.

While large-scale NbS interventions are crucial, their implementation in South American countries remains limited. This is primarily due to their complexity and the need for robust state organisation and transparency. Nevertheless, the prevalence of small-scale green interventions, such as urban gardens and similar initiatives, with strong communal involvement and non-induced participation,[1] presents a valuable opportunity for exploration to identify viable solutions to the current Climate Change that cities are facing (e.g., floods, heat waves and loss of biodiversity) (Castaldo et al., 2025). These initiatives, characterised by strong communal engagement and spontaneous participation, reflect a common form of human agency in public spaces—a distinctive expression in South American urban environments—and offer a foundation for advancing multi-species coexistence strategies. By acknowledging such approaches in small-scale NbS, it might be possible to promote participation and agency for both humans and other forms of biological diversity.

Finally, the evaluation method employed in this study demonstrates adaptability across various contexts within the Global South, providing a replicable framework for assessing green infrastructure through the perspective of multi-species coexistence. Nevertheless, it also highlights the necessity of critically re-evaluating how ecological integrity, accessibility, and cohabitation are substantively supported within both formal and informal urban landscapes.

[1] The term non-induced participation refers to the active involvement of communities in the planning, design, and maintenance of urban green spaces, often following a democratic process or communing. It differs from induced participation, which is often organised by external actors, including governments, non-governmental agencies, and other institutions that implement a specific participatory framework to engage citizens and enact empowerment (Wolff et al., 2022).

5 Conclusions

The findings presented in this study demonstrate that multi-species urbanism and the coexistence of small-scale nature-based solutions (NbS) can be implemented in everyday spaces located in both central and peripheral urban areas, across diverse contexts—for instance, in residential and commercial zones, as well as in neighbourhoods with varying socioeconomic and accessibility conditions. This evidence highlights the presence of alternative forms of life and interconnected spaces inhabited by non-human species, which are part of the urban infrastructure yet are commonly overlooked in architecture and urban planning.

This research invites us to rethink urban governance to support multi-species coexistence beyond dominant models of technologically driven planning and aesthetic-driven green infrastructure. While Singapore's structured approach excels in ecological metrics and data integration, it falls short in acknowledging more-than-human agency. Conversely, Manizales demonstrates that informal, grassroots-led NbS—despite lacking formal ecological planning—can produce rich multi-species entanglements when grounded in place-based knowledge, local stewardship, and inclusive spatial practices.

To advance multi-species urbanism, governance must shift from centralised, top-down logics towards participatory, polycentric models that acknowledge and redistribute spatial agency among human and non-human actors. This requires new institutional norms that legitimise bottom-up ecological interventions, expand ecological literacy among planners, and formally recognise informal actors as co-creators of urban ecosystems. Planning processes should be informed not only by performance metrics but also by relational and qualitative indicators that value ecological spontaneity, cultural significance, and the well-being of non-human entities.

Large-scale NbS in South America face institutional challenges, but small-scale interventions, such as urban gardens, should be recognised not as marginal greening efforts but as strategic ecological nodes capable of triggering larger urban transformations as well as mobilising local stakeholders. Their spontaneous nature offers insights into the need to enable spaces for adaptive responses and local representation of space. These initiatives, characterised by strong, autonomous community engagement, reflect everyday forms of agency and lay the groundwork for advancing multi-species coexistence. In Latin American contexts, where formal implementation of large-scale NbS is often constrained by institutional capacity, these bottom-up interventions represent viable, culturally embedded alternatives.

In this context, cities in the Global South can draw from Singapore's example, especially its effective governance and successful integration of nature into urban infrastructure. Building on grassroots efforts, South American cities can further develop multi-species design through inclusive planning, multidisciplinary collaboration, and active engagement with residents. This includes recognising diverse perceptions, ensuring clear communication, and sharing leadership with citizens, civil society groups, and other urban stakeholders.

We present the subsequent pragmatic recommendations derived from this study: (1) prioritising multi-species needs at various scales; (2) incorporating informal small-scale green interventions into formal urban planning without losing their bottom-up character or space; (3) funding small-scale, community-led nature-based solutions (NbS), particularly in under-resourced areas; (4) creating decentralised ecological governance platforms where civil society, institutions, and biodiversity experts collaborate on adaptive urban stewardship; (5) expanding ecological monitoring that combines citizen science with traditional methods; and (6) cooperation across urban planning and design sectors to expand the discussion of recognising multi-species coexistence and their agency in NbS.

Ultimately, advancing multi-species urban futures requires more than incremental green solutions; it demands a normative redefinition of urban life that challenges anthropocentric hierarchies and supports ecological justice. By recognising humans and non-humans as interdependent urban dwellers, planners, and policymakers can foster environments where coexistence becomes a principle—not a byproduct—of urban design. More exploratory research should be conducted in both the Global South and North. Especially noteworthy are the strategies that recognise the inherent right of natural biodiversity to coexist in urban and non-urban settings, as they may offer unique perspectives on designing multi-species urban futures. It's also important to acknowledge the differences in agency among species, as well as to understand spatial governance, policy instruments for the provision of ecosystem services (planning, management, financing, and evaluation), recognition and incentives for best practices and their influence on how nature occupies urban spaces.

References

Anguelovski, I. (2023). (in)justice in urban greening and green gentrification. In S. Villamayor-Tomas & R. Muradian (Eds.), *The Barcelona School of Ecological Economics and Political Ecology. Studies in ecological economics* (Vol. 8, pp. 235–248). Springer. https://doi.org/10.1007/978-3-031-22566-6_20

Anguelovski, I., Irazabal, C., & Connolly, J. J. T. (2019). Grabbed urban landscapes: Socio-spatial tensions in green infrastructure planning in Medellín: Grabbed urban landscapes. *International Journal of Urban and Regional Research, 43*(1), 133–156. https://doi.org/10.1111/1468-2427.12725

Apfelbeck, B., Snep, R. P., Hauck, T. E., Ferguson, J., Holy, M., Jakoby, C., MacIvor, J. S., Schär, L., Taylor, M., & Weisser, W. W. (2020). Designing wildlife-inclusive cities that support human-animal co-existence. *Landscape and Urban Planning, 200*, 103817.

Arias Maldonado, M. (2018). *Antropoceno: La política en la era humana* (Taurus/Penguin Random House Barcelona).

Atchison, J., Pickerill, J., Arnold, C., Gibbs, L., Gill, N., Hubbard, E., Lorimer, J., & Watson, M. (2024). Peopled landscapes: Questions of coexistence in invasive plant management and rewilding. *People and Nature, 6*(2), 458–473. https://doi.org/10.1002/pan3.10598

Benjumea, D. M., Kato, Y., & Chong, K. H. (2022). Participatory approaches to enact meaningful interconnectedness with the natural environment: A case study in Singapore. *Cities & Health*, 1134–1151. https://doi.org/10.1080/23748834.2022.2124012

Benjumea Mejía, D. M. (2025). *Beyond Human Realms: Exploring The Comprehensive Ecological Perspective of Spatial Agents and Their Contribution to Nature-Based Solutions*. Urban Futures-Cultural Pasts Sustainable Cities, Cultures & Crafts Volume 1: Sustainability & Participation. AMPS PROCEEDINGS SERIES 40.1 ISSN 2398–9467.

Bernal, G. A., Salgado-Gálvez, M. A., Zuloaga, D., Tristancho, J., González, D., & Cardona, O.-D. (2017). Integration of probabilistic and multi-Hazard risk assessment within urban development planning and emergency preparedness and response: Application to Manizales, Colombia. *International Journal of Disaster Risk Science, 8*(3), 270–283. https://doi.org/10.1007/s13753-017-0135-8

Boros, J. Z., Shevchenko, V., Cay, D., & Gualtieri, G. (2024). A framework for more-than-human Placemaking with data storytelling. *Journal of Digital Landscape Architecture, 2024*(9), 235–252. https://doi.org/10.14627/537752023

Bragança, L. S. (2025). MULTISPECIES PROJECT. A multispecies approach to communities in urban green spaces in times of the Anthropocene. In *Composing worlds: Humanities, health and wellbeing in the XXI century towards a more sustainable world* (pp. 11–28). Springer Nature Switzerland. https://doi.org/10.1007/978-3-031-87108-5_2

Brits, B. (2018). Brain trees: Neuroscientific metaphor and botanical thought. In G. Prudence & B. Brits (Eds.), *Covert plants: Vegetal consciousness and Agency in an Anthropocentric World (version 1, p. 266pp) [PDF]*. Punctum books. https://doi.org/10.21983/P3.0207.1.00

Buckley, Y. M., Austin, A., Bardgett, R., Catford, J. A., Hector, A., Iler, A., & Mariotte, P. (2024). The plant ecology of nature-based solutions for people, biodiversity and climate. *Journal of Ecology, 112*(11), 2424–2431. https://doi.org/10.1111/1365-2745.14441

Calvo, P., & Lawrence, N. (2023). *Planta sapience: Unmasking plant intelligence*. The Bridge Street Press.

Castaldo, A. G., Nocentini, M. G., Lemes De Oliveira, F., & Mahmoud, I. H. (2025). Nature-based solutions and urban planning in the global south: Challenge orientations, typologies, and viability for cities. *Land Use Policy, 150*, 107439. https://doi.org/10.1016/j.landusepol.2024.107439

Chamaidi, T., & Stavrakis, M. (2024). A Multispecies Interaction Design Approach: Introducing the Beings Activities Context Technologies (BACT) Framework. *Multimodal Technologies and Interaction, 8*(9), 77. https://doi.org/10.3390/mti8090077

Chardon, A.-C. (2002). *Un enfoque geográfico de la vulnerabilidad en zonas urbanas expuestas a amenazas naturales*. Editorial Centro de Publicaciones, Universidad Nacional de Colombia.

Chardon, A.-C., & Cifuentes Suarez, J. F. (2010). Resettle…, beyond four walls. An analysis from the sustainable habitat, in the city of Manizales. *Colombia. Bitácora Urbano Territorial, 1*, 11–34.

Cianchi, J. (2015). *Radical environmentalism: Nature, identity and more-than-human agency*. Palgrave Macmillan.

Coles, A. R., & Quintero-Angel, M. (2018). From silence to resilience: Prospects and limitations for incorporating non-expert knowledge into hazard management. *Environmental Hazards, 17*(2), 128–145. https://doi.org/10.1080/17477891.2017.1382319

Coskun, A., Cila, N., Nicenboim, I., Frauenberger, C., Wakkary, R., Hassenzahl, M., Mancini, C., Giaccardi, E., & Forlano, L. (2022). More-than-human Concepts, Methodologies, and Practices in HCI. *Conference on Human Factors in Computing Systems-Proceedings*. https://doi.org/10.1145/3491101.3516503

de Groot, R., Brander, L., van der Ploeg, S., Costanza, R., Bernard, F., Braat, L., Christie, M., Crossman, N., Ghermandi, A., Hein, L., Hussain, S., Kumar, P., McVittie, A., Portela, R., Rodriguez, L. C., ten Brink, P., & van Beukering, P. (2012). Global estimates of the value of ecosystems and their services in monetary units. *Ecosystem Services, 1*(1), 50–61. https://doi.org/10.1016/j.ecoser.2012.07.005

De Magalhães, C., & Carmona, M. (2009). Dimensions and models of contemporary public space management in England. *Journal of Environmental Planning and Management, 52*(1), 111–129. https://doi.org/10.1080/09640560802504704

de Manizales, A. (2017a). Documento Técnico de Soporte (DTS). In *Componente Urbano del Acuerdo Municipal 958 de 2017*. Plan de Ordenamiento Territorial.

Alcaldía de Manizales. (2017b). Plan de Ordenamiento Territorial 2017-2031. Acuerdo Municipal 958 de 2017.

Dempsey, N., & Burton, M. (2012). Defining place-keeping: The long-term management of public spaces. *Urban Forestry & Urban Greening, 11*(1), 11–20. https://doi.org/10.1016/j.ufug.2011.09.005

Depietri, Y., & McPhearson, T. (2017). Integrating the Grey, green, and blue in cities: Nature-based solutions for climate change adaptation and risk reduction. In N. Kabisch, H. Korn, J. Stadler, & A. Bonn (Eds.), *Nature-based solutions to climate change adaptation in urban areas: Linkages between science, policy and practice* (pp. 91–109). Springer International Publishing. https://doi.org/10.1007/978-3-319-56091-5_6

Dorst, H., Van Der Jagt, A., Raven, R., & Runhaar, H. (2019). Urban greening through nature-based solutions—Key characteristics of an emerging concept. *Sustainable Cities and Society, 49*, 101620. https://doi.org/10.1016/j.scs.2019.101620

Duerbeck, G., Schaumann, C., & Sullivan, H. (2015). *Human and Non-Human Agencies in the Anthropocene, 6*(1), 118–136. https://ecozona.eu/article/view/642

Elton, S. (2021). Growing methods developing a methodology for identifying plant agency and vegetal politics in the city. *Environmental Humanities, 13*(1), 93–112. https://doi.org/10.1215/220119198867219

Fahrig, L. (2017). Ecological responses to habitat fragmentation per se. *Annual Review of Ecology, Evolution, and Systematics, 48*(1), 1–23. https://doi.org/10.1146/annurev-ecolsys-110316-022612

Figueroa-Arango, C. (2024). *Integrating nature-based solutions in cities from the global south*. SELVAR.

Frantzeskaki, N. (2019). Seven lessons for planning nature-based solutions in cities. *Environmental Science & Policy, 93*, 101–111. https://doi.org/10.1016/j.envsci.2018.12.033

Geddes, I., Chatzichristou, C., Charalambous, N., & Ricchiardi, A. (2024). Agency within Neighborhoods: Multi-scalar relations between urban form and social actors. *Land, 13*(3), 269. https://doi.org/10.3390/land13030269

Grobman, Y. J., Weisser, W., Shwartz, A., Ludwig, F., Kozlovsky, R., Ferdman, A., Perini, K., Hauck, T. E., Selvan, S. U., Saroglou, S., Barath, S., Schloter, M., & Windorfer, L. (2023). Architectural multispecies building design: Concepts, challenges, and design process. *Sustainability, 15*(21), 15480. https://doi.org/10.3390/su152115480

Groutars, E. G., Kim, R., & Karana, E. (2024). Designing living artefacts for multispecies interactions: An ecological approach. *International Journal of Design, 18*(2), 59–78. https://doi.org/10.57698/v18i2.04

Huang, S. Y., Wang, Y., Llabres-Valls, E., Jiang, M., & Chen, F. (2024). Meta-connectivity in urban morphology: A deep generative approach for integrating human–wildlife landscape connectivity in urban design. *Land, 13*(9), 1397. https://doi.org/10.3390/land13091397

Haraway, D. (2010). When species meet: Staying with the trouble. *Environment and Planning D: Society and Space, 28*(1), 53–55. https://doi.org/10.1068/d2706wsh

Haraway, D. J. (2016). *Staying with the trouble: Making kin in the Chthulucene*. Duke University Press.

Hernandez-Santin, C., Amati, M., Bekessy, S., & Desha, C. (2022). A review of existing ecological design frameworks enabling biodiversity inclusive design. *Urban Science, 6*(4), 95. https://doi.org/10.3390/urbansci6040095

Jhagroe, S. (2024) Fences seeds and bees: The more-than-human politics of community gardening in Rotterdam. *Urban Studies, 61*(8), 1488–1507. https://doi.org/10.1177/00420980231208830

Kabisch, N., Frantzeskaki, N., & Hansen, R. (2022). Principles for urban nature-based solutions. *Ambio, 51*(6), 1388–1401. https://doi.org/10.1007/s13280-021-01685-w

Kohn, E. (2013). *How forests think: Toward an anthropology beyond the human*. University of California Press. https://doi.org/10.1525/california/9780520276109.001.0001

Kok, K. P. W., Loeber, A. M. C., & Grin, J. (2021). Politics of complexity: Conceptualizing agency, power and powering in the transitional dynamics of complex adaptive systems. *Research Policy, 50*(3), 104183. https://doi.org/10.1016/j.respol.2020.104183

Li, N. (2023). *A distinctive & delightful city Singapore urban design guidebook*. Urban Redevelopment Authority (URA).

Maller, C. (2021). Re-orienting nature-based solutions with more-than-human thinking. *Cities, 113*, 103155. https://doi.org/10.1016/j.cities.2021.103155

Marder, M. (2012). Plant intentionality and the phenomenological framework of plant intelligence. *Plant Signaling & Behavior, 7*(11), 1365–1372. https://doi.org/10.4161/psb.21954

Matos, F. A., & Roebeling, P. (2022). Modelling impacts of nature-based solutions on surface water quality: A rapid review. *Sustainability, 14*(12), 7381. https://doi.org/10.3390/su14127381

Mejía, M. A. (Ed.). (2016). *Naturaleza urbana: Plataforma de experiencias* (Instituto de Investigación de Recursos Biológicos Alexander von Humboldt).

Mosca, F., Calbi, M., Roccotiello, E., & Perini, K. (2025). A computational approach to assess the effects of ecological building envelopes on outdoor thermal comfort. *Sustainable Cities and Society, 120*, 106170. https://doi.org/10.1016/j.scs.2025.106170

Moyano, J., Dimarco, R. D., Paritsis, J., Peterson, T., Peltzer, D. A., Crawford, K. M., McCary, M. A., Davis, K. T., Pauchard, A., & Nuñez, M. A. (2024). Unintended consequences of planting native and non-native trees in treeless ecosystems to mitigate climate change. *Journal of Ecology, 112*(11), 2480–2491. https://doi.org/10.1111/1365-2745.14300

Myers, N. (2019). *From edenic apocalypse to gardens against eden: Plants and people in and after the anthropocene*. En K. Hetherington, Infrastructure, environment, and life in the anthropocene. Duke University Press. https://doi.org/10.2307/j.ctv121024s

Nesshöver, C., Assmuth, T., Irvine, K. N., Rusch, G. M., Waylen, K. A., et al. (2017). The science, policy and practice of nature-based solutions: An interdisciplinary perspective. *Sci. Total Environ., 579*, 1215–1227. https://doi.org/10.1016/j.scitotenv.2016.11.106

Parris, K. M., Amati, M., Bekessy, S. A., Dagenais, D., Fryd, O., Hahs, A. K., Hes, D., Imberger, S. J., Livesley, S. J., Marshall, A. J., Rhodes, J. R., Threlfall, C. G., Tingley, R., van der Ree, R., Walsh, C. J., Wilkerson, M. L., & Williams, N. S. G. (2018). The seven lamps of planning for biodiversity in the city. *Cities, 83*, 44–53. https://doi.org/10.1016/j.cities.2018.06.007

Pineda-Pinto, M., Kennedy, C., Collier, M., Cooper, C., O'Donnell, M., Nulty, F., & Castaneda, N. R. (2023). Finding justice in wild, novel ecosystems: A review through a multispecies lens. *Urban Forestry & Urban Greening, 83*(1), 127902.

Power, E. R. (2005). Human–Nature Relations in Suburban Gardens. Australian Geographer, 36(1), 39–53. doi:https://doi.org/10.1080/00049180500050847.

Prudence, G., & Baylee, B. (Ed.). (2018). Covert Plants: Vegetal Consciousness and Agency in an Anthropocentric World (Version 1, p. 266pp) [PDF]. Punctum books. https://doi.org/10.21983/P3.0207.1.00.

Qi, J., Mazumdar, S., & Vasconcelos, A. C. (2024). Understanding the relationship between urban public space and social cohesion: A systematic review. *International Journal of Community Well-Being, 7*(2), 155–212. https://doi.org/10.1007/s42413-024-00204-5

Randrup, T. B., & Jansson, M. (2020). Strategic management of urban open spaces. In *Urban open space governance and management* (pp. 190–203). Routledge, Taylor & Francis Group. https://doi.org/10.4324/9780429056109-3

Randrup, T. B., Svännel, J., Sunding, A., Jansson, M., & Sang, Å. O. (2021). Urban open space management in the Nordic countries. Identification of current challenges based on managers' perceptions. *Cities, 115*, 103225. https://doi.org/10.1016/j.cities.2021.103225

Rishbeth, C., & Rogaly, B. (2018). Sitting outside: Conviviality, self-care and the design of benches in urban public space. *Transactions of the Institute of British Geographers, 43*(2), 284–298. https://doi.org/10.1111/tran.12212

Robinson, J. M., Watkins, H., Man, I., Liddicoat, C., Cameron, R., Parker, B., Cruz, M., & Meagher, L. (2021). Microbiome-inspired green infrastructure: A bioscience roadmap for urban ecosystem

health. Arq. *Architectural Research Quarterly, 25*(4), 292–303. https://doi.org/10.1017/S1359135522000148
Russo, A., Esperon-Rodriguez, M., St-Denis, A., & Tjoelker, M. G. (2025). Native vs. non-native plants: Public preferences, ecosystem services, and conservation strategies for climate-resilient urban green spaces. *Land, 14*(5), 954. https://doi.org/10.3390/land14050954
Shimoyamada, S. (2019). Hybrid production of gardens: An actor–network analysis of heterogeneous power relations. *Tourism Geographies, 21*(2), 195–213. https://doi.org/10.1080/14616688.2018.1457075
Shingne, M. C. (2022). The more-than-human right to the city: A multispecies reevaluation. *Journal of Urban Affairs, 44*(2), 137–155. https://doi.org/10.1080/07352166.2020.1734014
Skodra, J., Zorita, S., Perez, I. G., & Moebus, S. (2020). Co-creating nature-based solutions for healthy and sustainable cities: Urban public health approach. *European Journal of Public Health, 30*(5), ckaa165.234. https://doi.org/10.1093/eurpub/ckaa165.234
Standish, R. J., & Parkhurst, T. (2024). Interventions for resilient nature-based solutions: An ecological perspective. *Journal of Ecology, 112*(11), 2502–2509. https://doi.org/10.1111/1365-2745.14377
Suárez Hormazábal, J. (2025). Andean Cosmotechnics: Notes for two defensive architectures in the San José Neighborhood in Manizales. *Colombia. Revista Diseña, 26*. https://doi.org/10.7764/disena.26.Article.5
Tabara, J. D., & Chabay, I. (2013). Coupling human information and knowledge systems with social-ecological systems change: Reframing research, education, and policy for sustainability. *Environmental Science and Policy, 28*, 71–81.
Tan, B. A., Gaw, L. Y.-F., Masoudi, M., & Richards, D. R. (2021). Nature-based solutions for urban sustainability: An ecosystem services assessment of plans for Singapore's first "Forest town". *Frontiers in Environmental Science, 9*. https://doi.org/10.3389/fenvs.2021.610155
URA (2017). *Updates to the Landscaping for Urban Spaces and High-Rises (LUSH) Programme: LUSH 3.0*. https://www.ura.gov.sg/Corporate/Guidelines/Circulars/dc17-06
Van Patter, L. E. (2023). Toward a more-than-human everyday urbanism: Rhythms and sensoria in the Multispecies City. *Annals of the American Association of Geographers, 113*(4), 913–932. https://doi.org/10.1080/24694452.2022.2134838
Villard, M.-A., & Metzger, J. P. (2014). Beyond the fragmentation debate: A conceptual model to predict when habitat configuration really matters. *Journal of Applied Ecology, 51*, 309–318. https://doi.org/10.1111/1365-2664.12190
Wesely, J. (2021). Towards understanding enabling environments for good practices in disaster risk management: An analysis of critical junctures in Manizales. *Colombia. Environment and Urbanization, 33*(2), 599–615. https://doi.org/10.1177/09562478211008873
Wolff, E., Rauf, H. A., Diep, L., Natakun, B., Kelly, K., & Hamel, P. (2022). Implementing participatory nature-based solutions in the global south. *Frontiers in Sustainable Cities, 4*. https://doi.org/10.3389/frsc.2022.956534

Diana M. Benjumea Mejia Core Learning, College of Interdisciplinary and Experiential Learning, Singapore University of Social Sciences, Singapore.

Tania Giraldo Ospina The National University of Colombia. Representation, communication, and Architecture Research Group, Manizales, Caldas, Colombia.

Integrating Nature-Based Solutions in Urban Planning. Challenges and Concepts of Regenerative Urbanism in Europe

Rafael Córdoba Hernández, Fernando Carmona Mateos, Javier Morán Uriel, and Federico Camerin

Abstract This chapter addresses the challenges and opportunities associated with the effective integration of nature-based solutions (NbS) into urban planning in Europe, affecting both urban policies and spatial plans, according to the notion of "regenerative urbanism." The work explores how NbS can serve as transformative tools to restore ecosystems, improve climate resilience, and promote socio-ecological well-being in urban environments. The main objective of the chapter is to propose a conceptual and methodological framework for integrating NbS into urban planning, emphasising multilevel governance, community engagement, and interdisciplinary collaboration. The study combines conceptual analysis with a review of current policy frameworks and case study evidence from European initiatives, particularly those funded under the Horizon 2020 and LIFE programmes (e.g., NATURVATION, EKLIPSE, and LIFE SeedNEB). Methodologically, it draws on policy review, spatial analysis and comparative urban research to identify barriers and enablers to NbS implementation in planning practice. The key findings of the research highlight persistent challenges in the attempt to integrate NbS into the realm of planning: fragmented governance, limited technical capacity, socio-spatial inequalities, and lack of robust monitoring and financing mechanisms. Despite these obstacles, the study highlights promising examples at both the building and community levels—such as green roofs, community gardens, and co-design processes—that demonstrate the potential of NbS to regenerate urban spaces while improving public health and social cohesion. By focusing on community-scale interventions and their integration into planning frameworks, this chapter aligns with the Nature and Climate Resilience section of the volume by two main contributions. Firstly, it proposes a novel perspective on operationalising regenerative urbanism through NbS by providing practical guidance for

R. Córdoba Hernández (✉) · F. Carmona Mateos · J. Morán Uriel
Department of Urbanística y Ordenación del Territorio, Universidad Politécnica de Madrid, Madrid, Spain
e-mail: rafael.cordoba@upm.es

F. Camerin
Department of Urbanismo y Representación de la Arquitectura, Universidad de Valladolid, Valladolid, Spain

A. Cheshmehzangi et al. (eds.), *Nature-based Urbanism*, Urban Sustainability,
https://doi.org/10.1007/978-981-95-6864-2_9

planners, policymakers and researchers. Secondly, it shows how NbS can help cities become more resilient, inclusive and sustainable by balancing environmental, social and spatial dimensions in the face of climate challenges.

Keywords Urban policy · Resilient urbanism · Spatial planning · Multidisciplinary approach · Ecosystem services · Environmental sustainability

1 Introduction

The twenty-first century has demonstrated that cities are both a significant source of global problems and a potential solution to these problems. This phenomenon is attributable to the escalating planetary urbanisation. Despite occupying a mere 2% of the Earth's land surface, cities are home to more than 50% of the global population, as reported in the *World Urbanisation Prospects: the 2018 Revision* (United Nations, 2019). Urbanisation signifies more than just population growth; it engenders a myriad of intricate challenges that impinge on all regions of the planet. Consequently, the elaboration of effective responses is imperative and of the utmost urgency.

A primary challenge confronting urban expansion is the dichotomy between urban growth and environmental sustainability. As urbanised land grows, cities encounter heightened demand for resources, infrastructure, and services, thereby exacerbating preexisting environmental concerns, including pollution, unsustainable energy consumption, waste generation, and ecosystem degradation. The capacity of cities to effectively address these challenges hinges on their ability to strike a balance between population growth and the conservation of natural resources, thereby ensuring sustainable development that can meet the needs of the present without compromising those of future generations.

Climate change introduces an additional layer of complexity. Cities are responsible for a considerable share of global greenhouse gas (GHG) emissions due to the concentration of industrial activities, transportation, and energy consumption (IPCC, 2022b). Concurrently, urban environments exhibit a high degree of vulnerability to the effects of climate change, including rising temperatures, increased frequency of extreme weather events, sea level rise, and droughts, owing to their physical characteristics and high population density. For instance, urban environments, due to factors such as extensive pavement and sparse vegetation density, experience heat island effects, which exacerbate the consequences of heat waves and increase the demand for cooling energy (IPCC, 2022a).

Conversely, as the global population continues to migrate towards urban settings, the pressure on urban mobility systems is escalating (European Commission, 2023b). Transportation constitutes a significant challenge in contemporary European cities due to the proliferation of private vehicles and the saturation of road infrastructures, resulting in congestion, pollution, and diminished productivity. To mitigate these effects, cities should commit to a mix of uses, reducing travelling distances for a daily basis. Investing in efficient and sustainable public transportation, promoting active mobility (e.g., walking and cycling), and adopting emerging technologies that contribute to the decarbonisation of transport (e.g., electric vehicles or shared mobility solutions) are also recommended.

In addressing the adverse consequences of increasing urbanisation patterns, two synergistic approaches have come to the fore: resilient urbanism and regenerative urbanism. These approaches, though employing distinct strategies, converge their objectives to enhance city sustainability and liveability. Resilient urbanism emphasises a city's capacity to adapt and recuperate from crises, such as climate change or natural disasters, thereby ensuring swift recovery and a swift return to normalcy (Davoudi, 2014). Regenerative urbanism, in contrast, aims to proactively transform urban environments to activate processes of restoration and improvement of natural and social systems. This approach seeks to create cities that not only sustain themselves over time, but also revitalise the resources they consume (Thomson & Newman, 2020).

Resilient urbanism is an approach focused on the ability of cities to withstand challenges and to demonstrate resilience in the face of adversity. In contrast, regenerative urbanism adopts a more holistic approach, in which cities are designed to integrate nature into their spatial planning tools and to improve ecosystems and promote long-term well-being for urban inhabitants. Consequently, regenerative urbanism can be regarded as a progression of resilient urbanism. It proposes an urban model that not only responds to crises but also proactively enhances environmental and social conditions, searching for a harmonious and equitable city development.

2 Background Knowledge

In a context marked by the expansion of cities, environmental unsustainability, and the effects of climate change, regenerative urbanism should encompass not only the mitigation of environmental impacts but also the provision of innovative tools that facilitate the integration of solutions aimed at enhancing the climate resilience of cities. In this regard, the present chapter proposes a framework for the incorporation of Nature-based Solutions (NbS) into urban regeneration processes, exploring the potential of these strategies to be integrated into the planning and urban design phases.

Incessant urbanisation has accelerated the degradation of ecosystems due to the excessive growth of urban settings, without the application of sustainability parameters in their planning, coupled with the excessive consumption of natural resources and the gradual loss of natural habitats. This phenomenon is evidenced by the cases of Chinese ghost cities (Morán Uriel et al., 2024) and megacities, such as São Paulo or Mexico City (de la Luz Hernández-Flores et al., 2017), which have experienced explosive growth in recent decades. In such cases, the interaction between the impacts of climate change and urbanisation has given rise to novel challenges that society must confront to ensure its long-term survival (Górgolas Martín, 2019). In response to these challenges, regenerative urbanism has emerged as a pivotal strategy to reverse these adverse trends. This approach aims not only to mitigate the adverse effects of climate change but also to transform cities so that they can regenerate the resources they consume. It seeks to integrate natural processes from the earliest stages of urban planning.

Urban ecosystems are being recognised as pivotal components in safeguarding human settlements, facilitating a reconnection with nature, and encouraging environmentally sustainable behaviours and lifestyles (Zingoni de Baro, 2022). Conceptualised as intricate systems, cities should thus be in harmony with natural cycles to foster a sustainable and restorative environment. This is why a significant number of city administrations have initiated the redesign of their urban spaces, with the objective of promoting biodiversity and enhancing the relationship between humans and their environment. These urban redesigns leverage natural ecosystems as the foundation for a novel urbanism (Fig. 1).

Nevertheless, the languid pace of these transformations stands in stark contrast to the pressing urgency of climate change, thereby engendering a conspicuous temporal dissonance between the two processes. While cities are undergoing a gradual and long-term evolution, climate change is rapidly escalating, impeding their capacity to effectively adapt to its deleterious effects. While cities have been substantial contributors to climate change through greenhouse gas (GHG) emissions, they also serve as pivotal platforms for mitigating and adapting to climate change (Barton, 2009). Mitigation strategies prioritise reducing emissions by renewable energy, energy efficiency and conservation, as well as the generation of carbon sinks. However, the true impact of these strategies will only become evident in the long term. In the short to medium term, the damage has already been done, and extreme weather events are set to continue to affect the liveability of cities.

Fig. 1 Introducing nature within the Siemens Healthineers MedMuseum (Erlangen, Germany). *Source* Photo taken by the authors in September 2023

Fig. 2 Introducing climate regulation ecosystem services through a self-managed urban garden in the Village of Monte Francisco (Castro Marim, Portugal). *Source*: Photo taken by the authors in August 2024

Cities have the capacity to implement immediate changes, such as increased use of early warning systems to alert residents to extreme weather events or the adoption of NbS to mitigate the impacts of climate change, including excessive heat, flooding, and wildfires (Robbiati et al., 2023). However, a considerable discrepancy exists between complexity, cost, and time required to adapt to climate change and the rapid advancement of climate change itself. This creates a dilemma in which slow city transformations are pitted against the rapidly growing effects of climate change (Fig. 2).

This scenario necessitates a substantial transformation of the conventional city model, favouring an approach that not only mitigates damage but also fosters the regeneration of urban ecosystems and the territory as a whole, considering the ecosystemic contributions it can provide (Córdoba Hernández & Camerin, 2023). Contemporary strategies should prioritise not only the reduction of the ecological footprint of cities but also a more extensive rethinking that incorporates the surrounding territories, integrating rural, and natural landscapes into urban design. This approach necessitates a redesign of the everyday urban space to promote closer and more sustainable living, where reduced forced mobility favours local travel. The reactivation of local and circular economies, in conjunction with the promotion of local agriculture and food sovereignty policies, is pivotal to ensure territorial resilience (Morán Alonso et al., 2025). Moreover, the establishment of an urban and territorial metabolism characterised by closed energy, water, and material cycles is crucial to diminish reliance on external resources and to facilitate the regeneration of ecosystems that have been adversely affected by uncontrolled urban expansion.

The integration of this approach into urban planning is imperative for the restructuring of territories in the face of the challenges posed by urban sprawl and unstainable land use (Córdoba Hernández & Camerin, 2024). The transition towards a more efficient and regenerative management of the urban environment will not only enable better addressing of the current environmental and social crises but also the construction of cities capable of revitalising the natural resources they consume, ensuring a more equitable and resilient future.

In this context, the chapter focuses on three main aspects, trying to contribute, for each of them, to the wider discussion on NbS implementation in cities, serving as specific guidance and recommendations for planners, authorities, and other actors involved towards a more regenerative urbanism. Firstly, raising awareness of the NbS benefits solutions among planners and local communities in Europe. Secondly, considering the community participation and availability of financial and technical resources for a more effective planning and design of NbS. Thirdly, establishing systematic monitoring and evaluation mechanisms to ensure the long-term effectiveness of NbS, as well as to foster their scalability and replicability across the diverse urban contexts in Europe. The chapter takes into consideration the report *Towards an EU Research and Innovation policy agenda for Nature-Based Solutions & Re-Naturing Cities* (European Commission, 2015a), which aspires to provide guidance and recommendations for NbS implementation in cities from the European Union.

3 The Integration of NbS in Urban Planning. A Vision from the European Union

NbS are defined as actions supported by nature that concurrently provide environmental, social, and economic benefits in a sustainable and resilient manner. In the European context, these solutions have been increasingly promoted as a key tool to address urban challenges, especially in combating the effects of climate change and improving the quality of life in cities.

The growing recognition of NbS relevance for urban settings by the European Commission has underscored their aptitude to confront an array of environmental, social, and economic challenges across the continent. Consequently, NbS are being adopted with increasing frequency as instruments to restore ecological flows in cities and fortify the resilience of heterogeneous urban populations. The Horizon 2020 project's *Final Report of the Expert Group on Nature-Based Solutions and Re-Naturing Cities* identifies four objectives within which to frame future actions (European Commission, 2015b):

- The provision of a more sustainable and resilient urbanisation. This can be achieved through NbS that leverage natural areas within cities and their surroundings, aiming to capitalise on ecosystem services. This approach presents numerous strategic opportunities, which are categorised into three overarching interconnected challenges and trends:
  - The support of economic development in urban settings is contingent on the availability and quality of natural resources, such as water for sanitation or consumption. This solution has the potential to catalyse the emergence of

novel business models that decouple economic growth from resource depletion and inequitable distribution, underpinned by a circular economy. This, in conjunction with an augmented reliance on local resources, would result in enhanced efficiency in the utilisation of energy and materials.

- The implementation of NbS in urban planning holds the potential to offer opportunities for adaptation to climate change, increased resilience to risks such as droughts, floods, and heat waves, and to create opportunities for small-scale climate mitigation through increased carbon storage. These issues can also contribute to reducing pressure on more natural peripheral areas.
- The contribution to the social dimension of sustainable urbanisation is of paramount importance, as it is inextricably linked to public health. It has been widely demonstrated that there is a positive correlation between residential proximity to green spaces and a reduced incidence of various health problems, including cardiovascular disease, obesity, and depression. This correlation is particularly evident among the most vulnerable members of society, such as children and the elderly.

- Pushing for the restoration of degraded ecosystems. The drivers of ecosystem loss and degradation vary by ecosystem and location, but the assessment methodologies converged towards five kinds of pressures: habitat transformation, climate change, overexploitation of resources, introduction of invasive alien species, and pollution and nutrient enrichment. In the European context, these pressures are particularly salient in the context of agricultural intensification, loss of fertile soils, extensive agricultural uses, contamination of abandoned land, hydrological modifications of water bodies, intensification of forestry practices, urban sprawl, and climate change. The ways in which ecosystems are affected are varied and include issues such as water purification, protection against soil erosion, flood damage control, carbon sequestration, and the provision of habitable places and recreational opportunities that contribute to human well-being, economic stability, and physical security (Fig. 3).
- Developing climate change adaptation and mitigation measures. A significant challenge is the integration of grey infrastructure—comprised of sewerage networks and water treatment facilities—with green and blue infrastructure. This integration aims to augment traditional management methodologies by leveraging nature to enhance their efficacy in addressing climate change. Investing in these solutions, such as the development of low-cost and low-maintenance products or the implementation of low-carbon solutions for climate change mitigation, has been demonstrated to enhance the cost-effectiveness of responses to social, environmental, and potentially economic challenges.
- Improving risk management and territorial resilience. NbS implementation presents significant opportunities to mitigate the frequency and intensity of various hazards. Consequently, NbS should be incorporated into a comprehensive suite of integrated risk management measures and actions, as they have the potential to yield greater benefits compared to conventional methods. Ecosystemic

Fig. 3 Restoration of the Manzanares River as it passes through Madrid. *Source*: Photo taken by the authors in December 2020

services, such as water regulation and retention, climate regulation, and biomass production, have the potential to enhance resilience in urban settings.

The mounting body of knowledge concerning NbS demonstrates their aptitude for fine-tuning to specific social contexts and their capacity to engender manifold benefits that have been shown to surpass the efficiency and cost-effectiveness of more conventional approaches (Nesshöver et al., 2017). Notwithstanding this, and in accordance with the aforementioned objectives, the *Final Report of the Expert Group on Nature-Based Solutions and Re-Naturing Cities* proposes seven priorities in terms of nature-based research and innovation initiatives to address the societal challenges outlined in the four objectives:

- Encouraging urban regeneration. Changes in land use, abandoned land, and disused areas represent a significant challenge for cities. While this issue is already being addressed by urban regeneration strategies, the incorporation of NbS in this field has the potential to offer numerous ecological benefits, including the reduction of energy and resource demands and costs, decreased air pollution, enhanced flood control, and increased leisure opportunities.
- Improving well-being in urban settings. This measure seeks to accompany the improvement of the health and well-being of citizens, while providing other ecological and economic goods and services. A growing body of evidence demonstrates the positive effects of access to high-quality green spaces and landscapes on health, well-being, social cohesion, and community support. NbS implementation enhancing neighbourhood spaces has been demonstrated to stimulate healthy physical activity and promote the development of social bonds. This phenomenon occurs because citizens are drawn outdoors to use public spaces in greater numbers, feeling safer to move freely.

- Increasing the sustainability of materials and energy use. It has the potential to curtail resource demands and enhance energy and material efficiency. Its application in public and private wooded open space areas has already begun to demonstrate these effects. In cities such as Singapore, for instance, policies have been instituted that promote the conservation and expansion of urban green areas of various scales. These policies contribute to the cultural landscape and favour the reduction of energy demand for cooling by providing shading elements and reducing the heat island effect.
- Participating in multifunctional watershed management and ecosystem restoration. Restoring ecosystems such as grasslands, croplands and forests, as well as former industrial or vacant land, with NbS offers important benefits such as reducing erosion, flood risk or droughts, while improving water quality and quantity. The implementation of these actions can result in the creation of recreational areas and spaces within urban settings, offering cultural benefits in addition to regulatory and supply functions.
- NbS to enhance the insurance value of ecosystems. The value of ecosystems in reducing risk and maintaining resilience is of critical importance in disaster risk management. To this end, the development of methods and conceptual frameworks for assessing this value is essential. One approach involves collaboration with financial institutions and insurance companies to promote nature-based solutions and translate risk reduction capacity into economic benefits. This necessitates the creation of a legal framework that encourages the preservation and improvement of the insurance capacity of ecosystems, while also considering cultural aspects and human perceptions of risk and insurance.
- Developing carbon sequestration. This is a critical component in the broader context of climate change mitigation. Contemporary approaches within this domain place significant emphasis on biosequestration, which involves the strategic utilisation of living organisms as effective long-term carbon storage solutions.
- Increasing coastal resilience. In the context of rising sea levels, the subsidence of certain regions, and the escalating threat of flooding, there is a contentious interest in engineering solutions that are highly anthropogenic, while disregarding other potentially less detrimental environmental alternatives, such as the restoration of marshlands, which have demonstrated efficacy in mitigating the risk of flooding and coastal erosion. For instance, the Canary Islands' Environmental Promotion Plan "*Adapta Costas Canarias*" (Gobierno de Canarias, 2022) suggests that as many as 147 beaches could be lost by 2050 due to coastal erosion, with profound consequences for both the environment and the islands' economy (Fig. 4). It is estimated that the Canary Islands would suffer an annual GDP loss of approximately 11%, amounting to more than 1000 million euros by the specified date. Projections indicate that by the year 2100, the number of affected beaches will rise to 153, with a 45.2% loss of total area and an economic impact of 4.5 billion euros per year.

Fig. 4 Bollullo Beach (Villa de la Orotova, Tenerife). *Source*: Photo taken by the authors in November 2023

To sum up, in contrast to non-urban environments, the EU vision on NbS tends to prioritise climate change adaptation over direct mitigation of its causes. This approach encompasses climate regulation, enhancement of air quality, and mitigation of the heat island effect through strategies such as the implementation of green roofs or permeable surfaces. Additionally, NbS promote the development of cultural and recreational ecosystem services, which can exert a positive influence on urban planning.

Beyond the scope of climate resilience, the implementation of NbS in European urban settings has yielded notable co-benefits, including the enhancement of urban biodiversity, the promotion of public health and the well-being of citizens, social cohesion, and the empowerment of local communities. The European NATURVATION project is a notable example in this regard, having identified numerous NbS interventions across various European cities. These interventions demonstrate the potential to transform urban environments into more sustainable and liveable spaces (ICLEI, 2021). The EKLIPSE project further underscores the value of collaboration between experts and communities in facilitating the integration of NbS into urban planning. This collaboration enhances our understanding of the impacts of these innovative approaches, thereby driving their adoption in cities across Europe to address current environmental, social, and economic challenges (Watt et al., 2019).

4 Discussion and Implications

Notwithstanding the numerous advantages it offers, NbS integration into urban planning in European cities still poses the following challenges.

The first challenge regards the fact that research findings indicate that these solutions are more cost-effective and efficient than traditional approaches; however, urban governance remains a substantial impediment (Mahmoud et al., 2025). Urban planning is frequently characterised by rigid practices and regulations, which hinders the incorporation of innovative solutions such as NbS. Addressing this "governance gap" necessitates enhanced collaboration among diverse stakeholders, as emphasised by Frantzeskaki et al. (2020), encompassing local authorities, communities, and experts. The incorporation of such actions into urban planning (and along with policy frameworks) is considered a fundamental factor in ensuring their effective implementation within urban governance, constituting an essential step for their acceptance and consolidation. Urban policies and spatial plans should play a crucial role in guiding NbS implementation, ensuring that they remain adaptable to changes arising from tacit knowledge, previous experiences, and continuous learning during their implementation and environmental management (Kabisch et al., 2022).

The second challenges regard the NbS implementation and optimisation within the framework of regenerative urbanism, which necessitates a continuous, robust, and multilevel dialogue between strategies, decision-makers, and citizens. This contrasts with conventional urban development, which follows a linear and hierarchical process. Regenerative urbanism advocates a more holistic and dynamic approach. This approach should be conceptualised as a process rather than a discrete action, emphasising adaptability and the integration of future requirements. As indicated in a recent European Commission report (2023a), p. 44), the transition to a truly regenerative urban model requires the facilitation of effective collaboration by decision-makers at all stages of the process, encompassing the co-design, co-implementation, and co-monitoring phases. While ensuring the social and environmental equity of NbS is the responsibility of decision-makers, this approach also offers an opportunity to transform traditional governance models, opening the way to new possibilities within participatory planning.

The third challenge is the fact that NbS implementation in urban settings is confronted with numerous barriers that impede their widespread adoption, despite their considerable potential to address the challenges posed by climate change and enhance the resilience of cities. A pervasive reluctance to adopt climate adaptation measures has been observed (Abellán López, 2021), which also impacts NbS for several key reasons: uncertainty regarding their effectiveness (Chausson et al., 2020), challenges in mobilising the necessary investment (Kabisch et al., 2016), and constraints stemming from governance (Pattberg et al., 2022). However, despite these challenges, significant opportunities exist for transforming cities into more sustainable spaces by fostering systemic approaches that, in addition to strengthening environmental resilience, involve local communities and promote social innovation.

The fourth challenge regards one of the most notorious obstacles in the NbS implementation, i.e. the need for space, as these solutions necessitate larger areas in comparison to conventional grey infrastructure. However, not all urban spaces equally suited to NbS due to constraints related to topography, structure, or demographics, which limits the available options for their integration into cities. This situation may result in competition with traditional alternatives that possess more experience and financial resources. Additionally, densely populated cities encounter heightened challenges in identifying available space, thereby necessitating the exploration of innovative solutions such as brownfield redevelopment or the incorporation of these strategies into existing infrastructure, including green roofs or vertical gardens. Alternatively, larger-scale actions, entailing the demolition and relocation of existing buildings, should be considered. Furthermore, it is imperative that cities develop land-use prioritisation strategies that promote NbS implementation, particularly in urban settings vulnerable to climate impacts, such as coastal zones or flood-prone areas.

The fifth issue is that, despite their considerable promise, NbS are frequently regarded as having less potential than grey infrastructures. The lack of robust data demonstrating NbS tangible benefits remains a critical challenge, limiting their acceptance and application at the policy and technical levels (Bockarjova et al., 2020). Consequently, there is a pressing need to bolster scientific research that validates the multifaceted benefits of NbS, expanding beyond environmental considerations to incorporate economic and social dimensions. A comprehensive approach to evaluating these projects is essential to demonstrating their long-term cost-effectiveness and their capacity to generate positive impacts on the citizens' health and social cohesion. Additionally, continuous monitoring of these interventions and the collection of data on their effectiveness are essential to ensure their success and replicability in other urban contexts.

The sixth challenge is the scarcity of knowledge and technical capacity among certain municipal administrations, underscoring the need for the promotion of training and knowledge exchange among diverse governmental levels and professional disciplines. This initiative should prioritise the establishment of collaborative networks, thereby facilitating the integration of NbS within urban policies. The promotion of city-to-city cooperation initiatives and the implementation of digital platforms facilitating access to data and studies on the benefits and feasibility of NbS in different urban contexts are crucial measures for institutional capacity building. The implementation of NbS projects necessitates a systemic and interdisciplinary approach, which is often hindered by specialisation and a lack of cross-cutting vision in single municipalities. This underscores the importance of strong institutional support and the need for NbS to be effectively integrated into urban planning mechanisms at multiple scales (Longato et al., 2023).

Finally, it is critical to address the social challenges that may arise with NbS implementation, including the potential displacement of local communities and the phenomenon of "greentrification" (Anguelovski et al., 2022), the possibility of adverse cultural influences (Diep et al., 2022), and the absence of substantial community involvement in decision-making processes (Puskás et al., 2021). To mitigate these risks, it is crucial that NbS policies adopt an inclusive approach that ensures the participation of local communities in the design, planning, and implementation of these projects. This approach ensures that NbS not only responds to environmental needs but also to social demands, promoting a more equitable and just urban development. In addition, the creation of compensation and support mechanisms for affected communities can help avoid social exclusion and negative impacts derived from the revaluation of intervened areas. These factors require careful attention to ensure that NbS contribute to inclusive and equitable urban development.

5 Concluding Remarks

The efficacy of NbS in addressing environmental challenges and fostering urban sustainability is acknowledged at the institutional level as evidenced by the European case, even though NbS do not represent a novel concept. Indeed, NbS have been implemented for decades in various cities, but their application has evolved considerably in recent years. Initially, these solutions focused on the creation of parks and the restoration of urban ecosystems, such as rivers and wetlands. However, more complex interventions have emerged, including sustainable water management, the integration of eco-friendly street furniture, and the use of green roofs and facades. NbS aim not only to mitigate the effects of climate change but also to enhance the resilience of cities. It is thus essential to learn from the challenges highlighted in this analysis and to work for improving the framework within which NbS are planned, implemented, and managed on the grounds of the following recommendations.

Firstly, it is necessary to address the complex coordination among multiple stakeholders, ranging from local authorities to communities and civil organisations, constitutes a major issue. Overcoming this challenge necessitates a holistic approach, adapted to the specific characteristics of each urban setting, considering the local environmental, social, and cultural context. The political willingness to establish a robust ecological structure within cities can also facilitate the formulation of sustainable planning policies. In fact, an effective NbS implementation is contingent upon the integration of territorial and legal constraints within the planning process. NbS incorporation into spatial plans necessitates the establishment of collaborative networks at various levels, thereby enabling the integrated management of natural resources and a higher degree of adaptation to the unique characteristics of each territory.

Secondly, the efficacy of the more NbS-oriented planning process hinges upon the incorporation of local communities in the design, co-implementation, and monitoring of NbS, alongside the establishment of long-term, stable management bodies. This approach ensures the adaptation of interventions to the social and cultural features of the environment, enhancing their effectiveness. Additionally, it fosters a sense of ownership and engagement among citizens in urban regeneration initiatives. Community participation is imperative to ensure that NbS not only respond to environmental challenges but also to social demands for equity and justice, avoiding undesirable processes such as greentrification, in which environmental development displaces the most vulnerable populations.

Thirdly, despite the considerable potential of NbS, concerns have been raised regarding the possibility that their implementation may result in a diversion of attention from other pressing environmental issues, such as emissions reduction and the preservation of intact ecosystems. A NbS-based approach should balance environmental and social challenges in a comprehensive manner. The incorporation of NbS within the framework of sustainable development necessitates the implementation of adaptive policies and planning processes that are capable of continual adjustment to lessons learned and evolving knowledge, thereby ensuring the maximisation of their long-term benefits.

Fourthly, the scarcity of available space in densely populated areas, coupled with the perception that NbS are less efficacious than traditional infrastructure, constitutes significant impediments to their implementation. To overcome these challenges, it is imperative to mobilise investments and develop innovative strategies that prioritise urban resilience. Urban management agencies can play a pivotal role in facilitating cooperation between local actors, integrating technical expertise, and mobilising financial resources to surmount these barriers.

Eventually, the establishment of stable, multidisciplinary management bodies dedicated to regenerative urbanism and the integration of NbS is urgent for ensuring inclusive and equitable urban development. These bodies should function as permanent coordination platforms, facilitating continuous collaboration between policymakers, technicians, and local communities, which should lead to ongoing dialogue and more informed and participatory decision-making processes. These bodies should also ensure the effective integration of NbS into urban planning mechanisms at multiple scales, ranging from territorial management to local action. By adopting a systemic and long-term approach, these bodies can address the challenges posed by climate change while promoting the resilience and sustainability of cities. Their role encompasses the coordination of strategies, the mobilisation of resources, and the articulation of the participation of various actors, including governments at different levels, civil society, technical experts, and the private sector.

In conclusion, establishing permanent governance structures specialised in regenerative urbanism facilitates the transformation of NbS into operational and adaptive tools that evolve in accordance with urban needs and dynamics. By consolidating a holistic vision of sustainable development, these governance bodies can contribute meaningfully to the development and implementation of urban policies and spatial plans that acknowledge the significance of biodiversity and the balance between the built and the natural environment. Moreover, their efforts should engender transparency and accountability, which are pivotal factors in preserving public trust and ensuring the sustainability of policies over the course of successive political cycles, thereby surmounting the impediments imposed by brief political cycles. The institutionalisation of these management bodies can establish a solid foundation for the integration of NbS in urban planning, creating synergies between different sectors and levels of government. This, in turn, can facilitate the transformation of cities into more sustainable, resilient and just spaces, where urban regeneration becomes a permanent and evolving practice, capable of responding effectively to current and future environmental and social challenges.

Acknowledgements Rafael Córdoba HernándezEliminate has carried out this research in the framework of the "LIFE SeedNEB Project (grant agreement no. 101148064): Beautifying cities through nature: implementation of biodiversity friendly NBS under the New European Bauhaus in 3 pilot neighbourhoods in Spain, Italy and Hungary (LIFE23-ENV-ES-LIFE SeedNEB)", approved under the call for proposals of the EU Environment and Climate Action Programme (LIFE), 2023.

Fernando Carmona Mateos has participated as co-author as part of his research *"Planeamiento urbanístico en contexto de vulnerabilidad ambiental. Mecanismos de gestión y financiación de actuaciones sobre el medio urbano"* within his PhD programme in *Sostenibilidad y regeneración urbana* at the Universidad Politécnica de Madrid.

Federico Camerin has participated as coauthor in the frame of the research project "*Las propiedades militares en Europa: Estrategias de Regeneración y sus efectos sobre Procesos de Producción Socio-Espaciales*" funded by the *Ministerio de Ciencia, Innovación y Universidades, la Agencia Estatal de Investigación* through the *Plan Estatal de Investigación Científica, Técnica y de Innovación 2021-2023*, and co-funded by the ESF+.

Authors' Contribution: Roles played by contributors according to CRediT taxonomy: Rafael Córdoba Hernández (RCH): 25%; Fernando Carmona Mateos (FCM): 25%; Javier Morán Uriel (JMU): 25%; Federico Camerin (FC): 25% to research otputs. Conceptualisation: RCH; Data curation: FCM; Formal analysis: FCM; Funding acquisition: FCM; Investigation: RCH; Methodology: RCH; Project administration: FC; Resources: JMU; Software: JMU; Supervision: FC; Validation: FC; Visualisation: JMU; Writing—original draft: RCH, FCM, JMU & FC; Writing— review & editing: RCH, FCM, JMU, & FC.

References

Abellán López, M. Á. (2021). El cambio climático: negacionismo, escepticismo y desinformación. *Tabula Rasa, 37*, 283–301. https://doi.org/10.25058/20112742.n37.13

Anguelovski, I., Connolly, J. J. T., Cole, H., Garcia-Lamarca, M., Triguero-Mas, M., Baró, F., Martin, N., Conesa, D., Shokry, G., del Pulgar, C. P., Ramos, L. A., Matheney, A., Gallez, E., Oscilowicz, E., Máñez, J. L., Sarzo, B., Beltrán, M. A., & Minaya, J. M. (2022). Green gentrification in European and North American cities. *Nature Communications, 13*(1), 3816. https://doi.org/10.1038/s41467-022-31572-1

Barton, J. R. (2009). Adaptación al cambio climático en la planificación de ciudades-regiones. *Revista de Geografía Norte Grande, 43*, 5–30. https://doi.org/10.4067/S0718-34022009000200001

Bockarjova, M., Botzen, W. J. W., van Schie, M. H., & Koetse, M. J. (2020). Property price effects of green interventions in cities: A meta-analysis and implications for gentrification. *Environmental Science and Policy, 112*. https://doi.org/10.1016/j.envsci.2020.06.024

Chausson, A., Turner, B., Seddon, D., Chabaneix, N., Girardin, C. A. J., Kapos, V., Key, I., Roe, D., Smith, A., Woroniecki, S., & Seddon, N. (2020). Mapping the effectiveness of nature-based solutions for climate change adaptation. *Global Change Biology, 26*(11). https://doi.org/10.1111/gcb.15310

Córdoba Hernández, R., & Camerin, F. (2023). Assessment of ecological capacity for urban planning and improving resilience in the European framework. *Cuadernos de Investigación Geográfica, 49*(2), 119–142. https://doi.org/10.18172/cig.5638

Córdoba Hernández, R., & Camerin, F. (2024). The application of ecosystem assessments in land use planning: A case study for supporting decisions toward ecosystem protection. Futures, 161, 103399. doi:https://doi.org/10.1016/J.FUTURES.2024.103399.

Davoudi, S. (2014). Climate change, securitisation of nature, and resilient urbanism. *Environment and Planning C: Government and Policy, 32*(2), 360–375. https://doi.org/10.1068/c12269

Gobierno de Canarias. (2022). *PIMA ADAPTA COSTAS Evaluación del riesgo frente al cambio climático en las costas de Canarias*. https://opendata.sitcan.es/upload/medio-ambiente/pima/20220531IP_ResumenPIMA.pdf

de la Luz Hernández-Flores, M., Otazo-Sánchez, E. M., Galeana-Pizaña, M., Roldán-Cruz, E. I., Razo-Zárate, R., González-Ramírez, C. A., Galindo-Castillo, E., & Gordillo-Martínez, A. J. (2017). Urban driving forces and megacity expansion threats. Study case in the Mexico City periphery. *Habitat International, 64*. https://doi.org/10.1016/j.habitatint.2017.04.004

Diep, L., Parikh, P., dos S., D.,. B. P., Bourget, A. F., Dodman, D., & Martins, J. R. S. (2022). "It won't work here": Lessons for just nature-based stream restoration in the context of urban informality. *Environmental Science and Policy, 136*. https://doi.org/10.1016/j.envsci.2022.06.020

European Commission. (2015a). *Towards an EU Research and Innovation policy agenda for Nature-Based Solutions & Re-Naturing Cities. Final Report of the Horizon 2020 Expert Group on 'Nature-Based Solutions and Re-Naturing Cities.'* https://ec. europa.eu/newsroom/horizon2020/document.cfm?doc_id=10195.

European Commission. (2015b). Towards an EU research and innovation policy agenda for nature-based solutions & re-Naturing cities. https://doi.org/10.2777/765301

European Commission. (2023a). *Guidelines for co-creation and co-governance of nature-based solutions: insights form EU-funded projects* (I. Ferreira, G. Lupp, & I. Mahmoud (eds.)). Publications Office of the European Union. https://data.europa.eu/doi/10.2777/157060

European Commission. (2023b). *Horizon Europe Work Programme 2023–2024 8. Climate, Energy and Mobility towards climate neutrality. August 2021*, 1–528.

Frantzeskaki, N., Vandergert, P., Connop, S., Schipper, K., Zwierzchowska, I., Collier, M., & Lodder, M. (2020). Examining the policy needs for implementing nature-based solutions in cities: Findings from city-wide transdisciplinary experiences in Glasgow (UK), Genk (Belgium) and Poznań (Poland). Land Use Policy, 96, 104688. doi:https://doi.org/10.1016/j.landusepol.2020.104688.

ICLEI. (2021). Making Urban Nature bloom. Report, 53(9). https://naturvation.eu/sites/default/files/news/files/making_urban_nature_bloom.pdf

IPCC. (2022a). Climate change 2022: Impacts, Adaptation and Vulnerability. Contribution of Working Group II to the Sixth Assessment Report of the Intergovernmental Panel on Climate Change. doi:https://doi.org/10.1017/9781009325844.

IPCC. (2022b). Climate change 2022: Mitigation of climate change. Contribution of Working Group III to the Sixth Assessment Report of the Intergovernmental Panel on Climate Change. doi:https://doi.org/10.1017/9781009157926.020.

Kabisch, N., Frantzeskaki, N., & Hansen, R. (2022). Principles for urban nature-based solutions. *Ambio, 51*(6), 1388–1401. https://doi.org/10.1007/s13280-021-01685-w

Kabisch, N., Frantzeskaki, N., Pauleit, S., Naumann, S., Davis, M., Artmann, M., Haase, D., Knapp, S., Korn, H., Stadler, J., Zaunberger, K., & Bonn, A. (2016). Nature-based solutions to climate change mitigation and adaptation in urban areas: Perspectives on indicators, knowledge gaps, barriers, and opportunities for action. *Ecology and Society, 21*(2), 342. https://doi.org/10.5751/ES-08373-210239

Longato, D., Cortinovis, C., Balzan, M., & Geneletti, D. (2023). A method to prioritize and allocate nature-based solutions in urban areas based on ecosystem service demand. *Landscape and Urban Planning, 235*, 104743. https://doi.org/10.1016/j.landurbplan.2023.104743

Mahmoud, I., Dubois, G., Liquete, C., & Robuchon, M. (2025). Understanding collaborative governance of biodiversity-inclusive urban planning: Methodological approach and benchmarking results for urban nature plans in 10 European cities. *Urban Ecosystems, 28*(2), 17. https://doi.org/10.1007/s11252-024-01656-5

Martín, G. (2019). Del «urbanismo expansivo» al «urbanismo regenerativo» directrices y recomendaciones para reconducir la herencia territorial de la década prodigiosa del urbanismo español (1997–2007). Aplicación al caso de estudio del litoral andaluz coast. *Ciudad y Territorio Estudios Territoriales, 51*(199), 81–100. https://recyt.fecyt.es/index.php/CyTET/article/view/76716

Morán Alonso, N., Viedma Guiard, A., Simón Rojo, M., & Córdoba Hernández, R. (2025). Agricultural land suitability analysis for land use planning: The case of the Madrid region. *Land, 14*(1), 134. https://doi.org/10.3390/land14010134

Morán Uriel, J., Camerin, F., & Córdoba Hernández, R. (2024). Urban horizons in China: Challenges and opportunities for community intervention in a country marked by the Heihe-Tengchong line. In G. Siew, Z. Allam, & A. Cheshmehzangi (Eds.), *Diversity as catalyst: Economic growth and urban resilience in global cityscapes* (urban Sust, pp. 105–125). Springer. doi:https://doi.org/10.1007/978-981-97-8776-0_6.

Nesshöver, C., Assmuth, T., Irvine, K. N., Rusch, G. M., Waylen, K. A., Delbaere, B., Haase, D., Jones-Walters, L., Keune, H., Kovacs, E., Krauze, K., Külvik, M., Rey, F., van Dijk, J., Vistad,

O. I., Wilkinson, M. E., & Wittmer, H. (2017). The science, policy and practice of nature-based solutions: An interdisciplinary perspective. In *Science of the Total Environment* (Vol. 579, pp. 1215–1227). doi:https://doi.org/10.1016/j.scitotenv.2016.11.106.

Pattberg, P., Kaiser, C., Widerberg, O., & Stripple, J. (2022). 20 years of global climate change governance research: Taking stock and moving forward. *International Environmental Agreements: Politics, Law and Economics, 22*(2), 295–315. https://doi.org/10.1007/s10784-022-09568-5

Puskás, N., Abunnasr, Y., & Naalbandian, S. (2021). Assessing deeper levels of participation in nature-based solutions in urban landscapes—A literature review of real-world cases. *In Landscape and Urban Planning, 210*. https://doi.org/10.1016/j.landurbplan.2021.104065

Robbiati, F. O., Natalia, C., Gustavo, B., Gustavo, O., Jim, C. Y., Mario, S., Emmanuel, H., Rubio, E., Leonardo, G., & Lelia, I. (2023). Vegetated roofs as a nature-based solution to mitigate climate change in a semiarid city. Nature-Based Solutions, 3, 100069. doi:https://doi.org/10.1016/j.nbsj.2023.100069.

Thomson, G., & Newman, P. (2020). Cities and the Anthropocene: Urban governance for the new era of regenerative cities. *Urban Studies, 57*(7), 1502–1519. https://doi.org/10.1177/0042098018779769

United Nations. (2019). World Urbanization Prospects: The 2018 Revision. In *World Urbanization Prospects: The 2018 Revision*. https://doi.org/10.18356/b9e995fe-en

Watt, A., Ainsworth, G., Balian, E., Cojocaru, G., Darbi, M., Dicks, L., Eggermont, H., Furman, E., Goudeseune, L., Huybrecht, P., Kelemen, E., Koch, F., Konstantinou, Z., Livoreil, B., Locher, K., Lux, A., Mehring, M., Nesshoever, C., Paloniemi, R., et al. (2019). EKLIPSE: Engaging knowledge holders and networks for evidence-informed European policy on biodiversity and ecosystem services. *Evidence & Policy, 15*(2), 253–264. https://doi.org/10.1332/174426418X15314036194114

Zingoni de Baro, M. E. (2022). An integrated framework for designing regenerative sustainable urban environments. In *Cities and nature: Vol. part* (pp. 97–116). Springer, Cham. https://doi.org/10.1007/978-3-030-90559-0_5

Nature as Everyday Practice and Place-Making

Circle: Migrations and Mundane Mobilities of Childhood

Jennifer L. Kent

Abstract The focus of this chapter is one school's celebration of a migratory bird: the Bar-tailed Godwit. Its intention is to draw attention to the ways nature might be incorporated into cities in a way that is more intentional and effective, with a focus on children's day-to-day travel. The godwit is famous for the world's longest non-stop migration—flying 9000 kilometres each spring from the Arctic Circle to specific mudflats along Australia's east coast, including those in Abbotsford, a suburb of Sydney. At Abbotsford's local school, godwits are honoured through art and stories, highlighting their journey and Indigenous significance. The message is global, respectful and communal—a joyful acknowledgement of the wonder of nature in cities. Juxtaposed with this message are the day-to-day 'migratory' practices of families in Abbotsford. Although the suburb's urban fabric provides an ideal environment for children to access school by walking and cycling, most come by car, reflecting the preferences and practices of mobility in Australian cities more generally. In this conceptual piece, an exploratory case-study methodology is used to highlight the simultaneous contrasts and parallels between the flight of the godwit and children's experiences of daily transport. First, a social practice lens is used to position nature-based urban imagery as a shaper of practice with potential to provoke transitions to more sustainable ways of living in cities. The chapter progresses to explore themes and examples of ways both nature and day-to-day mobility in cities are presented to children. Using the case of the godwits in Abbotsford, the chapter reveals a disconnect—between children's everyday mobility and their understandings of the awe and wonder of the godwit. The chapter concludes by exploring opportunities to better connect urban images of nature with everyday practices, particularly for children. It then articulates lessons for city planners and scholars, outlining the importance for both groups to notice frictions between the intent articulated through imagery, and the way people live in cities.

Keywords Urban imagery · Nature in schools · Children and nature · Active transport to school · Social practice theory · Sydney · Godwits

J. L. Kent (✉)
School of Architecture, Design and Planning, The University of Sydney, Sydney, NSW, Australia
e-mail: Jennifer.kent@sydney.edu.au

A. Cheshmehzangi et al. (eds.), *Nature-based Urbanism*, Urban Sustainability,
https://doi.org/10.1007/978-981-95-6864-2_10

1 Introduction

The focus of this chapter is one school's celebration of a migratory bird: the Bar-tailed Godwit. The Bar-tailed Godwit is a relatively plain-looking bird, with a migratory pathway that scales the globe. It is famous for its ability to fly longer distances than any other bird in the world. In spring, flocks of come from the Arctic Circle to a small selection of bays along the Australian east coast. Over the summer, they recover from their voyage and feed on mudflats in preparation to fly again in Autumn.

Abbotsford is a suburb situated on a bay in Sydney, Australia, where godwits come each Spring to feed. At the local public school, the birds are celebrated and supported, with their journey, precarity, and place in Indigenous stories incorporated into the iconography surrounding the school. The school's message celebrates nature in cities—global in scope, respectful in tone, and shared in spirit.

In contrast, everyday transport choices for school travel in Abbotsford reflect a broader national reliance on private cars. The peninsula suburb provides an ideal environment for children to access school by walking and cycling, yet at school drop off times the curb sides temporarily adorned with family cars, which eject uniformed children, school bags and packed lunches before driving off, or turning to head back home. Although Abbotsford's streets are quiet, characterised by reduced speed limits, marked bike lanes, adequate footpaths and grid-like designs, only some students walk, accompanied by parents and carers, and very few access the school independently. Parents and children, it seems, are either unaware or accepting of the negative impacts of cars on the same environment that is cherished for the journey of the godwit.

In this conceptual and theoretical chapter, attention is drawn to the simultaneous contrasts and parallels between the flight of the godwit and children's experiences of mundane mobilities. The intention is to explore an example of how nature in cities is presented to children, and how it might be extended to encourage children to reflect on their own attitudes and practices. In Abbotsford, the children are inspired through the school by the godwit's unique ability to travel independently and actively, and in this way the bird is elevated to the realm of needing care and concern. The connection, however, between what the children do each day to arrive and depart from school and what the Godwits do annually is missed. The imagery fails to call upon the children to consider their own routines of mobility against that of the Godwits. The circle between the tale of the godwit and the reality of children's mobility is interrupted by routines, roads, and role modelling, which are all defined by the private car.

The chapter opens by positioning existing travel trends in many cities, particularly those associated with children, as problematic. Focussing on the need for cities to transition away from private car dependence, the chapter progresses to review the various elements which shape day-to-day travel. Social practice theory is used to highlight the way socio-spatial ordering in cities—including through travel—can be understood as a complex interplay between the built environment and psycho-social/cultural dynamics. The role of imagery in sculpting the emotions and cultural scripts that determine travel attitudes and patterns, particularly for parents and children, is highlighted. The chapter then introduces one aspect of imagery used by a school in Sydney, Australia, to instil within children respect for nature, understandings of environmental precarity, and wonder at the day-to-day natural elements present in the urban environment. This imagery is focussed on a migratory bird who nests near the school—the bar-tailed godwit. This is then juxtaposed to the everyday realities of children, which are dominated by unsustainable practices, including practices of mobility. The chapter concludes that in fostering awe at the godwit, yet failing to connect its way of being in the world to the practices of children, an opportunity is lost to nurture children's appreciation for more sustainable and respectful day-to-day travel. More broadly, the chapter recommends a more holistic and intentional appreciation of the benefits of nature in cities, which extend to inspire a more respectful approach to day-to-day life.

2 Background

2.1 Cities, Cars and Children's Mobility

Governing institutions in most global north urban contexts now articulate intentions to decrease reliance on the private car for day-to-day transport (Evans, 2023). Often this is through the promotion of alternatives (for example, walking and cycling), and the pursuit of reductions in the distances between uses in cities (primarily through densification). Despite these stated intentions, over the last ten years private car ownership—an accurate predictor of car use—has continued to rise globally, reflecting increased incomes and degrees of urbanisation. The majority of this growth has occurred in non-OECD countries aspiring to emulate the levels of privacy, convenience and comfort enjoyed by the automobile global north (Haarstad et al., 2023). Vehicle ownership in wealthier contexts, however, also continues to grow, albeit at a slower pace (Saeidizand et al., 2022).

Unsurprisingly, trends in the way children travel mirror global trends. Most scholarship in this space has its focus on school travel, examining determinants of the variations in the uptake of children's active transport to school from context to context. While not universal, many nations have experienced dramatic declines in children walking to school since the 1970s (Ji et al., 2022). In Australia, for example, 70% of students walked or cycled to school in 1974, compared to 30% by 2010. Most are now driven to school, particularly in primary school (Garrard, 2011). These declines continue, despite the substantial scholarship recognising the benefits of children's active transport to school, including engagement with physical activity, socialisation, enhanced academic performance and better mental health (for example, Ramanathan et al., 2014; García-Hermoso et al., 2017; Hassevoort et al., 2016).

Outside of the school journey, families play a significant role in sustaining the dominance of the car. In Australia's two largest cities—Sydney and Melbourne—motor vehicle ownership is highest among households with children (The Australian

Bureau of Statistics, 2021). In Melbourne, young families (defined as one or two parents with all children under the age of 10) are among the least likely to use public transport. In 2010, only 5% of trips by couples with children and 4% of trips by sole-parent households were made using public transport, compared to 14% of trips by other household types (The Victorian Department of Transport, 2010). Caring for children generates a substantial amount of travel, with mothers of children under 10 making more trips per day than any other demographic group—and they are also the least likely to use public transport. Children aged 0–10 make an average of 2.5 trips per day, with over 90% of these trips taken by private car (The Victorian Department of Transport, 2018). In Sydney, around 50% of trips are made for purposes related to domestic responsibilities and caregiving, with trips serving passengers and those to school and childcare comprising the majority (The Bureau of Transport Statistics, 2021).

This dominance of the car in parenting practices is unfortunate given the fact children have the most to benefit from decreased car-dependence. From the impacts of climate change to the heightened vulnerability of children's bodies to road trauma and respiratory illness (outlined below), it is children who absorb the problem of car-dependency in the present and the future. Furthermore, childhood is a time where attitudes about transport are developed, with various studies demonstrating that children who are exposed to less car-dependent practices and appreciations are likely to grow into less car-dependent adults (for example, Baslington, 2008; Baslington, 2009).

The ongoing failure to curb growth in private car ownership, and subsequent dependency on the car, generates and perpetuates many of the problems faced by contemporary cities. From the perspective of human health, the devastating toll of trauma and death from car collisions is the most obvious harm (James et al., 2020). Cars are also implicated in a series of more chronic conditions, including respiratory illnesses associated with car emissions (Lelieveld et al., 2015), and the impacts of noise on mental health (Hegewald et al., 2020). More subtle still is the way cars perpetuate inactive lifestyles (Frederick et al., 2018) and the stress of 'just-in-time' ways of living undeniably associated with access to a car (Urry, 2004). Traffic congestion caused by cars is also associated with stress, lost time and income (for example, BITRE 2015). And finally, cars remain key contributors to climate change—in 2018 the International Energy Agency reported that the transport sector accounts for a quarter of global CO2 emissions, of which 74% is caused by the ever-growing subsector of road transport (The International Energy Agency, 2018). Only some of this will be addressed by the slow transition to electric vehicles (Zhao et al., 2023).

Increased awareness of the problem with private cars and the benefits of alternative modes has stimulated intense attention to the determinants of car-ownership and use in an effort to shift mobility towards a more sustainable and healthier regime. Literature now abounds on ways to prompt transitions towards more sustainable travel practices such as walking, cycling, and public transport use, including for children, as well as how we can reduce demand for travel at all. Most recognise that shifting established transport practices in cities is enduringly complex, requiring theoretical and conceptual frameworks emphasising the interplay of material and institutional structures with psycho-social paradigms. This chapter draws from one such theoretical frame—social practice theory.

2.2 *Introducing Social Practice Theory*

Theories of practice (Reckwitz, 2002; Schatzki et al., 2001) especially critique the view that behaviours are the result of either individualised and linear decision-making processes or overbearing and unavoidable forces of material and institutional structure (Urry, 2012). Instead, routine human actions—or practices—are understood as influenced by interactions between the wider environment, socialised preferences and individual processes of deliberation (Watson, 2012). The way a child travels to school, for example, is a practice shaped not only by urban form but also influenced by emotional and cultural dimensions of parenting.

While social practice theory deploys a series of tools and concepts, its most ubiquitous concept is that practices are construed of elements. This concept provides a practical scaffold from which to deconstruct the multiple shapers of spatially and temporally complex practices such as those related to travel (Kent, 2021). Elements are the material, structural, psycho-social and cultural preconditions that interact to comprise practices. For the purposes of this chapter, the elements of focus are labelled 'materials and structures', 'feelings and emotions', and 'cultural scripts'.

Materials and structures, particularly infrastructure and other technical provisions, are generally well researched and well known by those seeking to promote sustainable transport in cities. For example, higher residential and employment densities decrease the distances between uses (reviewed by Kent et al. 2023a). Mixed uses will also support walkable access instead of car reliance (Fonseca et al., 2021; Panter et al., 2016; Panter et al., 2019). Design features, such as separated and connected bike lanes and footpaths, and adequate signage (Le Gouais et al., 2021; Mölenberg et al., 2019), will also enable some car trips to be replaced by walking and cycling. Public transit that is predictable, regular, clean, and planned to connect the places people need to go is linked to increased ridership (Boisjoly et al., 2018; Ingvardson & Nielsen, 2018; Ingvardson & Nielsen, 2019) and decreased modal share by car (Mulalic & Rouwendal, 2020). With specific reference to families and children, designing apartment buildings in higher-density areas such that children are welcome (Kerr et al. 2021, Kent et al. 2023b), reducing school drawing areas to shorten school travel distances (Yan et al., 2019), and traffic calming measures around schools (Nicholson et al., 2014) are all structural interventions designed to support less car-reliant childhoods.

While structural determinants of travel practices remain dominant in research, it is difficult to dismiss that emotions are also invariably at play when it comes to ways of travelling day-to-day in cities. Although less prolific than literature on transport's structural determinants, there has been research on the ways bodily sensations (feeling hot, cold, tense, or relaxed (Bean et al., 2021)), accompanied by emotional states (feeling empowered, free, safe, constrained, or angry (Kent, 2014)) might preference the autonomy and privacy of the private car. Consider, for example, new parents' appreciation of the sensory cocoon of the car, and its provision of welcome emotional and sensory relief from parenting a newborn (Sheller, 2004). For parents of school-aged children, imagine the way the terror of a speeding car might prevent or delay decisions to allow children to walk to school alone (Crawford et al., 2017; Foster et al., 2014).

Complementing the elements of structures and emotions is an additional layer of influence pertaining to culture. This is the element of know-how, inclination, skill, and acceptance of the way things are, referred to here as 'cultural scripts'. For example, in many car-dependent contexts there is a prevailing acceptance that 'if you have kids, you need a car' (Freudendal-Pedersen & Kesselring, 2016, 580). In these same contexts there is cultural aversion to the idea of raising a family in an apartment (more likely to have access to public transport) rather than a house (Kerr et al., 2018).

In the context of understanding private car use, the key benefit of a practice approach is the ability to see prerequisites for behaviour change as multifaceted (Kent, 2021). When viewed through the lens of interacting elements, it seems obvious that examinations of the material built environment alone will fail to elucidate true understandings of travel practices. Emotional and cultural elements, and the ways these things interact, must also be considered. And nature can be considered a component of all three.

2.3 Practice Theory, Parenting, and the Private Car

Having established the problems of car dependency and the possibilities of social practice theory in understanding this phenomenon, attention now turns to determinants of children's travel, and the role of urban imagery, particularly images of nature.

As outlined in Sect. 2.1 above, children have the most to gain from reduced car-dependency, and exposure to alternative modes of travelling. Of particular relevance is that development of an appreciation of carless ways of being mobile can develop into changed travel behaviour as the child comes of age. Children who are taught to appreciate and accept sustainable travel modes as 'normal', for example, are more likely to express positive attitudes to these modes (Lunke, 2025) and even self-select into built environments where car dependence is unnecessary (Van Acker et al., 2019).

To understand the determinants of children's present and future transport behaviour, most research defers to the structural approach, and this is reflected in dominant policy paradigms. For example, there is a mountain of research emphasising the way reduced distances between home and school, and safer environments around schools, are associated with children's active transport to school (Chica-Olmo et al., 2018; Macdonald et al., 2019; Rodríguez-López et al., 2017). Other studies focus on the role of the parent as the gatekeeper of children's travel, with particular attention paid to the way parents' perception of safety acts as a deterrent to children's active transport (Rothman et al., 2021).

This chapter has its focus on one particular and under-explored opportunity to enlist children themselves as shapers of transport practices—whether it be in the present or the future—urban imagery. Its specific focus is the use of imagery of nature in school environments. The following section goes on to propose a conceptual framework for how imagery, including images of nature in schools, might fit within the elements of social practice theory to understand the influence of imagery on children's travel behaviour.

2.3.1 The Role of Imagery in the Travel Practices of Children

The term 'urban imagery' has a long history in urban geography and can be used to describe the spectrum that sits somewhere between visual symbolic elements *within* urban environments (such as public architecture and art), and visual representations *of* urban environments (such as depictions of cities in texts and maps) (Domosh, 1992). This chapter has its focus on the former with a particular focus on artistic imagery in and around school environments.

Imagery is akin to a language, capable of eliciting emotion and communicating meaning. Taking a practice perspective, it is possible to see how structural imagery (for example, signs and posters) might reflect dominant cultural scripts and aspirations (for example, expectations for recycling), as well as evoke emotions (for example, pride). As such, urban imagery can be viewed as a shaper of practice, and its potential to provoke transitions to more sustainable ways of living in cities cannot be dismissed as individualistic or trivial.

Children's emotions and their understandings of social norms are particularly open to engagement through both storytelling and art (Guo et al., 2020), as well as more the mundane icons and symbols peppered throughout their realm (Na et al., 2021; Siu et al., 2015). Various studies have shown the ways children can be engaged through urban imagery to develop awareness and appreciation of natural elements, with school environments proving particularly influential (as reviewed by Guo et al. (2020)). Schools too have assumed responsibility for educating children to become environmentally aware and healthy citizens for many years, and imagery of nature is often used to pull this narrative together (Elegbede et al., 2023).

Other studies have demonstrated the way imagery is used to support children walking and cycling to school. These visual campaigns often use vibrant and engaging imagery to appeal to children and their parents. Posters and advertisements featuring happy, active children walking or cycling in safe and scenic environments attempt to foster positive associations with active transport and often emphasise the health benefits (Carlin et al., 2016; Villa-González et al., 2018; Ward et al., 2007).

While it is beyond the scope of this piece to comprehensively review or evaluate, compare and contrast the use of imagery in the promotion of nature, care for the environment and active transport in school environments, a cursory review prompts reflection on an opportunity.

The way animals navigate their world with little impact on those around them can provide particularly valuable vehicles to foster children's understandings of the importance of sustainable travel practices and promote awareness of what a sustainable transport system looks like. There is a notable disconnect, however, in how nature and mobility are visually represented in school-based programs promoting environmental awareness. Environmental education often uses imagery to foster emotional connections to nature (Gough, 2013; Payne, 2010), yet it rarely includes depictions of everyday mobility practices like walking or cycling to school or other destinations. Conversely, active transport campaigns tend to rely on urbane or cartoonish visuals, omitting ecological themes (Jones et al., 2019). This separation reinforces what Malone (2007) critiques as the 'urban-nature disconnect', where nature is taught as something distant from children's lived experiences. This is surprising given the role of mobility and travel in natural systems—from the importance of pollen transfer to species migration. Nature needs travel and provides inspiring stories of ways to travel with little impact. Yet, the way we travel in modern cities and the travel habits we socialise our children into are destroying nature. This absence of integrated imagery suggests a missed opportunity to frame sustainable travel as part of a broader environmental ethic, reflecting a tendency to under-estimate children's capacities to think holistically about their environment, including the impact of their day-to-day routines on that environment.

There are some example case studies illustrating ways this gap might be bridged. Sweden's "Grönare Skolvägar" initiative uses nature-themed signage along walking routes to school (Gericke et al., 2020), while Suwon's EcoMobility World Festival in South Korea featured school murals linking active transport with clean air and biodiversity (ICLEI, 2013). These examples demonstrate ways combining nature and day-to-day mobility in educational contexts can foster deeper ecological understanding and engagement, especially in urban contexts where transport systems are a major contributor to environmental degradation.

Taking this observation and these examples as inspiration, this chapter progresses to explore an opportunity to patch these two together in a different context, highlighting how nature might be used to support sustainable transport transitions in cities. The following section introduces an exploratory case study (after Yin, 2013) to demonstrate how children's urban imagery of nature might be broadened to promote nature's protection and cities that are healthier and more sustainable. Let's start with the Bar-tailed Godwit.

3 Case Study: Godwits, Schools, and Children's Mobility

3.1 The Wonder of the Bar-Tailed Godwit

The Bar-tailed Godwit (godwit) is a relatively large bird for a wader bird, but small and slight when foraging on the mudflats amongst the Silver Gulls, pelicans and ibis that also call Australia and New Zealand's coastal estuaries home from time to time. From afar it isn't a particularly striking bird. Godwits are mainly mottled brown above and a lighter buff colour below. It has dull white underwings and a long, slightly upturned bill (Partenko, 1936). Like so many birds, on closer inspection, its beauty and grace become more apparent. And like so many unassuming pieces of the puzzles that make up the ecosystems around us, its true wonder is in the quiet accomplishment of feats we'd never dare to dream of Fig. 1.

The godwit performs the longest non-stop migratory flight in the world (Alerstam et al., 2003). In the southern hemisphere's Spring, flocks of godwits fly from the Arctic Circle to a small selection of bays and inlets along the Australian east coast and the coasts of New Zealand. Over the summer, they recover from their voyage and feed on mudflats in preparation to fly again in Autumn. When the time is right, they fly non-stop from Australia and New Zealand to China's Yellow Sea and its rich tidal flats, where they refuel briefly before the return flight to Alaska to breed. They are extremely proficient at this risky endeavour; the survival rate is more than 90 percent (Beehler, 2025).

What is epic about the godwits' flight is that it is long and continuous. The southbound flight, from Alaska to Australia and New Zealand, covers at least 8000 kilometres and lasts from eight to 10 days and nights through pounding rain, high

Fig. 1 Godwit in Abbotsford *(source: Simon Gatward, SPG Photography)*

winds, and other perils (Beehler, 2025). And the godwits don't stop. They don't rest, touch down, feed or drink for the entire flight. In fact, unlike other long-haul birds, they don't even use dynamic soaring to travel—their mode of transport requires that they flap their brown speckled wings over and over again.

The known continuous distance record for a godwit migration is 13,560 kilometres. It was set in October 2023 by a five-month-old adult male Bar-tailed Godwit with a tag code of 234,684—colloquially known as Bruce. Bruce, and the flock he was flying with, encountered inclement weather on their way to New Zealand. They veered off course to a more distant landing in Tasmania. By the time Bruce touched down, he had flapped his wings for 266 hours, or just over 11 days, non-stop.

How the birds do this has been the subject of intense scrutiny. We now know that the internal organs of godwits shrink and grow throughout the year, positioning the body to be as aerodynamic as possible. Their bodies have an extraordinary capacity to store fat. And their lungs are designed to expand and contract to support their efforts. They live, work and travel in groups, guided by consensus as to where and when to go. They rely on several cues for navigation, particularly the sun and the stars, to find their way back to the same specific sites at the end of their flight. They do this for each of the 15 to 20 years of their lives. Incredibly, three-month-old godwit juveniles fly their nonstop maiden voyage without adult supervision. It was likely that Bruce was flying as part of a flock of younger godwits, left alone by their parents who would have departed a couple of months earlier. The younger ones stay behind longer to take every possible opportunity to fatten themselves up before they set out to travel so far.

Like so many species around the world, the godwits are threatened. They are resilient, but their numbers are dwindling as their feeding grounds are handed over to development, their migratory pathways warm and the terrains where they breed shrink. The way we live, with our carbon and our waste and our intrusion, threatens their hardy but sensitive survival.

One of the bays where godwits land is in the suburb of Abbotsford in inner-urban Sydney, Australia.

3.2 The Ordinariness of Abbotsford, Sydney, Australia

Abbotsford is a peninsula suburb, with its western shoreline forming part of an inlet off the Parramatta River known as Hen and Chicken Bay. Each Spring, godwits appear on the bay's mudflats. They are shrunken and tired from their migration.

Away from the water, Abbotsford is a suburb that reflects various urban elements and eras of development typical of its host city, Sydney. Sydney is in the southern hemisphere but part of the colonised global north, enjoying relatively high per capita income and providing home for over six million people. The city enjoyed rapid growth throughout the 1950s, 60 s, and 70 s—a time at which car ownership was also becoming the norm. Sydney's development, therefore, was predicated on widespread private car ownership, resulting in a sprawling, low-density urban area covering

12,500 square kilometres, sprawling all directions from its eastern origins. Since the late twentieth Century, there have been attempts to introduce higher-density, mixed-use developments near transit services in middle ring 'greyfield' suburbs and the inner city. Low-density residential areas, however, still dominate, particularly in the outer suburbs (Randolph & Freestone, 2012), and a detached house on a sizeable block remains a cultural aspiration with a long history (Bunker et al., 2005, Kerr et al. 2021). This poses significant challenges for alternative transport planning (Glazebrook, 2009; Legacy et al., 2017). There are substantial distances between destinations, making walking and cycling impractical and often unsafe (Ellison & Greaves, 2011; Pucher et al., 2010). Although fixed and light rail, bus, and ferry services are available, public transport accessibility is inconsistent (Tsai et al., 2014). Non-work trips are often hampered by unreliable connections, limited off-peak schedules, and difficulties in transporting loads and travelling with children (Kent & Mulley, 2023). This structural backdrop merges with cultural narratives that position the car as the normal, necessary and natural way to traverse the city, particularly with children. The bodily sensations of heat while walking in summer, or wet on a rainy day are other things to avoid rather than accept as the price of a less car-dependent environment. In short, for a confluence of structural, cultural and emotional reasons, car travel remains the predominant mode of transport for all trip purposes in Sydney as a city, mirroring trends in other major cities in the colonies (NSW Bureau of Transport Statistics, 2020).

While the Sydney urban agglomeration sprawls over 90 km to the north, south and west, Abbotsford is located just 10 kilometres from Sydney's eastern CBD, and considered part of the city's inner-urban sphere. It is relatively well serviced with buses departing regularly to key destinations, and a ferry that trips along the Parramatta River to the city, on the half hour. Abbotsford's streets are on the quiet side, with even its one single busier road characterised by reduced speed limits and well-marked bike lanes. All streets all have footpaths on both sides that stretch along in a grid-like shape; residential densities are relatively mixed, as are uses, with small-scale shops and services scattered throughout and within walkable distance to most. In short, the suburb essentially satisfies the dreams of urban designers and transport planners seeking sustainable solutions, hosting the destinations, diversities., and distances associated, on paper, with active transport use and a shift away from the private car (as reviewed in Sect. 2.1).

Although structural provisions for alternatives are relatively good, Abbotsford remains a suburb, in a city, in a nation, shaped by private car use. Over 87% of Abbotsford households own a car, and 44.6% own at least two cars (The Australian Bureau of Statistics, 2021). Of those working and travelling to work, around 60% go by car (The Australian Bureau of Statistics, 2016). Data is scarce on the mode share for other trips (such as trips for shopping, leisure and school), however there are various indicators suggesting cars dominate the travel practices for Abbotsford's residents. Traffic congestion around the area, for example, is recorded as intensifying (Transport for New South Wales, 2025), and the key complaint addressed by the local Council relates to the lack of car parking available at shopping and service precincts (City of Canada Bay, 2021).

Fig. 2 Godwit iconography at Abbotsford Public School *(source: the author, 2025)*

The suburb is home to just over 7000 people living in around 3000 households, with slightly less than 40% of these being households with dependent children. Socio-economically, the area is diverse, hosting more high-income households than Greater Sydney, but also a high proportion of households with lower incomes. Approximately 500 of Abbotsford's 7000 residents are primary school-aged children, and more than 80% of these kids attend the local primary school—Abbotsford Public School (The Australian Bureau of Statistics, 2021).

At Abbotsford Public, the godwits are celebrated and supported. Their journey, precarity and place in indigenous stories is incorporated into the iconography surrounding the school. Its logo for example, and other signage around the school, all feature the godwit image, and children are taught of the godwit's feat (Fig. 2). The school has contributed to public art displayed around the locality to draw attention to the importance of the mudflats because they provide the food that fuels the godwit migration (Fig. 3.).

The message is global, respectful, and communal. It is a celebration of the godwit based on the amazing ability the bird has to be mobile. It is consistently a message that, somehow, we need to protect the role the bay that is these children's backyard plays in making this mobility be.

Sitting alongside this message are the day-to-day transport practices of families in Abbotsford, which are ostensibly car-dependent. While seemingly providing an ideal *structural* environment for less cars, the suburb echoes the practices and preferences of the city that hosts it. At school drop off time the kerb sides lining the perimeter of the public school are temporarily adorned with SUVs—a steady march of cars eject uniformed children, school bags and packed lunches before driving off, or turning to head back home—parents who are likely either too concerned about safety or busy to encourage their children to walk to school. Although data on the mode share for the journey to school in Abbotsford does not exist, there is no reason to think it would contradict norms for the Greater Sydney region, with over 70% of the suburbs' primary school-aged children driven to school in a private car on a daily basis (Garrard, 2011).

Fig. 3 Drawings used by school students to protect godwits *(source: the author, 2025)*

In the face of contradictions—between urban form (which supports children's active school transport but hosts car-dependency) imagery (which promotes the godwit), and aspirations (which seek to both protect the godwit and the child)—it is interesting to draw simultaneous contrasts and parallels between the flight of the godwit and children's experiences of mundane mobilities.

The godwits have an undeniable rhythm about them, which echoes the lives of children, who thrive on predictability and routine. The idea of an unquestioned annual practice would appeal to children, whose lives are reassuringly punctuated with rituals—the school week, the school holidays, birthdays and Christmas. As with children, movement is also an integral component of the godwit's survival. Children also need and like to move about—to connect, to explore, to do the play that is their work. While a 500 metre walk to the park may not be a 9000 kilometre flight around the world, it is still a practice with capacity to lull a child into a state of flow, distracted from thought by the simple act of being mobile and in pursuit of a destination. And, like children and the mobility regimes their parents enrol them in, the godwits migration is seemingly unquestioned by the godwits in flight, positioned as entirely necessary to survive in the current environment in which they find themselves. The godwit must be mobile to feed and breed. The modern-day child must be mobile to be safe, and attend the tutoring, violin lessons and sport that will make them worthy citizens. Again, we are reminded that the need, appreciation and desire to be mobile are universalising experiences—across ages, cultures, and species.

And now to contrasts. There is the obvious—with an average life expectancy between 15 and 20, a healthy godwit will make their first unaccompanied migration at age 3 months. The age at which a child is granted their independent mobility licence—or ability to travel without an adult—varies between cultures but in Australia it is between 9 and 12 (Schoeppe et al., 2016). Our mobility with and for children, it seems, is imbued with a fear not felt by the godwits. Then picture the flock flying together, aware yet responding to intuition, travelling extreme distances on an ancient route, simply so that they can accommodate the shifting availabilities of resources around the globe. Bending themselves around what is available, rather than taking until there is nothing left to take. They adapt their bodies and practices, rather than force their surroundings to adapt. There is something ancient and magical about what these birds do. They defy what we can understand as humans, and if we can sit in that space of not quite understanding, we can sense a certain enormity. Their flight is not only carbonless. Its only impact is absorbed by the stealth-like bodies of these birds, which are built to do what is done.

4 Discussion

The focus of this chapter is the way mobility and transport are presented to people, particularly children, who are forming ideas about what is right, wrong, normal, and odd, including when it comes to day-to-day ways to travel. In the background sect. I established that there is an element of co-production to urban imagery as it merges with structural provisions, emotions and cultural scripts to shape the way we travel in cities. It is therefore possible that the scales and subjects used to entrain children to different images of mobility actually do influence practices of mobility. As a result, it is important that we teach children about monumental feats of mobility—like the flight of the godwit—and compare and contrast these to the way we travel day-to-day. The chapter proceeds to detail an exploratory case study as a way to demonstrate the missed opportunity in one Australian primary school exploring one exceptional migratory bird.

Of course, the regime that is private car use across the globe will not be challenged through images of migratory birds alone. Similarly, the entrainment of children to appreciations of more respectful mobility will not occur simply through the careful presentation of imagery. This exploratory case study suggests it needs to be made more explicit. The children are inspired to awareness of the godwit's unique ability to travel, and in this way the bird is elevated to the realm of needing care and concern. The connection, however, between what the children do each day to arrive and depart from school, and what the godwits do annually, and the way they do it, is missed. The imagery does not call upon the children to consider their own routines of mobility against that of the godwits, nor to consider how their routines might be impacting the godwits' ability to do what they do. The point is that the co-production of the imagined and the lived in transport practices is not necessarily a natural progression, and more can be done.

The circle, it seems, between the tale of the godwit and the reality of children's mobility, is incomplete—disrupted by the realities of parenting and travelling in cities by car. As such, the magic of the godwit sits in one realm of the minds of children and their parents, and the reality of day-to-day sits in another—forever disconnected.

5 Conclusion

This conceptual and theoretical chapter has drawn attention to the simultaneous contrasts and parallels between the flight of the godwit and children's experiences of mundane mobilities. Highlighting the important role of urban imagery in using nature as a subject, the chapter has highlighted how images and stories about nature can be used to entrain children not only to awe and respect, but also to different ways of living in cities, including ways of travelling day-to-day. But still, while Abbotsford Public's celebration of the godwit as a mobile subject is a lost opportunity to provoke consideration of reflections on our own ways of doing mobility, it is not a hopeless endeavour. The bird is amazing, and its feat undoubtedly inspires children.

In conclusion, attention is drawn to what we can learn as city planners and scholars by noticing frictions between the intent articulated through imagery and the way people live in cities. These tensions reflect aspir against mundane realities. They may be eye-rollingly frustrating in their presentation, yet they are actually also tiny beacons of hope. Hope that we do believe in the magic of nature, including its ability to travel lightly. It is time now for us to use our fascination by making more explicit the links between our routine migrations and those of the various species remaining. In this sense we may be prompted to reflect on how, comparatively, what we do daily is so resource intensive, so scarring, so disrespectful to those around us, and so costly to the world that hosts us.

References

Alerstam, T., Hedenström, A., & Åkesson, S. (2003). Long-distance migration: Evolution and determinants. *Oikos, 103*(2), 247–260.

Baslington, H. (2008). Travel socialization: A social theory of travel mode behavior. *International Journal of Sustainable Transportation, 2*(2), 91–114.

Baslington, H. (2009). Children's perceptions of and attitudes towards, transport modes: Why a vehicle for change is long overdue. *Childrens Geographies, 7*(3), 305–322.

Bean, R., Pojani, D., & Corcoran, J. (2021). How does weather affect bikeshare use? A comparative analysis of forty cities across climate zones. *Journal of Transport Geography, 95*, 103155.

Beehler, B. M. (2025). *Flight of the godwit: Tracking epic shorebird migrations.* Smithsonian Books.

Boisjoly, G., Grisé, E., Maguire, M., Veillette, M.-P., Deboosere, R., Berrebi, E., & El-Geneidy, A. (2018). Invest in the ride: A 14 year longitudinal analysis of the determinants of public transport ridership in 25 north American cities. *Transportation Research Part A: Policy and Practice, 116*, 434–445.

Bunker, R., Holloway, D., & Randolph, B. (2005). The expansion of urban consolidation in Sydney: Social impacts and implications. *Australian Planner, 42*(3), 16–25.

Carlin, A., Murphy, M. H., & Gallagher, A. M. (2016). Do interventions to increase walking work? A systematic review of interventions in children and adolescents. *Sports Medicine, 46*, 515–530.

Chica-Olmo, J., Rodríguez-López, C., & Chillón, P. (2018). Effect of distance from home to school and spatial dependence between homes on mode of commuting to school. *Journal of Transport Geography, 72*, 1–12.

City of Canada Bay (2021). "Mayor calls for parking scheme around metro sites to support local business and residents." Retrieved February 4, 2025.

Crawford, S. B., Bennetts, S. K., Hackworth, N. J., Green, J., Graesser, H., Cooklin, A. R., Matthews, J., Strazdins, L., Zubrick, S. R., D'Esposito, F., & Nicholson, J. M. (2017). Worries, 'weirdos', neighborhoods and knowing people: A qualitative study with children and parents regarding children's independent mobility. *Health & Place, 45*, 131–139.

Domosh, M. (1992). Urban imagery. *Urban Geography, 13*(5), 475–480.

Elegbede, I., Matti-Sanni, R., Moriam, O., & Emily Osa, I. (2023). *Sustainability education and environmental awareness* (pp. 1–9). Springer.

Ellison, R. B., & Greaves, S. (2011). Travel time competitiveness of cycling in Sydney, Australia. *Transportation Research Record: Journal of the Transportation Research Board, 2247*(−1), 99–108.

Evans, C. (2023). "Revealed: Colonial rule nearly doubles UK's historical contribution to climate change." Retrieved December 19, 2023 from https://www.carbonbrief.org/revealed-colonial-rule-nearly-doubles-uks-historical-contribution-to-climate-change/.

Fonseca, F., Ribeiro, P. J., Conticelli, E., Jabbari, M., Papageorgiou, G., Tondelli, S., & Ramos, R. (2021). Built environment attributes and their influence on walkability. *International Journal of Sustainable Transportation, 16*(7), 660–679.

Foster, S., Villanueva, K., Wood, L., Christian, H., & Giles-Corti, B. (2014). The impact of parents' fear of strangers and perceptions of informal social control on children's independent mobility. *Health & Place, 26*, 60–68.

Frederick, C., Riggs, W., & Gilderbloom, J. H. (2018). Commute mode diversity and public health: A multivariate analysis of 148 US cities. *International Journal of Sustainable Transportation, 12*(1), 1–11.

Freudendal-Pedersen, M., & Kesselring, S. (2016). Mobilities, futures & the City: Repositioning discourses–changing perspectives–rethinking policies. *Mobilities, 11*(4), 575–586.

García-Hermoso, A., Esteban-Cornejo, I., Olloquequi, J., & Ramírez-Vélez, R. (2017). Cardiorespiratory fitness and muscular strength as mediators of the influence of fatness on academic achievement. *The Journal of Pediatrics, 187*, 127–133.

Garrard, J. (2011). *Active travel to school literature review*. ACT Health.

Gericke, N., Manni, A., & Stagell, U. (2020). The green school movement in Sweden–past, present and future. In A. Gough, L. Chi-Kin, & P. K. Tsang (Eds.), *Green schools globally: Stories of impact on education for sustainable development* (pp. 309–332). Springer International Publishing.

Glazebrook, G. (2009). *Developing a thirty year public transport plan for Sydney*. University of Technology.

Gough, A. (2013). The emergence of environmental education research: A "history" of the field. In R. B. Stevenson, M. Brody, J. Dillon, & A. E. J. Wals (Eds.), *International handbook of research on environmental education* (pp. 13–22). Routledge.

Guo, D., McTigue, E. M., Matthews, S. D., & Zimmer, W. (2020). The impact of visual displays on learning across the disciplines: A systematic review. *Educational Psychology Review, 32*(3), 627–656.

Haarstad, H., Grandin, J., Kjærås, K., & Johnson, E. (2023). *Haste: The slow politics of climate urgency*. UCL Press.

Hassevoort, K. M., Khan, N. A., Hillman, C. H., & Cohen, N. J. (2016). Childhood markers of health behavior relate to hippocampal health, memory, and academic performance. *Mind, Brain, and Education, 10*(3), 162–170.

Hegewald, J., Schubert, M., Freiberg, A., Romero Starke, K., Augustin, F., Riedel-Heller, S. G., Zeeb, H., & Seidler, A. (2020). Traffic noise and mental health: A systematic review and meta-analysis. *International journal of environmental research public health, 17*(17), 6175.

IICLEI – Local Governments for Sustainability. (2013). EcoMobility World Festival 2013: Suwon, South Korea, Retrieved August 01, 2025 from https://www.ecomobilityfestival.net/home/suwon-2013/.

Ingvardson, J. B., & Nielsen, O. A. (2018). How urban density, network topology and socio-economy influence public transport ridership: Empirical evidence from 48 European metropolitan areas. *Journal of Transport Geography, 72*, 50–63.

Ingvardson, J. B., & Nielsen, O. A. (2019). The relationship between norms, satisfaction and public transport use: A comparison across six European cities using structural equation modelling. *Transportation Research Part A: Policy and Practice, 126*, 37–57.

James, S. L., Lucchesi, L. R., Bisignano, C., Castle, C. D., Dingels, Z. V., Fox, J. T., Hamilton, E. B., Liu, Z., McCracken, D., & Nixon, M. R. (2020). Morbidity and mortality from road injuries: Results from the global burden of disease study 2017. *Injury Prevention, 26*(Suppl 2), i46–i56.

Ji, X., Guan, H., Lu, M., Chen, F., & Qin, W. (2022). International research Progress in school travel and behavior: A. *Literature Review and Bibliometric Analysis., 14*(14), 8857.

Jones, R. A., Blackburn, N. E., Woods, C., Byrne, M., van Nassau, F., & Tully, M. A. (2019). Interventions promoting active transport to school in children: A systematic review and meta-analysis. Preventive medicine, 123, 232–241.https://www.ecomobilityfestival.net/home/suwon-2013/

Kent, J. L. (2014). Still feeling the car–the role of comfort in sustaining private car use. *Mobilities, 10*(5), 726–747.

Kent, J. L. (2021). The use of practice theory in transport research. *Transport Reviews*, 1–23.

Kent, J. L., Crane, M., Waidyatillake, N., Stevenson, M., & Pearson, L. (2023a). Urban form and physical activity through transport: A review based on the d-variable framework. *Transport Reviews*, 1–29.

Kent, J. L., Mason, M., Virgona, N., & Ryan, H. (2023b). Parenting in proximity to others: The importance of transitions and trajectories. *Urban Policy and Research*, 1–14.

Kent, J. L., & Mulley, C. (2023). Travel with dogs: The need to accommodate "messy trips" in healthy and sustainable transport transitions. *Journal of Transport & Health, 28*, 101559.

Kerr, S. M., Gibson, C., & Klocker, N. (2018). Parenting and neighbouring in the consolidating city: The emotional geographies of sound in apartments. *Emotion Space and Society, 26*, 1–8.

Kerr, S.-M., Klocker, N., & Gibson, C. (2021). From backyards to balconies: Cultural norms and parents' experiences of home in higher-density housing. *Housing Studies, 36*(3), 421–443.

Le Gouais, A., J. R. Panter, A. Cope, J. E. Powell, E. L. Bird, J. Woodcock, D. Ogilvie, L. Foley and c. iConnect (2021). "A natural experimental study of new walking and cycling infrastructure across the United Kingdom: The Connect2 programme." Journal of Transport and Health 20.

Legacy, C., Curtis, C., & Scheurer, J. (2017). Planning transport infrastructure: Examining the politics of transport planning in Melbourne, Sydney and Perth. *Urban Policy and Research, 35*(1), 44–60.

Lelieveld, J., Evans, J. S., Fnais, M., Giannadaki, D., & Pozzer, A. (2015). The contribution of outdoor air pollution sources to premature mortality on a global scale. *Nature, 525*(7569), 367–371.

Lunke, E. B. (2025). Does adolescent experience influence mobility later in life? A propensity score matching approach. *Journal of Transport Geography, 123*, 104129.

Macdonald, L., McCrorie, P., Nicholls, N., & Olsen, J. R. (2019). Active commute to school: Does distance from school or walkability of the home neighbourhood matter? A national cross-sectional study of children aged 10-11 years, Scotland, UK. *BMJ Open, 9*(12).

Malone, K. (2007). The bubble-wrap generation: Children growing up in walled gardens. *Environmental Education Research, 13*(4), 513–527. https://doi.org/10.1080/13504620701581612

Mölenberg, F. J. M., Panter, J., Burdorf, A., & van Lenthe, F. J. (2019). A systematic review of the effect of infrastructural interventions to promote cycling: Strengthening causal inference from observational data. *International Journal of Behavioral Nutrition and Physical Activity, 16*, 1–31.

Mulalic, I., & Rouwendal, J. (2020). Does improving public transport decrease car ownership? Evidence from a residential sorting model for the Copenhagen metropolitan area. *Regional Science and Urban Economics, 83*, 103543.

Na, L., R. Ibrahim and S. Z. Abidin (2021). "The Influence of Public Artworks with Children's Painting Characteristics on Urban Culture and Public Space." *Enviro Sci Poll Res And Mang: ESPRM* 113.

Nicholson, L. M., Turner, L., Slater, S. J., Abuzayd, H., Chriqui, J. F., & Chaloupka, F. (2014). Developing a measure of traffic calming associated with elementary school students' active transport. *Transportation Research Part D: Transport and Environment, 33*, 17–25.

Panter, J., Guell, C., Humphreys, D., & Ogilvie, D. (2019). Title: Can changing the physical environment promote walking and cycling? A systematic review of what works and how. *Health & Place, 58*, 102161.

Panter, J., Heinen, E., Mackett, R., & Ogilvie, D. (2016). Impact of new transport infrastructure on walking, cycling, and physical activity. *American Journal of Preventive Medicine, 50*(2), e45–e53.

Partenko, L. (1936). The Bar-tailed godwit and its races. *The Auk*, 194–197.

Payne, P. G. (2010). Moral spaces, the struggle for an intergenerational environmental ethics and the social ecology of families: An "other" form of environmental education. *Environmental Education Research, 16*(2), 209–231. https://doi.org/10.1080/13504620903580545

Pucher, J., Garrard, J., & Greaves, S. (2010). Cycling down under: A comparative analysis of bicycling trends and policies in Sydney and Melbourne. *Journal of Transport Geography*.

Ramanathan, S., O'Brien, C., Faulkner, G., & Stone, M. (2014). Happiness in motion: Emotions, Well-being, and active school travel. *Journal of School Health, 84*(8), 516–523.

Randolph, B., & Freestone, R. (2012). Housing differentiation and renewal in middle-ring suburbs: The experience of Sydney, Australia. *Urban Studies February*, 1–19.

Reckwitz, A. (2002). Basic elements of a theory of social practices—a perspective in social theory. *Zeitschrift Fur Soziologie, 32*(4), 282–301.

Rodríguez-López, C., Salas-Fariña, Z. M., Villa-González, E., Borges-Cosic, M., Herrador-Colmenero, M., Medina-Casaubón, J., Ortega, F. B., & Chillón, P. (2017). The threshold distance associated with walking from home to school. *Health Education and Behavior, 44*(6), 857–866.

Rothman, L., Hagel, B., Howard, A., Cloutier, M. S., Macpherson, A., Aguirre, A. N., McCormack, G. R., Fuselli, P., Buliung, R., HubkaRao, T., Ling, R., Zanotto, M., Rancourt, M., & Winters, M. (2021). Active school transportation and the built environment across Canadian cities: Findings from the child active transportation safety and the environment (CHASE) study. *Preventive Medicine, 146*.

Saeidizand, P., Fransen, K., & Boussauw, K. (2022). Revisiting car dependency: A worldwide analysis of car travel in global metropolitan areas. *Cities, 120*, 103467.

Schatzki, T., K. Knorr Cetina and E. von Savigny, Eds. (2001). The practice turn in contemporary theory..

Schoeppe, S., Tranter, P., Duncan, M. J., Curtis, C., Carver, A., & Malone, K. (2016). Australian children's independent mobility levels: Secondary analyses of cross-sectional data between 1991 and 2012. *Children's Geographies, 14*(4), 408–421.

Sheller, M. (2004). Automotive emotions—feeling the car. *Theory Culture & Society, 21*(4–5), 221–242.

Siu, K. W. M., Lam, M. S., & Wong, Y. L. (2015). Designing signs for children: A study of children's drawings for safety signs. *Communication Design, 3*(2), 106–123.

The Australian Bureau of Statistics (2016) Census of population and housing—census data by location..
The Australian Bureau of Statistics (2021) Census of population and housing—census data by location..
The Bureau of Transport Statistics. (2021). *NSW 2010–2019 household travel survey*. Bureau of Transport Statistics, Transport for NSW.
The International Energy Agency. (2018). *The world energy outlook*. IEA Publications.
The Victorian Department of Transport (2010) *Victorian integrated survey of travel and activity* (VISTA)..
The Victorian Department of Transport (2018) *Victorian integrated survey of travel and activity* (VISTA)..
Transport for New South Wales (2025). "Traffic Statistics and Volume, Sydney." Retrieved February 3, 2025 from https://www.transport.nsw.gov.au/operations/roads-and-waterways/corporate-publications/statistics/traffic-statistics/traffic-volume.
Tsai, C. H., C. Mulley and G. Clifton (2014). The spatial interactions between public transport demand and land use characteristics in the sydney greater metropolitan area. Australasian Transport Research Forum, ATRF 2012—Proceedings.
Urry, J. (2004). The 'system' of automobility. *Theory Culture & Society, 21*(4–5), 25–39.
Urry, J. (2012). Changing transport and changing climates. *Journal of Transport Geography, 24*, 533–535.
Van Acker, V., Mulley, C., & Ho, L. (2019). Impact of childhood experiences on public transport travel behaviour. *Transportation Research Part A: Policy and Practice, 130*, 783–798.
Villa-González, E., Barranco-Ruiz, Y., Evenson, K. R., & Chillón, P. J. P. M. (2018). *Systematic review of interventions for promoting active school transport., 111*, 115–134.
Ward, D. S., Linnan, L., Vaughn, A., Neelon, B., Martin, S. L., & Fulton, J. E. (2007). Characteristics associated with US walk to school programs. *International Journal of Behavioral Nutrition and Physical Activity, 4*, 1–10.
Watson, M. (2012). How theories of practice can inform transition to a decarbonised transport system. *Journal of Transport Geography, 24*, 488–496.
Yan, Y., M. Burke and A. Leung (2019). Travel behaviour differences between private and public-school students in South East Queensland. Australasian Transport Research Forum, ATRF 2019-Proceedings.
Yin, R. K. (2013). *Case study research: Design and methods*. Sage publications.
Zhao, X., Hu, H., Yuan, H., & Chu, X. (2023). How does adoption of electric vehicles reduce carbon emissions? Evidence from China. *Heliyon, 9*(9).

Designing Ecocentric Urban Spaces for Human-Nature Connection: A Case Study of Orchard Hill Station Precinct

Ahmad Fariz Al Hazmi

Abstract The human-nature connection is critical for sustaining ecological conservation, since people's ability to access and interact with nature can promote ecological stewardship. However, urbanisation has become a major threat to ecological integrity in many cities, including the Western Parkland City in the Greater Sydney Area. This loss of human-nature interaction can hinder effective ecological conservation and accelerate environmental degradation. Orchard Hills, a priority precinct in the Greater Penrith to Eastern Creek area, is expected to anticipate significant growth with the new Metro station. This offers an opportunity to adopt an ecocentric approach, aiming to protect and enhance existing ecosystems while still accommodating growth. However, this opportunity also poses a threat to the existing riparian ecosystem near Wianamatta South Creek. Therefore, this study aims to develop urban design recommendations that foster human-nature connections within the precinct, utilising an ecocentric urbanism approach. This study employs a design practice-led research methodology, drawing on contemporary urban design literature to establish a theoretical framework based on Biodiversity Sensitive Urban Design (BSUD), Nature-based Solutions (NbS), and Nature-Driven Urbanism (NDU). This study finds that applying NDU in the early design process can optimally ensure landscape resilience by identifying vulnerabilities and key ecosystems for protection. Subsequently, implementing NbS as a foundational strategy for designing green and blue infrastructure can strengthen ecosystem resilience. BSUD provides a valuable methodology for designing habitats tailored to specific species. This research thus offers insights into an ecological approach for master planning a new precinct around a transit station. It suggests that a thoughtfully integrated ecocentric design can balance urban development with environmental resilience by supporting biodiversity, enhancing ecological connectivity, and fostering meaningful human-nature interactions.

Keywords Ecocentric Urbanism · Nature-based Solution · Biodiversity Sensitive Urban Design · Nature-driven Urbanism · Human-nature Connection

A. F. Al Hazmi (✉)
School of Architecture, Design and Planning, The University of Sydney, Sydney, NSW, Australia
e-mail: fariz.hzmi@gmail.com

A. Cheshmehzangi et al. (eds.), *Nature-based Urbanism*, Urban Sustainability,
https://doi.org/10.1007/978-981-95-6864-2_11

1 Introduction

Urbanisation is one of the leading drivers of ecological loss in cities worldwide. However, the standard ecological offsetting practices, which typically relocate nature areas outside cities, only reduce the opportunities for urban residents to interact with nature in their everyday lives (Colding et al., 2020). This human-nature connection is important to sustain ecological conservation, as people's ecological stewardship can be better developed when they have daily interaction with nature (Colding et al., 2020; Garrard et al., 2018; Kabisch et al., 2022; Kirk et al., 2021). To foster this connection in cities, the ecosystem and biodiversity must first be available in the cities (Colding et al., 2020). However, urban development objectives frequently conflict with ecological goals; hence, a balance is essential for urban development growth and the preservation of functioning ecosystems (Garrard et al., 2018; Kirk et al., 2021). Therefore, this necessitates an 'ecocentric urbanism' approach that places significance on meeting ecological objectives in urban development, aiming to protect and build on the existing ecosystem while still accommodating growth (Garrard et al., 2018; Kabisch et al., 2022; Roggema, 2020b).

2 Case Study Context and Study Aim

Urbanisation remains one of the key planning challenges in Greater Sydney (Greater Cities Commission, 2022), with Western Parkland City (WPC) projected to exceed a population of 1.5 million by 2056 (Greater Cities Commission, 2022; Greater Sydney Commission, 2018a). This growth will intensify the urban development and place pressure on the region's extensive protected natural areas critical for ecological conservation within the metropolitan region (Greater Cities Commission, 2022; Greater Sydney Commission, 2018b).

The Cumberland Plain natural area supports rich native biodiversity in WPC region. However, ongoing land-use activities have led to over-clearing beyond 70% of its original extent, resulting in habitat fragmentation, reduced ecological connectivity, and heightened biodiversity loss risk (NSW Department of Planning and Environment, 2022; Tozer et al., 2015). Meanwhile, Wianamatta South Creek (see Fig. 1) is a key waterway corridor in the Cumberland Plain, with 80% of the WPC region planned within its catchment to support population growth (Greater Sydney Commission (GSC), 2018a). By 2015, urban areas occupied 20% of the catchment, degrading the creek's ecosystem from untreated stormwater discharge (Birtles, 2020; Greater Sydney Commission (GSC), 2018b). As urbanisation expands to 80% by 2056, stormwater discharge is projected to increase, exacerbating pollution and biodiversity loss (Birtles, 2020).

Orchard Hills (see Fig. 1), located under the jurisdiction of Penrith City Council, is a priority precinct for rural-to-urban transformation under the Greater Penrith to Eastern Creek strategic framework. Capitalising from the new Metro station on the

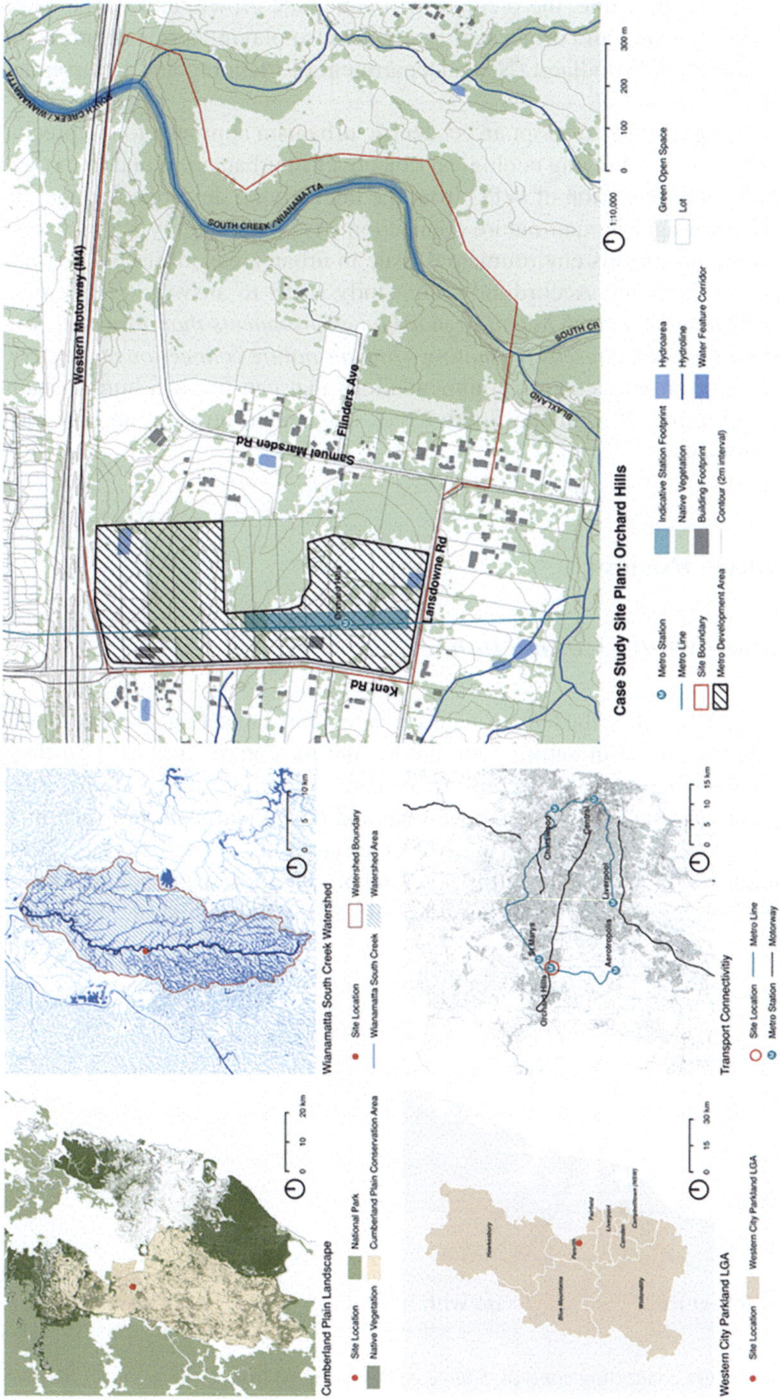

Fig. 1 Case study site location. *Source* Author

Western Sydney Airport line, the precinct will anticipate growth and is planned to facilitate diverse housing and urban services while also playing a role in protecting the Cumberland Plain Woodland (NSW Department of Planning and Environment (DPE), 2023).

There is an opportunity to adopt an ecocentric urbanism approach to the precinct design that prioritises achieving ecological objectives in urban areas and to mitigate further ecological degradation of WPC from the intensifying development and ultimately foster a stronger human-nature connection. As elaborated by Kabisch et al. (2022), the solution for this environmental issue in urban areas must also be found in the urban environment. Accordingly, this study seeks to answer, '*How can an ecocentric urbanism approach help design urban environments that protect ecology while facilitating growth, thereby promoting a human-nature connection in the city?*'. It aims to develop urban design recommendations that can foster a human-nature connection within the Wianamatta South Creek catchment, utilising an ecocentric urbanism approach.

3 Literature Review

3.1 Human-Nature Connection

Colding et al. (2020) define the human-nature connection as a sustainable relationship that allows people 'to be in nature, with nature and for nature' (see Fig. 2). Being *in* nature involves access to and comfort in nature (Colding et al., 2020). Being *with* nature entails direct physical engagement and ecological learning, such as by facilitating '*cues to care*' (Colding et al., 2020; Garrard et al., 2018; Kirk et al., 2021). Being *for* nature demonstrates resulting potential pro-environmental behaviours and stewardship (Colding et al., 2020; Garrard et al., 2018; Kabisch et al., 2022; Kirk et al., 2021).

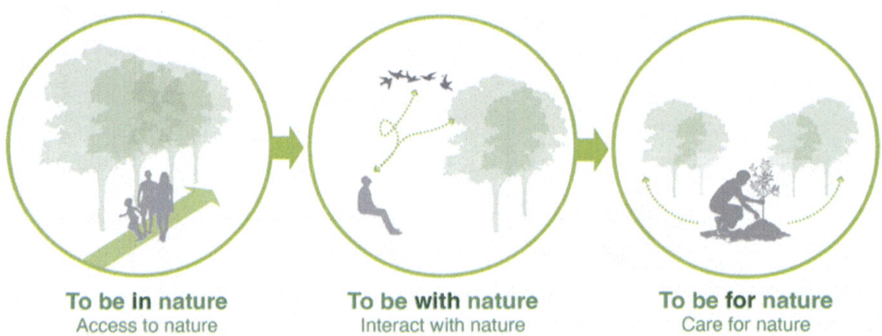

Fig. 2 Human-nature connection concept. *Source* Author, adapted from Colding et al. (2020)

Regular interaction with nature in urban settings is vital to maintain ecological awareness and biodiversity conservation, while also offering health and well-being benefits, including reduced mortality and improved mental health (Bosch & Sang, 2017), and fostering social cohesion and a sense of place (Kabisch et al., 2022; Kirk et al., 2021). Therefore, ensuring human-nature connection in cities is essential not only for sustaining the health and resilience of ecosystems but also for enhancing the well-being of urban residents.

3.2 Nature-Based Solution

Nature-based solutions (NbS) challenge conventional technical approaches to urban issues by using natural elements to address societal and climate challenges in cities (Kabisch et al., 2022). They are underpinned by the conviction that both biodiversity and human well-being rely on healthy, resilient ecosystems. Thus, this approach promotes the sustainable conservation and management of natural or modified ecosystems to maintain the delivery of ecosystem services (Cohen-Shacham et al., 2019). Sijmons (2020) and Roggema (2021) further define NbS as a 'contact' method that brings nature into closer interaction with urban environments by using the properties, features, and complex systems of nature. Kabisch et al. (2022) outline five key NbS principles: (1) *Systematic Understanding*, which promotes ecosystem conservation at all spatial scales; (2) *Benefitting People and Biodiversity*, which seeks to balance human use and ecological protection; (3) *Inclusivity*, which integrates local cultural knowledge to meet community needs; (4) *Context Consideration*, which ensures solutions are tailored to local environmental and cultural conditions; and (5) *Communication and Learning*, which aims to reconnect people with nature to enhance environmental stewardship.

3.3 Biodiversity Sensitive Urban Design (BSUD)

Biodiversity Sensitive Urban Design (BSUD) opposes the conventional conservation model that responds to biodiversity loss from urban development by offsetting habitats outside cities. Instead, BSUD recognises cities as potential biodiversity hotspots, where on-site conservation can support on-site ecosystem services (Garrard et al., 2018). BSUD thus aims to reintegrate nature into the urban fabric, enabling direct delivery of its services to the residents while fostering human-nature interactions and ecological stewardship. However, there are conflicts between the human need for urban development and the maintenance of thriving ecosystems, which BSUD seeks to navigate (Garrard et al., 2018; Kirk et al., 2021).

The BSUD framework, developed by Garrard et al. (2018) and expanded by Kirk et al. (2021) in their Fisherman's Bend case study, involves documenting biodiversity

Table 1 BSUD principles (Garrard et al., 2018; Parris et al., 2018)

5 BSUD Principles (Garrard et al., 2018)	7 Lamps of Planning for Biodiversity (Parris et al., 2018)
Maintain and introduce habitat	Protection
	Construction
	Novelty
Facilitate dispersal	Connectivity
Minimise threats and anthropogenic disturbance	Benevolence
Facilitate natural ecological processes	Cycles
	Interactions
Improve potential positive human-nature interaction	–

values, selecting target species, using ecological knowledge to inform BSUD objectives and actions, and assessing these BSUD recommendations to decide which meet both biodiversity and development objectives.

Garrard et al. (2018) outline five core BSUD principles, aligned with the *Seven Lamps of planning for biodiversity* (Parris et al., 2018), as summarised in Table 1. These include protecting ecologically valuable areas, creating new habitats in urban settings, and acknowledging the value of novel ecosystems. Ensuring ecological connectivity is vital to counter habitat fragmentation. Benevolent or 'softer' urban form or structure is equally important to minimise worsened anthropogenic impacts on the urban ecology. BSUD also promotes the facilitation of natural ecological processes and encourages urban designs that enable positive human-nature interactions to address human-nature conflicts.

3.4 Nature-Driven Urbanism

Nature-driven urbanism (NDU) is an approach that switches the perspective from 'adding nature into the cities' to embedding cities into nature, guided by natural processes to enhance resilience and adaptability; hence, the city is a part of nature (Roggema, 2020c, 2021). The overarching principle is to allow nature to guide spatial design, in which natural landscape and ecological system processes can dictate where and how urban form and function can occur (Roggema, 2020b, 2023). Therefore, as shown in Fig. 5, the natural landscape and ecosystem should be placed at the forefront of the design process, including mapping the landscape (e.g., topography), ecology and potential vulnerabilities (e.g., flood) across different scales. This will determine where urban development should be located to maintain its ecological integrity and resilience. This process is followed by the design of green and ecological options before embedding urban forms and functions (Roggema, 2020a, 2020b).

4 Methodology

4.1 Theoretical Framework

The notion of improving the human-nature connection criticises the standard ecological conservation approach of offsetting biodiversity in areas far from where the loss occurs, often outside urban fringes, which restricts urban dwellers' access to nature and reduces their interaction with nature (Colding et al., 2020; Kirk et al., 2021; Maron et al., 2016). If the ability to be in nature is the key attribute of human-nature connection, then the existence of ecosystems within cities becomes the prerequisite of human-nature connection (Colding et al., 2020). Therefore, this requires an approach in designing urban areas oriented towards the ecosystem, namely 'ecocentric urbanism'.

Several existing urbanism frameworks that signify achieving ecological objectives, previously discussed, contribute to the development of the methodological framework of this study, as shown in Fig. 5. A literature study identifies two distinct approaches of ecocentric urbanism, offering unique perspectives between '*embedding nature in cities*' and '*embedding cities in nature*' (Fig. 3). The NbS and BSUD are similarly underpinned by the conception of bringing nature back to the city (Kabisch et al., 2022; Roggema, 2021; Sijmons, 2020). In essence, BSUD is part of the NbS but specifically underscores biodiversity objectives as the overarching goal (Garrard et al., 2018). On the other hand, NDU is an optimistic approach that emphasises the role of nature in guiding the design (Roggema, 2020b, 2023). However, it is argued that NDU is most applies to greenfield areas or those with a smaller urban footprint, where natural ecosystems remain intact and can guide urban development. Hence, it can be challenging to adopt NDU in urban areas where nature might no longer exist or be limited.

The existing rural character of the case study area, with patches of native habitat found predominantly along the creek corridor, and the anticipated development intensification from the metro station (refer to Fig. 1) present an opportunity for integrating the two ecocentric perspectives in the design process. The NDU can guide the design process in ensuring that the new development 'embedded' in the natural landscape can protect the existing ecosystem. Meanwhile, the NbS and BSUD can guide the design of urban ecology to improve and build on the existing ecosystem. Therefore, this research integrates NDU, NbS, and BSUD into a comprehensive framework to guide both analysis and design recommendations. Each individual component of the site analysis is drawn from these conceptual theories, as in Fig. 4.

4.2 Methodological Framework

This study employs a design practice-led research methodology, aiming to bridge theory and practice by developing practical tools for real-world applications. It

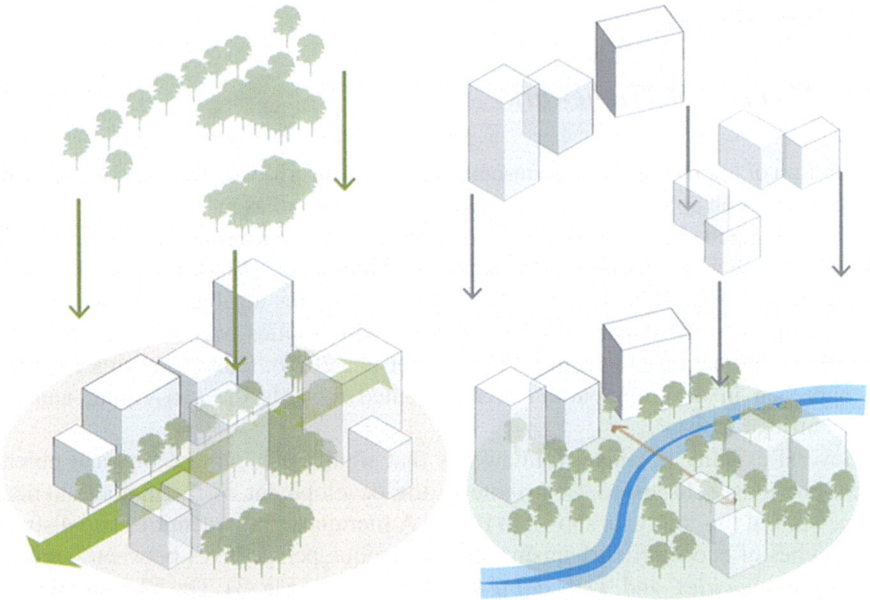

Fig. 3 Two distinct perspectives of ecocentric urbanism. *Source* Author

explores challenges in urban design while encouraging design solutions that go beyond market demands to influence both practice and academic discussions (Carmona, 2020; Firth, 2021; Martins, 2021a, 2021b). As Carmona (2020) highlights that theoretical work can inform practice, while practice-led research can contribute to academic discourse. Therefore, this research draws on contemporary academic literature in urban design to develop a theoretical framework that informs the design process and defines important site analysis criteria relevant to the topic. Ultimately, this research aims to provide design recommendations specific to the case study area.

Figure 5 outlines the methodological approach to achieve the two main objectives. Each objective is associated with specific sub-questions. The key method of this research involves site analysis, a review of planning policies, and synthesis of findings to identify opportunities and constraints, which, alongside literature insights, inform the final design outcomes. A literature review has been conducted to identify criteria for analysis, and a mix of qualitative and quantitative data analysis is appropriate for this study. Additionally, a key limitation is the absence of local ecological knowledge from Aboriginal and First Nations people—an important criterion in nature-based solution frameworks—due to time constraints and lack of accessible secondary data. Future research should incorporate this knowledge to strengthen the design and research process.

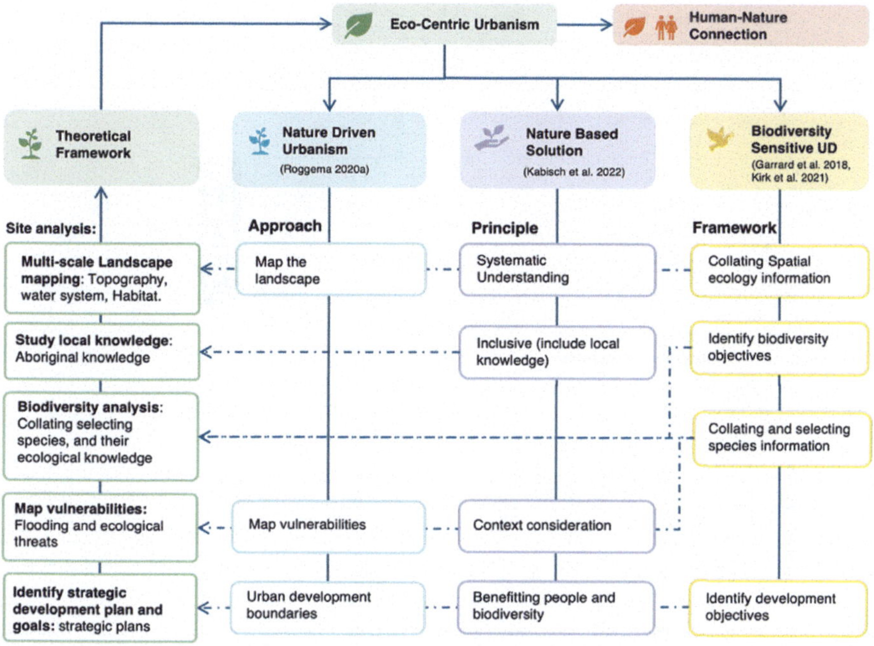

Fig. 4 Theoretical framework. *Source* Author, adapted from Garrard et al. (2018), Kirk et al. (2021); Roggema (2020b)

5 Site Analysis and Design Recommendation

5.1 Site Analysis

A. Biophysical Characters—Water Landscape and Topography

To accurately identify the biophysical features of the site, a cross-scale analysis has been conducted. At the macro level, the water landscape map (Fig. 1) demonstrates that the site is situated within the watershed of the Wianamatta South Creek, directly west of the main creek. At the micro level, the topographic map (Fig. 6) reveals that the site's contours range from 15 to 54 mAHD. The eastern half of the site consists of low-lying terrain, while the western half shows a steeper rise in elevation with the highest local point, reaching 82 mAHD, lying west of the site. Consequently, it can be understood that water flows from the highest elevation in the west towards the low-lying eastern area and eventually into the creek.

B. Landscape Vulnerabilities—Flooding and Urban Heat Island

The flood study of Wianamatta South Creek by the State of NSW Government (2022), shows that the site's low-lying eastern portion lies within a floodplain, with the 1% and 5% AEP models impact existing open spaces and vacant lots, while the probable

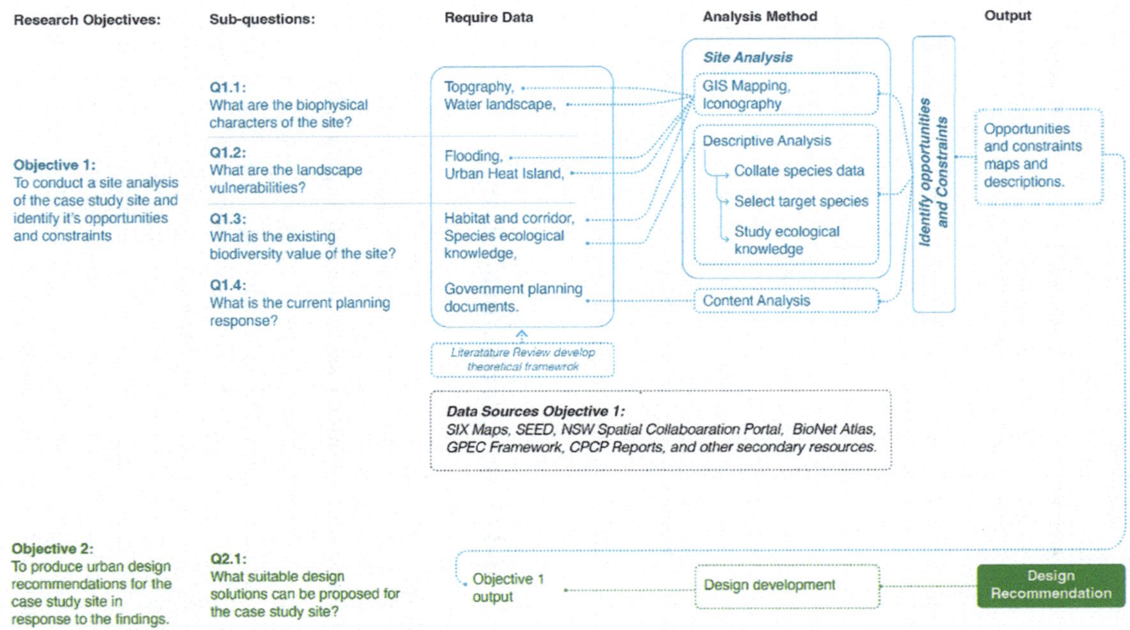

Fig. 5 Methodological Framework. *Source* Author

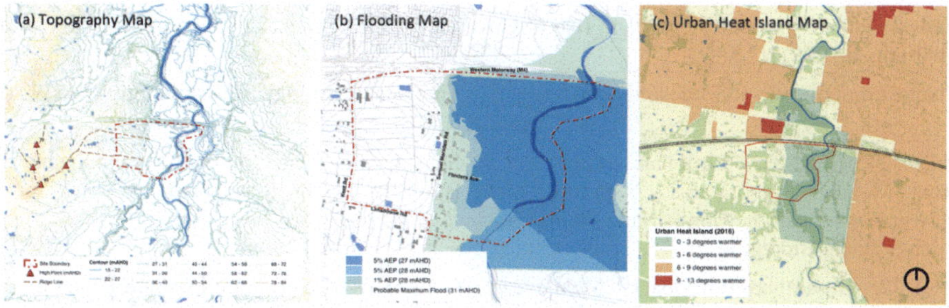

Fig. 6 Topography Map (a), Flooding Map (b) and Urban heat island Map (c). *Source* Author, obtained from State of NSW Government (2022) and State of NSW & DCEEW (2019)

maximum flood (PMF) extends to nearby residential areas (Fig. 6). Local catchment runoff is the main cause of flooding, highlighting the need for future development to reduce direct runoff into the creek.

Additionally, the urban heat island map (Fig. 6) reveals higher temperatures in the developed western half of the site, compared to the cooler eastern half near the creek and green areas (State of NSW & DCEEW, 2019). This temperature contrast underscores the importance of natural features—such as waterways and vegetation—in mitigating urban heat. Rising temperatures may also stress biodiversity, including the Grey-headed Flying Fox (State of NSW Department of Environment and Climate Change, 2008), which indicates the need for integrated green and blue infrastructure in future development.

C. **Existing Biodiversity—Flora**

The site hosts 4 types of Plant Community Types (PCTs) which house threatened ecological communities (Table 2). The four communities share similarities in their structural composition, typically consisting of canopy, shrub, and grass ground layers, which potentially provide habitats for diverse fauna (NSW Government Department of Planning, Industry and Environment, 2021; State Government of NSW & NSW Department of Planning, Housing and Infrastructure, 2022). Several species within these communities, including Broad-leaved ironbark, Forest Red Gum, and Narrow-leaved Ironbark, flower during winter, while others bloom in summer (Fig. 7). This indicates that the site can provide food resources and habitat for biodiversity throughout the year. For example, the Broad-leaved ironbark provides food for Grey-headed Flying Fox (State of NSW Department of Environment and Climate Change, 2008), while the nectar-rich Narrow leaved ironbark attracts bees (iNaturalist, 2024b).

The NSW Government Department of Planning, Industry, and Environment (2021) identifies that most PCTs in the area are in a thinned condition, except for PCT 725, which remains largely intact (Fig. 8). A thinned condition reflects a partially cleared canopy and more open structure, while intact conditions indicate healthier ecosystems with continuous canopy and structural diversity. Although PCT 725 is mostly intact, its communities are fragmented and isolated, especially from others

Table 2 Plant community type data (NSW Government Department of Planning, Industry, and Environment, 2021)

Plant Community Type Code and Name	Threatened Ecological Communities (NSW Status)
PCT 724: Broad-leaved Ironbark—Grey Box—Melaleuca decora grassy open forest on clay/gravel soils of the Cumberland Plain, Sydney Basin Bioregion	Shale Gravel Transition Forest in the Sydney Basin Bioregion (Endangered)
PCT 725: Broad-leaved Ironbark—Melaleuca decora shrubby open forest on clay soils of the Cumberland Plain, Sydney Basin Bioregion	Cooks River/Castlereagh Ironbark Forest in the Sydney Basin Bioregion (Endangered)
PCT 835: Forest Red Gum—Rough-barked Apple grassy woodland on alluvial flats of the Cumberland Plain, Sydney Basin Bioregion	River-flat Eucalypt Forest on Coastal Floodplains of the New South Wales North Coast, Sydney Basin and South East Corner Bioregions (Critically Endangered)
PCT 849: Grey Box—Forest Red Gum grassy woodland on flats of the Cumberland Plain, Sydney Basin Bioregion	Cumberland Plain Woodland in the Sydney Basin Bioregion (Critically endangered)

situated north of the Wianamatta South Creek catchment. PCT 724 also shows sparse and scattered distribution across the site.

In contrast, PCT 835, though largely in thinned condition, is more spatially connected, following the Wianamatta South Creek water corridor and forming a crucial habitat link. Better-quality intact communities of PCT 835 are found in the northern part of the catchment, while PCT 849 is mostly concentrated within the Wianamatta Regional Park (north) and the core areas to the southwest. The Wianamatta South Creek is recognised as a BioMap regional biodiversity corridor, connecting core habitat areas surrounding the site (see Fig. 8). Therefore, PCT 835 along the Wianamatta South Creek corridor plays a critical role as a habitat corridor, linking the two biodiversity core areas where PCT 849 is mostly concentrated. Accordingly, future development could prioritise the protection of the creek corridor habitat while enhancing connectivity among all PCTs on the site.

D. **Existing Biodiversity—Fauna**

Several threatened fauna species listed as both vulnerable and endangered in New South Wales are recorded around the site, as shown in Fig. 9 and Table 3. The following analysis is based on the *Cumberland Plain Assessment Report* by NSW Government Department of Planning, Industry and Environment (2021).

The Yellow-bellied Glider *Petaurus australis*, which requires an extensive habitat corridor for dispersal, has been identified along the creek, where it connects to the larger habitats, such as Wianamatta Regional Park to the north (refer to Fig. 9). Similarly, the Black Bittern *Botaurus flavicollis*, a water bird found along the creek, relies on riparian corridors for shelter, foraging, and dispersal. These records reinforce the role of Wianamatta South Creek as a critical ecological network connecting

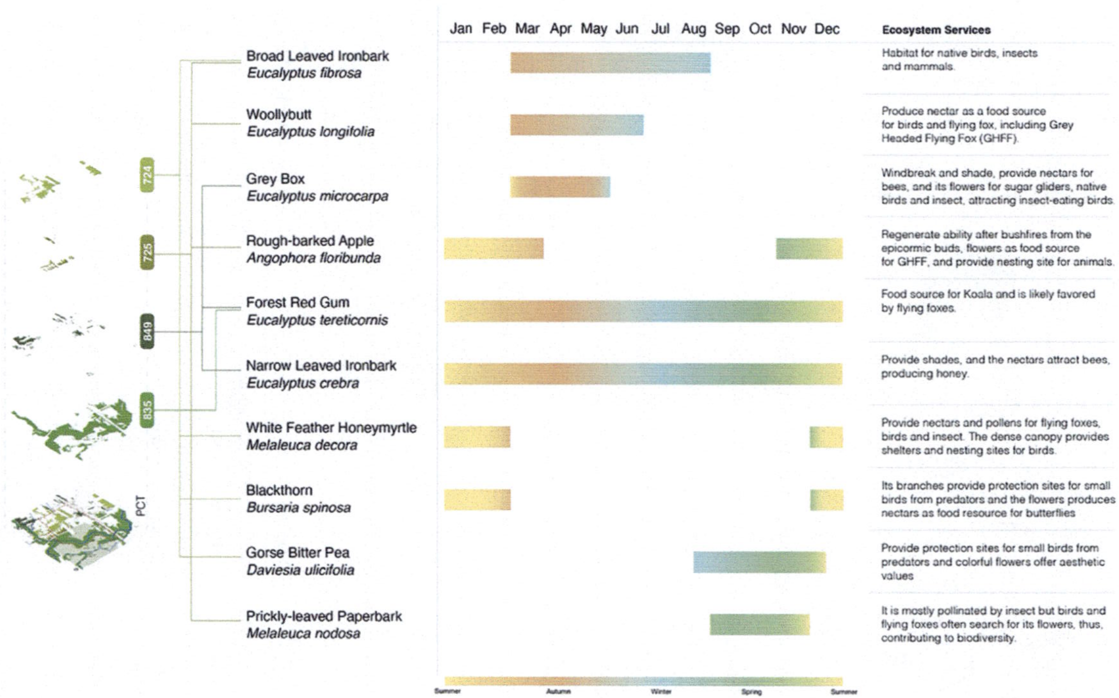

Fig. 7 Phenology and ecosystem services of native vegetations on-site. *Source* Author, obtained from NSW Department of Primary Industries and Regional Development (2010), iNaturalist (2024d), Australian Government & Natural Resources South East (n.d.), Greening Australia (2017), iNaturalist (2024c), Howes (2021), iNaturalist (2024a), iNaturalist (2024b), Growing Illawarra Natives (n.d.), Australian Native Plants Society (Australia) (2024a, 2024b), and Howes (2022)

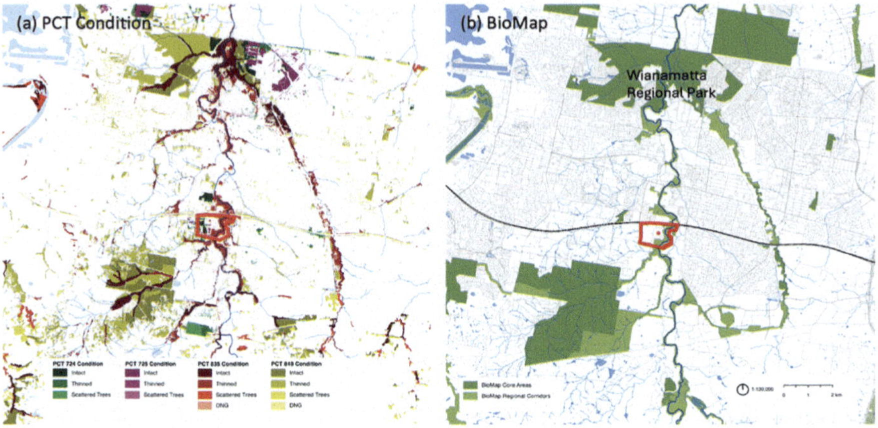

Fig. 8 PCT Condition (a) and BioMap (b). *Source* Author, obtained from NSW Government Department of Planning, Industry and Environment (2021)

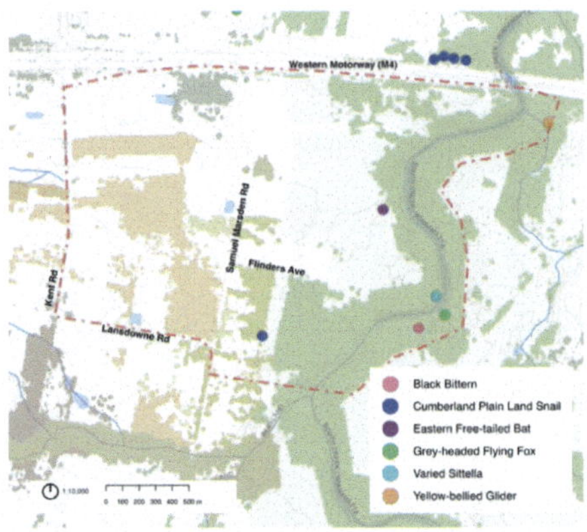

Fig. 9 Threatened Fauna Species Sighting. *Source* Author, obtained from Atlas of Living Australia (n.d.), NSW Environment and Heritage (2023), NSW Government Department of Planning, Industry and Environment (2021)

Table 3 Threatened Fauna Species On-site (Atlas of Living Australia (n.d.), NSW Environment and Heritage (2023), NSW Government Department of Planning, Industry and Environment (2021)

Species	NSW Conservation Status
Yellow-bellied Glider (*Petaurus australis*)	Vulnerable
Black Bittern (*Ixobrychus flavicollis*)	Vulnerable
Grey-headed Flying Fox (*Pteropus poliocephalus*)	Vulnerable
Eastern Freetail-bat (*Ozimops ridei*)	Vulnerable
Varied Sittella (*Daphoenositta chrysoptera*)	Vulnerable
Cumberland Plain Land Snail (*Meridolum corneovirens*)	Endangered

larger habitats, particularly for species requiring extensive corridors for movement and those relying on riparian ecosystems.

Moreover, flying animals, including the Grey-headed Flying Fox, Eastern Freetail-bat, and Varied Sittella, are primarily recorded around the creek in less fragmented habitats. However, habitat fragmentation does not hinder their movement; hence, scattered and degraded native vegetation communities in urban areas remain suitable for them. Lastly, the Cumberland Plain Land Snail, listed as an endangered species in NSW, has been identified as potentially supported by the existing plant community south of the site (PCT 725). Therefore, this analysis shows that the site's diverse habitats have the potential to enable on-site conservation of various threatened animals in future urban areas, necessitating early consideration in the design process.

E. **Target Species—Grey-Headed Flying Fox**

Grey-headed Flying Fox is selected as the primary target species in this study mainly for its role in the ecosystem, adaptability to live in the urban landscape, and the existing habitat on the site. The species primarily feeds on nectar and fruits of native vegetation that can be found on-site (complete list refer to Table 4). On top of that, Grey-headed Flying Foxes can travel in search of food up to 40 km in one night with an average distance of 10.9 km from their camping sites that are found within a 10–30 km radius around the site (Fig. 10), and their movement is not constrained by habitat fragmentation. Because they feed on nectar and fruit ability to forage over a long distance, they play a vital role in pollination and seed dispersal in the ecosystem. This supports more native vegetation growth, especially across the urban landscape, where habitats become more fragmented and degraded (Commonwealth of Australia, 2021; State of NSW Department of Environment and Climate Change, 2008).

Therefore, with the presence of native ecological communities providing year-round food resources for the Grey-headed Flying Fox and the proximity to surrounding camping spots, the site meets the criteria as '*a habitat critical to the survival of the Grey-headed Flying Fox*' in the National Recovery Plan for the Grey-headed Flying Fox (Australian Government, 2021, p. 15); thus, it has the potential as their foraging habitat. The species' habitat requirements are summarised in Table 5,

Table 4 On-site vegetation as food source for Grey-headed Flying Fox (Commonwealth of Australia, 2021; State of NSW Department of Environment and Climate Change, 2008)

Plant Name	Important Winter and Spring Plant
• Cabbage Gum (*Eucalyptus amplifolia*) • Sydney Blue Gum (*Eucalyptus saligna*) • Forest red gum (*Eucalyptus tereticornis*) • Smooth-barked apple (*Angophora costata*) • Rough-barked apple (*Angophora floribunda*)	• Blackbutt (*Eucalyptus pilularis*) • Narrow-leaved Ironbark (*Eucalyptus crebra*) • Broad-leaved ironbark (*Eucalyptus fibrosa*)

Fig. 10 Grey-headed flying fox camp sites map. *Source* Author, obtained from State of NSW & NSW Department of Climate Change, Energy, the Environment and Water (2010)

which highlights the practical integration of habitat provision into urban design strategies. However, the site presents potential threats to the species, including habitat loss, noise disturbance, heat stress, and climate change, as summarised in Table 6.

F. Current Planning Response

The incoming Metro service in Orchard Hills, providing efficient public transport across the metropolis, will undoubtedly unlock the development potential in the area. Table 7 reviews and refers to the current planning responses relevant to ecological and landscape issues in the Greater Penrith to Eastern Creek strategic plan by the NSW Department of Planning and Environment (2023) (Fig. 11).

Table 5 Habitat requirements for the Grey-headed Flying Fox

Requirement	Description
Diverse Vegetation	The species requires a diversity of vegetation types as food sources, distributed across extensive areas. Accordingly, conserving foraging habitat for the species solely within designated conservation reserves is insufficient, necessitating integrated spatial planning to include foraging habitat across a broader landscape (Commonwealth of Australia, 2021)
Vegetation Cover & Year-Round Food Supply	Although the likelihood of Grey-headed Flying Fox foraging declined as it gets distant from their campsites, the probability rises with increasing vegetation cover and food availability (Mcdonald-Madden et al., 2005). Therefore, continuous, year-round food availability is critical to support foraging needs
Preferred Foraging Sites with Reduced Disturbance	Parks are preferred over streets due to lower noise and light disturbances (Mcdonald-Madden et al., 2005). Accordingly, a more benevolent urban environment with fewer anthropogenic disturbances (e.g., noise, light) supports better foraging conditions
Revegetation Efforts	Revegetation of degraded habitats to mimic original ecosystems is important, especially for addressing food needs and mitigating heat stress (State of NSW Department of Environment and Climate Change, 2008)

Table 6 Potential threats to the Grey-headed Flying Fox (Commonwealth of Australia, 2021; State of NSW Department of Environment and Climate Change, 2008)

Threat	Description
Foraging Habitat Loss	Urbanisation and agriculture contribute to excessive clearing of native vegetation, leading to food shortages and increased mortality
Noise Disturbance	Traffic from the M4 motorway and noise from construction or anthropogenic activities can disturb Grey-headed Flying Fox movement, potentially causing mothers to drop their young
Electrocution Risk	The presence of a 500 kV power line on-site poses a serious risk of electrocution
Heat Stress	Elevated temperatures from the urban heat island effect (3–6 °C warmer) increase stress and mortality among the population. Consequently, future development should incorporate design strategies that contribute to cooling the site's environment
Climate Change	Alters flowering cycles of vegetation, reducing the continuity of food resources and threatening long-term survival

Table 7 Current planning response (AECOM, 2018; NSW Department of Planning and Environment (DPE), 2023)

Summary	Relevant Actions in Greater Penrith to Eastern Creek
Rezoning to mixed-use centred, approximately 30 dwellings per hectare	2.1.2, 3.1.9, 2.2.1, and 2.6.2
Align with Cumberland Plain Conservation Plan Strategy, designating land for urban development and biodiversity conservation	2.1.2
New transport corridor (Outer Sydney Orbital (OSO)) along Wianamatta South Creek will be constructed on a via duct to minimise footprint (AECOM, 2018)	–
Ecologically focused actions to protect and enhance ecological systems and biodiversity through restoration, revegetation, and improve habitat corridors	2.1.2, 2.3.1, 2.3.5, 2.3.7(i and v), 5.1.2, and 6.3.7 (vi)
Development to respond with the topography, water landscapes, and hazard risk	2.3.3
Using sustainable and nature-based solution for water and hazard risk management (including flooding and urban heat), and avoid high-flood risk areas	5.1.2, 6.1.2, 6.2.4 (ii), 2.3.7 (vii), 6.2.4, and 6.3.1
Include green corridors and open spaces to integrate with natural features and active transport	2.3.7(iv), 6.1.4(ii)

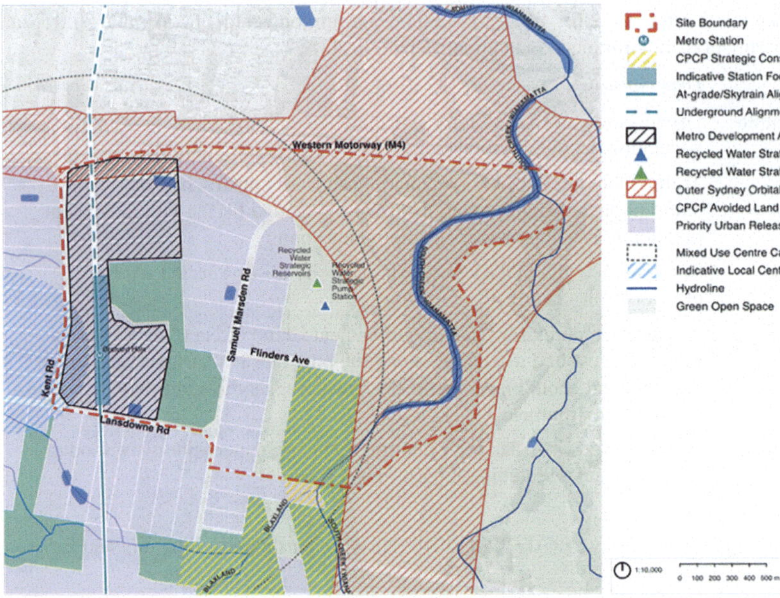

Fig. 11 Greater Penrith to Eastern Creek Plan. *Source* Author, obtained from NSW Department of Planning and Environment (2023)

G. **Opportunities and Constraints**

The following Table 8 and Fig. 12 outline the opportunities and constraints based on the site analysis findings, aiming to provide a foundation for formulating design recommendations.

5.2 Design Recommendation

A. **Design Strategies**

Five design strategies, as in Fig. 13, are developed to respond to the identified opportunities and constraints, as previously discussed.

a. **Creek Corridor Strategy**. This strategy aims to protect the remaining habitat of PCT 835 as a significant connectivity corridor and enhance it further through revegetation of adjacent land areas to mimic habitat characteristics while integrating low-impact recreational activities to facilitate access to nature.
b. **Green Strategy**. The four PCTs found across the site demonstrate a diverse habitat, with various native vegetation providing year-round food resources for native wildlife, particularly the Grey-headed Flying Fox. However, the majority is in a thinned condition, indicating past clearing events. The Metro construction site

Table 8 Site opportunities and constraint

Opportunities	Constraint
1. Proximity of the metro station to natural areas of Wianamatta South Creek becomes an opportunity to implement an ecocentric approach 2. Integrating water-sensitive design to enhance ecology and resilience by providing habitat corridor, mitigating flooding, and reducing urban heat island 3. Existing habitat provides year-long food resources for native fauna, including Grey-headed Flying Fox 4. Protecting Plant Community Type 835 along the creek as key corridor connecting core habitats 5. Current strategic planning and conservation plans (CPCP and Greater Penrith to Eastern Creek) support ecological focused development	1. Flooding restricts development 2. Urban heat island impacts fauna, which can be worsened after alteration of the landscape due to development 3. Vegetation clearing, in which construction of the Metro has resulted in the clearing of native vegetation, and land categorised as priority urban release in Greater Penrith to Eastern Creek can further lead to vegetation loss 4. Transport corridor (OSO) threatens riparian ecosystem although it is proposed to be elevated on a viaduct 5. Existing 500kV power line restricts ecological connectivity and pose electrocution risk to Grey-headed Flying Foxes

Source Author

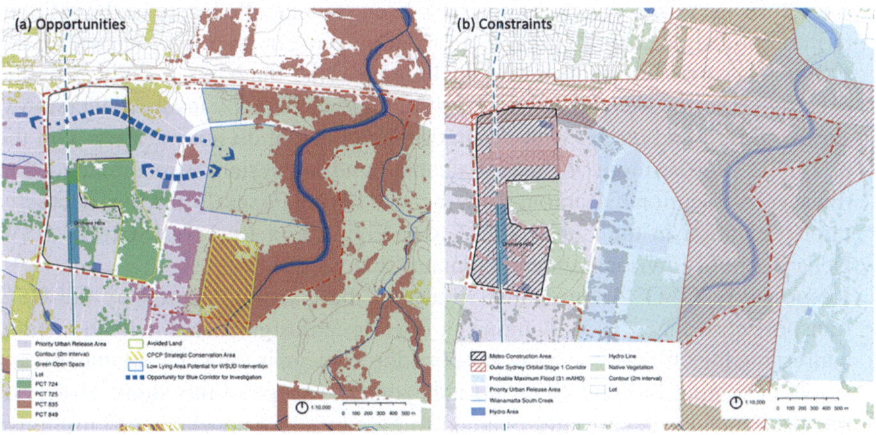

Fig. 12 Site opportunities (a) and constraints (b). *Source* Author

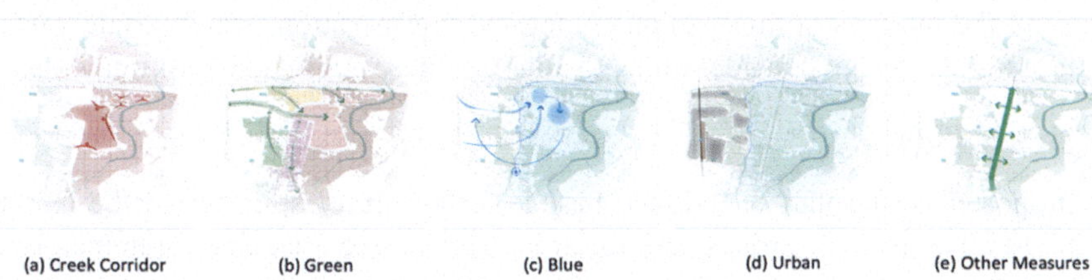

Fig. 13 Key design strategies. *Source* Author

has caused habitat removal, and future mixed-use development can potentially exacerbate this loss. Consequently, it necessitates a strategy to protect the remnant habitats while enhancing the habitat connectivity by creating a green corridor connecting future urban areas with the natural areas towards the creek, using tree species that provide food for Grey-headed Flying Foxes.

c. **Blue Strategy**. The low-lying area to the west of the site, adjacent to the creek, is flood prone, and occupies nearly half of the site. Therefore, this area presents an opportunity for integrating nature-based water management infrastructure, such as a combined system of wetlands and bioretention swales along the green corridor. This approach can mitigate flooding by reducing and treating stormwater runoff into the creek, addressing the primary cause of flooding in the Wianamatta South Creek, which is local catchment runoff (State of NSW, 2022). Moreover, combined with the green corridor, this strategy can contribute to cooling the environment and reducing urban heat, which is a significant threat to Grey-headed Flying Fox (Commonwealth of Australia, 2021; State of NSW Department of Environment and Climate Change, 2008).
d. **Urban Strategy**. Following the Nature-Driven Urbanism framework (Roggema, 2020a, 2020b), the areas suitable for urban development have been identified after recognising the vulnerable landscape and areas of ecological significance. Therefore, the proposed urban areas have been strategically located to be resilient to flooding, while ensuring a protection and improvement of ecological areas.
e. **Other Measures**. Other measures include ensuring the habitats divided by the existing powerline remain connected through lower canopy layers (understorey, shrub, and ground cover) with vegetation heights that do not interfere with the powerline. Furthermore, the use of a cable bundling system is recommended to reduce electrocution risk among Grey-headed Flying Foxes, following successful implementation in the Cabramatta Creek Flying Fox Reserve (State of NSW Department of Environment and Climate Change, 2008).
f. **Response to OSO**. Additionally, regarding the constraint posed by the OSO transport corridor, it is challenging to develop a design recommendation fully addressing the OSO, as the transport corridor's design is not yet available. Meanwhile, designing for the OSO itself is beyond the scope of this study. However, the road corridor is currently planned to be elevated on a viaduct to minimise impact on the riparian corridor (AECOM, 2018). Therefore, instead of focusing on adapting to the OSO infrastructure, this study explores the potential of developing the site to optimise urban ecological conservation, which could inform future OSO design considerations to adapt to these ecological requirements.

B. Master Plan

The proposed masterplan is designed based on the future aspiration of the Greater Penrith to Eastern Creek strategy for the area to be a mixed-use development around Metro station, combined with the aim of this study to foster human-nature connection through ecocentric urbanism approach. The following sections will address the critical design elements of the masterplan.

Green–Blue Corridor

The green–blue corridor is a critical design element in the masterplan to establish habitat connection for urban wildlife (see Fig. 14). It is particularly designed to attract Grey-headed Flying Foxes to forage along this corridor, leading to the riparian habitat of Wianamatta South Creek. Therefore, the green corridor is designed to minimise anthropogenic disturbances. This includes reducing traffic noise by locating the traffic away from the corridor, limiting direct vehicle access and allowing only active transport networks adjacent to the corridor. The green corridor is also designed to restore habitat PCT 724 which has been cleared for Metro construction, such as by planting Broad-leaved Ironbark to provide food resources throughout winter and spring (see Table 4). Lastly, integrating bioretention swales into the green corridor as a nature-based water management solution aims to address potential heat stress to Grey-headed Flying Foxes from increasing urban heat temperature by contributing to cooling through natural processes, such as evapotranspiration and shading (World Bank, 2021).

These approaches align with the principles of BSUD as outlined by Garrard et al. (2018) and Parris et al. (2018). It facilitates connectivity, introduces the restoration of the previously cleared habitat, minimises anthropogenic threats, encourages benevolent design, and enabling ecological processes (refer to Table 1).

Wetland

A constructed wetland is proposed in the site's low-lying area near the creek to manage flooding and improve waterway health, incorporating a sediment basin, macrophyte zone, and open water zone in line with current standards (City of Ryde Council, 2015; Transport NSW, 2023) (Fig. 15). In addition to treating stormwater and creating aquatic habitats that may attract native wildlife, the wetland helps mitigate pluvial flooding by reducing stormwater runoff volume and enabling water reuse for non-potable purposes (World Bank, 2021). Occupying 6% of the total site area, the wetland is expected to remove 80% of total suspended solids, 40% of total nitrogen,

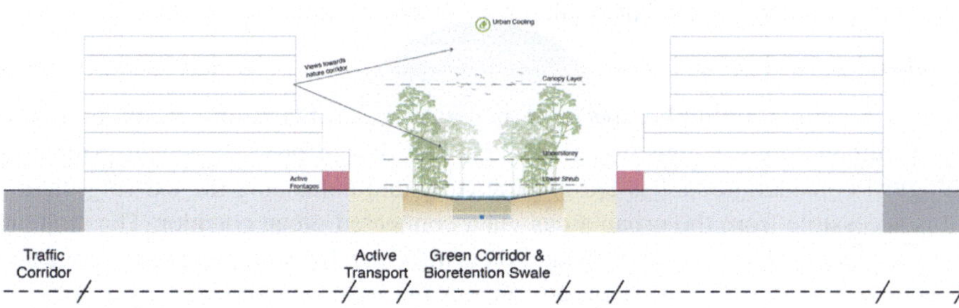

Fig. 14 Proposed green–blue corridor section. *Source* Author

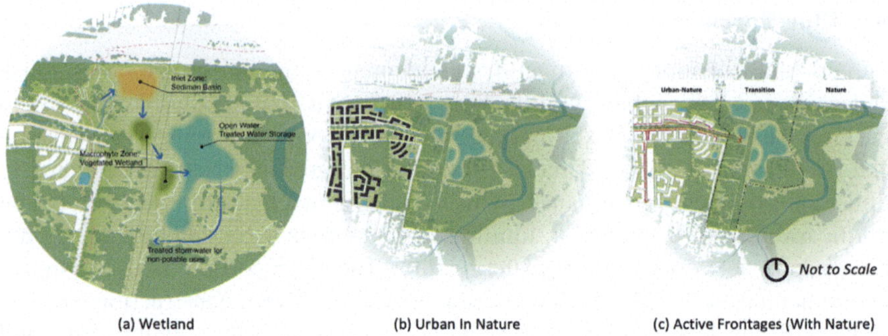

Fig. 15 Proposed Wetland (a), Design for '*In Nature*' human-nature connection principle (b), Design for 'With Nature' human-nature connection principle. *Source* Author

and 65% of total phosphorus from runoff, thus reducing pollution and supporting ecosystem health (City of Ryde Council, 2015).

C. **Design for Human-Nature Connection**

As discussed in the literature review, the three principles of human-nature connection are 'to be *in* nature, *with* nature and *for* nature' (Colding et al., 2020). The following will address how the proposed design promotes the first and second principles. Meanwhile, the third principle is an aspirational outcome of the first two, in which providing people access to nature and facilitating their interaction may contribute to fostering ecological stewardship and conservation efforts (Colding et al., 2020; Garrard et al., 2018; Kabisch et al., 2022; Kirk et al., 2021). While the relationship between nature exposure and environmental concern is complex and can be influenced by various factors, the third principle serves as an overarching goal and guiding principle in urban design, aiming to inspire sustainable practices to enhance human-nature connections in urban areas. This is in line with Colding et al.'s (2020) view that human-nature connection may contribute to sustainability transformations, while also recognising the knowledge-action gaps that challenge such assumptions.

In Nature

The first principle highlights the importance of accessible nature as the prerequisite of human-nature connection (Fig. 15). The design process, which prioritises ecological considerations, has led to protecting and enhancing the existing habitat that is accessible from the urban areas via a connected green corridor. The strategy on conserving the Grey-headed Flying Fox is informed by ecological knowledge and incorporated into the design. These efforts may increase local biodiversity as they are efficient pollinators and seed dispersals (State of NSW Department of Environment and Climate Change, 2008).

The buildings are oriented along the contour lines to integrate with the natural topography, while still maximising northern orientation for maximum sunlight. Additionally, the proposed bioretention swale and wetland guide movement towards the creek, both of which aim to evoke the sense of natural landscape. This approach embeds the city within its natural surroundings, allowing ecological features to thrive in urban areas.

With Nature

The second principle emphasises facilitating direct interactions between people and nature. Once people have access to nature, fostering their interaction is essential. In this proposal, the site is divided into three zones with varying levels of interaction (Fig. 15).

The first zone, 'urban-nature zone', aims to establish human-nature interaction in daily activities by orienting active frontages of mixed-use buildings and community facilities from the station area towards the green corridor, and eventually leading them into the more natural habitat areas. Furthermore, the 'transition zone', located where most revegetation will occur, is a zone for transitioning from a more urbanised areas into the natural habitat areas. This zone is designed for recreational activities with minimal environmental impact, such as water-based playgrounds, active transport routes, and picnic areas, to support periodic interactions. Lastly, the "nature zone" along the creek serves as a protected area, that still permits human access, such as by providing low impact infrastructure, including boardwalks and interactive information boards as a 'cue to care', where visitors can learn about the local biodiversity. This principle is vital, as urban nature experiences can potentially foster ecological stewardship, ultimately contributing to biodiversity conservation in cities (Colding et al., 2020; Garrard et al., 2018; Maron et al., 2016).

6 Conclusion

This study proposes that adopting an ecocentric approach can guide urban design and master planning efforts that aim to foster positive human-nature connections in cities. This study integrates design processes from several ecocentric frameworks, including Nature-based Solutions, Biodiversity Sensitive Urban Design, and Nature-driven Urbanism. Applying NDU early in the design process, focusing on embedding cities in nature, can help ensure landscape resilience by requiring early identification of landscape vulnerabilities and ecosystems for protection (Roggema, 2020a, 2020b).

Subsequently, incorporating NbS as the first step in designing green and blue solutions can further strengthen the resilience of the landscape. Moreover, BSUD offers a valuable methodology for designing habitats tailored to target species, contributing to broader ecosystem benefits (Garrard et al., 2018; Kirk et al., 2021).

This study thus provides insight into an ecological approach for master planning a new precinct around a transit station. It suggests that a thoughtfully integrated ecocentric design can balance urban development with environmental resilience, supporting biodiversity, enhancing ecological connectivity, and fostering meaningful human-nature interactions. In this way, ecocentric principles offer a pathway toward sustainable urban growth, enabling cities that are not only resilient but also enriching for both people and nature.

References

AECOM. (2018, March). *Outer Sydney Orbital Transport Corridor DRAFT STRATEGIC ENVIRONMENTAL ASSESSMENT*.

Atlas of Living Australia. (n.d.). *Explore your area*. https://www.ala.org.au

Australian Government & Natural Resources South East. (n.d.). *Grey Box Grassy Woodlands*. https://cdn.environment.sa.gov.au/landscape/docs/lc/grey-box-fact-sheet.pdf

Australian Native Plants Society (Australia). (2024a). *Bursaria spinosa*. https://anpsa.org.au/plant_profiles/bursaria-spinosa/

Australian Native Plants Society (Australia). (2024b). *Daviesia ulicifolia*. https://anpsa.org.au/plant_profiles/daviesia-ulicifolia/

Birtles, P. J. (2020). South Creek in Far Western Sydney: Opportunities for a new waterway focused city. In R. Roggema (Ed.), *Nature driven urbanism* (pp. 209–224). Springer International Publishing. https://doi.org/10.1007/978-3-030-26717-9_10

van den Bosch, M., & Sang, A. O. (2017). Urban natural environments as nature-based solutions for improved public health – A systematic review of reviews. *Environmental Research, 158*, 373–384. https://doi.org/10.1016/j.envres.2017.05.040

Carmona, M. (2020). To truly 'live', urban design needs accessible interdisciplinary research. *Journal of Urban Design, 25*(1), 5–9. https://doi.org/10.1080/13574809.2019.1706319

City of Ryde Council. (2015). *Water sensitive urban design guideline*.

Cohen-Shacham, E., Andrade, A., Dalton, J., Dudley, N., Jones, M., Kumar, C., Maginnis, S., Maynard, S., Nelson, C. R., Renaud, F. G., Welling, R., & Walters, G. (2019). Core principles for successfully implementing and upscaling Nature-based Solutions. *Environmental Science & Policy, 98*, 20–29. https://doi.org/10.1016/j.envsci.2019.04.014

Colding, J., Giusti, M., Haga, A., Wallhagen, M., & Barthel, S. (2020). Enabling relationships with nature in cities. *Sustainability, 12*(11). https://doi.org/10.3390/su12114394

Commonwealth of Australia. (2021). *National recovery plan for the grey-headed flying fox*.

Firth, K. (2021). A fruitful coupling. *Urban Design Group Journal, 157*, 18–20.

Garrard, G. E., Williams, N. S. G., Mata, L., Thomas, J., & Bekessy, S. A. (2018). Biodiversity sensitive urban design. *Conservation Letters, 11*(2), 1–10. https://doi.org/10.1111/conl.12411

Greater Cities Commission (GCC). (2022). *Discussion paper the six cities region*.

Greater Sydney Commission (GSC). (2018a, March). *Greater Sydney region plan—A metropolis of three cities*. https://www.planning.nsw.gov.au/sites/default/files/2024-04/greater-sydney-region-plan.pdf

Greater Sydney Commission (GSC). (2018b, March). *Western city district plan*. https://www.planning.nsw.gov.au/sites/default/files/2024-04/western-city-district-plan.pdf

Greening Australia. (2017). *Eucalyptus microcarpa*. https://www.greeningaustralia.org.au/wp-content/uploads/2017/11/FACT-SHEET_Eucalyptus_microcarpa.pdf

Growing Illawarra Natives. (n.d.). *White feather honey-myrtle*. https://finder.growingillawarranatives.org/plants/plant/313

Howes, J. (2021). *Eucalyptus tereticornis Forest Red Gum, Red Irongum, also known as Queensland Blue Gum*. https://resources.austplants.com.au/plant/eucalyptus-tereticornis/

Howes, J. (2022). *Melaleuca nodosa Prickly-leaved Paperbark*.

iNaturalist. (2024a). *Forest Red Gum Eucalyptus tereticornis*. https://www.inaturalist.org/taxa/58208-Eucalyptus-tereticornis

iNaturalist. (2024b). *Narrow-leaved Ironbark Eucalyptus crebra*. https://www.inaturalist.org/taxa/162750-Eucalyptus-crebra

iNaturalist. (2024c). *Rough-barked Apple Angophora floribunda*. https://www.inaturalist.org/taxa/207930-Angophora-floribunda

iNaturalist. (2024d). *Woollybutt Eucalyptus longifolia*. https://inaturalist.ala.org.au/taxa/353878-Eucalyptus-longifolia

Kabisch, N., Frantzeskaki, N., & Hansen, R. (2022). Principles for urban nature-based solutions. *Ambio, 51*(6), 1388–1401. https://doi.org/10.1007/s13280-021-01685-w

Kirk, H., Garrard, G. E., Croeser, T., Backstrom, A., Berthon, K., Furlong, C., Hurley, J., Thomas, F., Webb, A., & Bekessy, S. A. (2021). Building biodiversity into the urban fabric: A case study in applying Biodiversity Sensitive Urban Design (BSUD). *Urban Forestry & Urban Greening, 62*, 1–14. https://doi.org/10.1016/j.ufug.2021.127176

Maron, M., Ives, C. D., Kujala, H., Bull, J. W., Maseyk, F. J. F., Bekessy, S., Gordon, A., Watson, E. M., Lentini, P. E., Gibbons, P., Possingham, H. P., Hobbs, R. J., Keith, D. A., Wintle, B. A., & Evans, M. C. (2016). Taming a wicked problem: Resolving controversies in biodiversity offsetting. *BioScience, 66*(6), 489–498. https://doi.org/10.1093/biosci/biw038

Martins, J. (2021a). Practice-led research in urban design. *Urban Design Group Journal, 157*, 17.

Martins, J. (2021b). Towards research-led dialogues between practice and academia. *Urban Design Group Journal, 157*, 39.

Mcdonald-Madden, E., Schreiber, E. S. G., Forsyth, D. M., Choquenot, D., & Clancy, T. F. (2005). *Factors affecting Grey-headed Flying-fox (Pteropus poliocephalus: Pteropodidae) foraging in the Melbourne metropolitan area, Australia, 30*(5). https://doi.org/10.1111/j.1442-9993.2005.01492.x

NSW Department of Planning and Environment (DPE). (2022, August). *Cumberland Plain Conservation Plan*. NSW Department of Planning and Environment. https://shared-drupal-s3fs.s3.ap-southeast-2.amazonaws.com/master-test/fapub_pdf/Lisa+Drupal+Documents/Cumberland-Plain-Conservation-Plan-202208.pdf

NSW Department of Planning and Environment (DPE). (2023). *Greater Penrith to Eastern Creek (GPEC) Investigation Area Strategic Framework*.

NSW Department of Primary Industries and Regional Development. (2010). *Broad-leaved red ironbark*. https://www.dpi.nsw.gov.au/__data/assets/pdf_file/0005/362795/Paddock-plants-Eucalyptus-fibrosa-Broad-leaved-Ironbark.pdf

NSW Environment and Heritage. (2023). *BioNet species sightings data*. https://www.environment.nsw.gov.au/topics/animals-and-plants/biodiversity/nsw-bionet/about-bionet-atlas/species-sightings-data

NSW Government Department of Planning, Industry and Environment. (2021). *Cumberland plain assessment report*.

Parris, K. M., Amati, M., Bekessy, S. A., Dagenais, D., Fryd, O., Hahs, A. K., Hes, D., Imberger, S. J., Livesley, S. J., Marshall, A. J., Rhodes, J. R., Threlfall, C. G., Tingley, R., Van Der Ree, R., Walsh, C. J., Wilkerson, M. L., & Williams, N. S. G. (2018). The seven lamps of planning for biodiversity in the city. *Cities, 83*, 44–53. https://doi.org/10.1016/j.cities.2018.06.007

Roggema, R. (2020a). Landscape first! nature-driven design for Sydney's third city. In R. Roggema (Ed.), *Nature driven urbanism* (pp. 81–109). Springer International Publishing. https://doi.org/10.1007/978-3-030-26717-9_5

Roggema, R. (2020b). Nature-driven urbanism. In R. Roggema (Ed.), *Nature driven urbanism* (pp. 1–8). Springer International Publishing. https://doi.org/10.1007/978-3-030-26717-9_1

Roggema, R. (2020c). The future of nature-driven Urbanism. In R. Roggema (Ed.), *Nature driven urbanism* (pp. 331–334). Springer International Publishing. https://doi.org/10.1007/978-3-030-26717-9_16

Roggema, R. (2021). From nature-based to nature-driven: Landscape first for the design of Moeder Zernike in Groningen. *Sustainability, 13*(4), 2368. https://doi.org/10.3390/su13042368

Roggema, R. (2023). The Eco-Cathedric city: Rethinking the human-nature relation in urbanism. *Land, 12*(8), 1501. https://doi.org/10.3390/land12081501

Sijmons, D. (2020). Contrast, contact, contract; pathways to pacify urbanization and natural processes. In R. Roggema (Ed.), *Nature driven urbanism* (pp. 9–42). Springer International Publishing. https://doi.org/10.1007/978-3-030-26717-9_2

State Government of NSW & NSW Department of Planning, Housing and Infrastructure. (2022). *Cumberland Plain Conservation Plan (CPCP).* The sharing and enabling environmental data portal. https://datasets.seed.nsw.gov.au/dataset/cpcp

State of NSW. (2022). *Wianamatta South Creek catchment flood study—Existing conditions.*

State of NSW, & DCEEW, N. D. of C. C., Energy, the Environment and Water. (2019). *NSW Urban Heat Island to Modified Mesh Block 2016*. The sharing and enabling environmental data portal. https://datasets.seed.nsw.gov.au/dataset/97815860-d840-4e67-8f73-1bfc665cb310. Accessed October 15, 2024.

State of NSW Department of Environment and Climate Change. (2008). *Best practice guideline for grey headed flying fox.*

State of NSW & NSW Department of Climate Change, Energy, the Environment and Water. (2010). *Grey-headed flying-fox camp boundaries.*

Tozer, M. G., Leishman, M. R., & Auld, T. D. (2015). Ecosystem risk assessment for Cumberland plain woodland, New South Wales Australia. *Austral Ecology, 40*(4), 400–410. https://doi.org/10.1111/aec.12201

Transport NSW. (2023). *Water sensitive urban design guideline.*

World Bank. (2021). A catalogue of nature-based solutions for urban resilience. *World Bank*. https://doi.org/10.1596/36507

Living with Nature: Vernacular Architecture and Socio-Ecological Landscape of Riverine Communities in Agusan Marsh, Philippines

Isidoro R. Malaque III and Rachel Mae P. Sitcharon

Abstract The twenty-first century presents pressing global challenges, notably rapid urbanisation and climate change. As urban living becomes increasingly central to human experience, the study of human settlements offers valuable insights into sustainable development, particularly through the lens of human-nature interactions. The human-nature relationship creating the built environment is evident in the vernacular architecture and landscape of indigenous communities. Shaped by indigenous knowledge system and practices, this chapter explores how vernacular architecture reflects adaptive responses to environmental conditions. Focusing on riverine communities in Agusan Marsh, Mindanao, Philippines, the study examines how local architectural forms embody resilience and ecological sensitivity. In Sabang Gibong riverine community, four typologies of floating houses—locally known as *hapa*, *hanger*, *hapa-hanger*, and *dos andanas—demonstrate* nuanced adaptations to fluctuating water levels. In La Flora riverine community, other than floating houses, the built environment includes stilt houses and hybrid structures combining floating and stilt technologies, which are shaped by the community's spatial and ecological context. The marsh ecosystem provides basic building materials such as bamboo for floaters and timber for main structure of the houses. Moreover, synthetic modern building materials were integrated in the construction of houses. With the marshland environmental conditions, these housing forms respond to seasonal and annual flooding through elevation strategies and buoyant design. Living with the waters, the vernacular architecture of Agusan Marsh illustrates how indigenous ingenuity fosters community resilience in the face of environmental variability. By foregrounding these practices, this chapter aims to contribute to broader discussions on climate adaptation and sustainable urbanism, offering lessons for similarly vulnerable cities and regions worldwide.

I. R. Malaque III (✉)
University of the Philippines Mindanao, Davao, Philippines
e-mail: irmalaqueiii@up.edu.ph

R. M. P. Sitcharon
Provincial Government of Agusan del Sur, Agusan del Sur, Philippines

A. Cheshmehzangi et al. (eds.), *Nature-based Urbanism*, Urban Sustainability,
https://doi.org/10.1007/978-981-95-6864-2_12

Keywords Indigenous knowledge · Vernacular architecture · Community resilience · Climate change · Environmental sustainability · Water urbanism · Agusan Marsh

1 Introduction

The twenty-first century presents multifaceted challenges that span both natural ecosystems and urban environments. Climate change, as one of the popular environmental challenges, among others, causes extreme weather conditions resulting to extreme flooding and, in the worst-case scenario, ending in disasters. As over half of the global population now resides in urban areas, cities—rather than being viewed solely as sources of environmental strain—can serve as dynamic platforms for generating innovative and context-sensitive solutions. Urbanisation, when critically examined, offers opportunities for resilience and adaptation. Since rapid increase in human population will have adverse impacts to the natural environment such as loss of biodiversity and alteration of the ecosystem accommodating agriculture, among other examples, limited resources must be properly managed to ensure sustainability.

While contemporary approaches to sustainability often emphasise technological innovation, an equally vital perspective lies in fostering a deeper understanding of the reciprocal relationship between humans and nature, as exemplified in vernacular environments. One example is to explore how the indigenous knowledge system and practices, also referred as IKSP, contributed to the creation of built environment as a product of man-nature relation. Specifically, this chapter examines the vernacular architecture of communities in Agusan Marsh, Philippines, where settlement patterns and housing typologies—such as floating and stilt houses—demonstrate adaptive strategies rooted in indigenous lifeways. In this context, this chapter aims to present and discuss the cases of two riverine communities and their vernacular architecture, evident in the floating and stilt houses, which sustain with the conditions of marshland environment. To frame this analysis, this chapter draws upon interdisciplinary literature encompassing indigenous knowledge, vernacular architecture, water urbanism, community resilience, and environmental sustainability. The socio-ecological principles evident in the architectural practices of Agusan Marsh communities offer valuable insights into how traditional knowledge can inform contemporary responses to climate change, disaster risk reduction, and adaptive design. As such, while the twenty-first century technology influenced the present design and construction of houses, it is also noted how the indigenous knowledge on community resilience could also be sourced as a solution for the twenty-first century challenges such as climate change adaptation and resiliency, and disaster risk reduction and management.

2 Living with Nature and the Waters in Marsh Landscape

The case studies on floating and stilt houses in Agusan Marsh, Philippines, are examined as manifestations of indigenous knowledge systems shaped by environmental conditions. These architectural forms, rooted in vernacular traditions, reflect the adaptive strategies of riverine communities, and contribute to a broader understanding of landscape-responsive design. The theoretical framework underpinning this analysis draws from key arguments in vernacular architecture and indigenous environmental adaptation. Specifically, the case of Sabang Gibong (Sitcharon & Malaque III, 2023a, 2023b) and the case of La Flora (Malaque III & Sitcharon, 2024) provide foundational insights into the socio-ecological dynamics of settlement formation and architectural expression. Synthesised from these studies, the theoretical perspectives inform the discussion presented in the subsequent sections of this chapter.

2.1 Indigenous Knowledge and Vernacular Architecture

Indigenous knowledge systems and practices have increasingly been recognised as vital contributors to sustainable development. The popular 1987 Brundtland Report that highlighted the concept of needs and limitations, it recognised the importance of traditional *"life-styles that can offer modern societies many lessons in the management of resources in complex forest, mountain, and dryland ecosystems"* (United Nations, 1987, pp. 27–28). In one of the pioneering and classical theories on vernacular architecture, Rudofsky's (1964) *"Architecture without Architects"* highlighted the relationship between human and nature in the shaping of the built environment.

Further explained, vernacular architecture is communal architecture as they are produced not by specialists but by the spontaneous and continuing activity of a whole people with a common heritage, acting within a community of experience (Rudofsky's, 1964). Consequently, indigenous knowledge provides an alternative paradigm to modern architectural and developmental frameworks. Its enduring presence in vernacular architecture exemplifies how traditional ecological wisdom and community-based design continue to inform sustainable practices in contemporary contexts.

The significance of vernacular architecture in shaping the built environment remains highly relevant in the twenty-first century, particularly within cultural and ecological discourses. Primarily, from the point of view of culture from Rapoport (1988), for example, acknowledged how vernacular architecture works well and how its aesthetics are based in the context of its own environmental quality. Emerging in the twenty-first century, theorists in vernacular architecture paid more attention to the people who shaped, use, alter, and whose behaviour or way of living is shaped by the form, instead of viewing vernacular architecture purely as objects or artefacts.

Also cited in Malaque III and Sitcharon (2024), referring from the same theorist Rapoport (2006), it is said that by moving from a natural-historical stage to a problem-oriented stage, it is possible to regard vernacular environments as a laboratory to interpret the wide range of human responses to various factors affecting their way of living. Moreover, this is strengthened by Asquith (2006, p. 129) as it is said that *"once the vernacular is seen not as static building form, but as constantly evolving, reacting to changes in the communities that shaped its form, it will become higher on the agenda in architectural education"*. Building on Rudofsky's (1964) foundational work, *"Architecture without Architects"*, and the theoretical contributions of Rapoport (2006) and Asquith (2006), it becomes evident that indigenous knowledge embedded in vernacular architecture offers profound insights into sustainable and context-sensitive design. These architectural expressions, often perceived as intuitively perfect within their natural and social environments, exemplify the harmonious integration of cultural heritage, ecological adaptation, and community resilience.

2.2 Emergence of Water Urbanism

Water urbanism has emerged as a critical framework for advancing climate change resilience, particularly in flood-prone and water-adjacent urban environments. This theme was prominently featured in the first Southeast Asian Urbanisms Seminar (SEAUS), held at the University of Architecture Ho Chi Minh City, Vietnam, in May 2023. During this conference, the case of Sabang Gibong (Sitcharon & Malaque III, 2023b), focusing on the management regimes and sustainable dimensions of floating houses, was presented. Consequently, this and other papers presented in the same conference eventually instituted the theoretical background on water urbanism and climate change resilience in the succeeding case study conducted in La Flora (Malaque III & Sitcharon, 2024). The papers on the theme of water urbanism were concerned with the vulnerability of urbanised environment brought by climate change inherent in the rising of overall flood water levels.

Also cited in Malaque III and Sitcharon (2024), examples of studies related to water urbanism presented at the conference include the following: water-oriented development and placemaking strategies for Bangkok (Udomcharoenchaikit, 2023); how slum dwellers survived with local indigenous measures in order to reduce risks of flood hazards in the coastal slum of Dharavi, Mumbai (Goyal & Silva, 2023); socio-ecological impacts of industrialisation and urbanisation of craft villages in the Red River Delta, Vietnam (Montresor, 2023); placemaking practices in riverine settlements after river modification projects in the case of Citarum River, Indonesia (Stassi, 2023); and integration of water in the infrastructure development in Ho Chi Minh (Trang et al., 2023).

In the case of Sabang Gibong (Sitcharon & Malaque III, 2023b), water urbanism is discussed with environmental dimension as one of the sustainable factors in floating houses, commonly for being flood resilient and adaptable to water level change. Dimensions of sustainable factors in floating houses, to include environmental, economic, and social, are adopted from Moon (2014) for advocacy for climate resilient architecture. Thus, the phenomenon of water urbanism, as illustrated through these diverse case studies, underscores the importance of adaptive strategies in vulnerable communities. It offers a compelling lens through which to examine disaster risk reduction, and to make cities and the people resilient to climate change.

2.3 Community Resilience and Environmental Sustainability

Floating and stilt houses as products of shared knowledge demonstrated the ability of communities to be resilient in the context of their environmental conditions, in this case, on the changing flood water levels as the nature in marshland landscape. Other than adopting the sustainable dimensions and factors identified by Moon (2014), this case study adopted the identified design principles from the classic literature of Ostrom (1990), which resulted from analysing long-enduring common-pool resources. Also known as CPRs, this *"refers to a natural or man-made resource system that is sufficiently large as to make it costly (but not impossible) to exclude potential beneficiaries from obtaining benefits from its use"* (Ostrom, 1990, p. 30). In the Philippine context, Ostrom (1990, pp. 82-88) cited the case of the historical *zanjera* irrigation communities of Ilocanos in Ilocos Norte, Philippines, derived from Spanish priests writing in 1630 (Lewis, 1980, as cited by Ostrom, 1990).

The project was established collectively by landowning farmers wanting to construct common irrigation systems and individuals organising themselves so as to acquire land. However, analysing various long-enduring CPR cases, Ostrom (1990, p. 88) found one similarity, and it is *"that all face uncertain and complex environments"*. Furthermore, the case study on river communities referred to this paper (Malaque III & Sitcharon, 2024; Sitcharon & Malaque III, 2023a, 2023b) was viewed from the socio-ecological model firstly developed by Bronfenbrenner (1979). In relation to human development, it was theorised that man/woman is influenced by his/her ecological environment of nested structures where he/she may or may not be directly in. As this theory was widely adopted other than in the social sciences, it birthed, among others, the socio-ecological approach to architecture and design.

Community resilience within vernacular and natural environments has been demonstrated across diverse contexts, particularly in regions vulnerable to climate-related hazards. For example, in the Philippines, when government provided housing as part of post-disaster recovery efforts after a devastating typhoon failed to meet environmental and living requirements, upon occupancy, the housing beneficiaries made incremental modifications and, in some cases, self-built their own housing units following the principles of vernacular architecture, such as elevating the main house on stilts to avoid being flooded (Malaque III & Golimlim, 2019). Another example in Thai-Myanmar border, the proposed settlement for refugees was sought to provide a community-based refugee camp experience that includes the building and occupation of vernacular architecture to help address alienation and build resilience (Aung & Pretty, 2022). Moreover, in the case of resettlement site of Kannagi Nagar in Chennai, the capital of Tamil Nadu State in India, being built on flood-prone marshlands where residents are exposed to the risks of flooding and water scarcity, one of the main findings of the research is the significant positive relationship between community resilience and adaptation behaviour (Haas et al., 2021). Assessed using five core dimensions to include trust, place attachment, collective efficacy, social networks, and social support, which significantly and positively influenced adaptation actions, it is implied that only when the inhabitants are supported by their social networks and have confidence in their community's capabilities, greater risk awareness which increase the number of adaptation measures can be taken. Collectively, these cases underscore the importance of integrating socio-ecological principles into resilience-building efforts. By recognising the interplay between social systems, ecological conditions, and vernacular practices, communities can develop context-sensitive strategies that not only reduce vulnerability but also sustain the integrity of their natural environments.

3 Case Study on the Riverine Communities in Agusan Marsh, Philippines

The settlement of human populations within marshland environments offers a valuable context for examining the socio-ecological frameworks employed by indigenous communities in their adaptation to and coexistence with natural systems. The presence of vernacular architecture in such landscape could provide evidence on how communities were able to relate and interact with their way of living within their natural environment. To explain this relationship, this section presents and discusses the case study on floating and stilt houses of riverine communities in Agusan Marsh, Philippines, highlighting how architectural forms reflect and reinforce sustainable living strategies within a wetland ecosystem.

3.1 Aim and Objectives

The case study provides preliminary understanding of the floating house phenomenon of Agusan Marsh was conducted in Sabang Gibong riverine community (Sitcharon & Malaque III, 2023a, 2023b). This was followed by another case study in La Flora, another riverine community where, aside from floating houses, stilt houses, and houses combining floating and stilt technologies are also found (Malaque III & Sitcharon, 2024). Referring to these conducted case studies, the aim of this paper is to present, discuss, and synthesise their findings with the environmental claims of the study towards achieving community resilience in the context of living with nature.

In case of further reading, details of the case studies on Sabang Gibong (Sitcharon & Malaque III, 2023a) and La Flora (Malaque III & Sitcharon, 2024) referred in this chapter are readily accessible on the Architectural Science Association (ANZAScA) website at https://archscience.org.

3.2 *Case Study Areas and Methodologies*

Agusan Marsh located in the province of Agusan del Sur stands as one of the most significant wetlands in the country, serving as catch basin of adjacent regions in Mindanao Island, the Philippines. It is estimated to cover a total land area of 110,000 hectares, including its floodplain spreading in six municipalities of Agusan del Sur province. Along the main Agusan River, two communities—also referred as *barangays* that means the smallest political unit in the Philippines, belonging to the municipality of Talacogon were chosen as case study areas. Firstly, the *barangay* of Sabang Gibong, at the junction of Gibong River and the main Agusan River, situates 100 per cent of its families residing in floating houses. With current statistics records, it has a population of 1013 individuals and a total of 248 total households. Secondly, the *barangay* of La Flora, also located along the main Agusan River with floating and stilt houses, has current population records of 1518 individuals and a total of 338 households. Almost 100 per cent of the total population in both *barangays* are identified as indigenous peoples, locally referring to themselves as *Talacognon* but statistically recorded in government records belonging to the *Manobo* ethnolinguistic group.

In addition to the frequent visits to Agusan Marsh, specific fieldwork was conducted for the two case studies discussed in this paper. For the Sabang Gibong case, the fieldwork was conducted from January to April 2023; and for the La Flora case, the fieldwork was conducted from April to May 2024. Basically, physical documentation of the houses as main primary data was collected through photo documentation, sketching, and direct observation, including immersion with the community. Moreover, to support the physical documentation, focus group discussions and interviews, including structured and key information, were also conducted. Primary data derived from focus group discussions, interviews, and direct observations were organised and reduced; interconnectedness was established; and themes were identified to ensure meaningful presentation and discussion of research results. Case study results, as originally reported to include the Sabang Gibong case (Sitcharon & Malaque III, 2023a, 2023b) and the La Flora case (Malaque III & Sitcharon, 2024), are referred to this chapter, presented, and discussed in the succeeding sections.

Fig. 12.1 Typologies of floating houses in Agusan Marsh, from left to right: *hapa*, *hanger*, *hapa-hanger* combination, and *dos andanas* (adopted from Sitcharon & Malaque III, 2023a, 2023b)

3.3 Architecture and Construction of Floating and Stilt Houses

Houses are locally referred as *ba'y*, which is closely associated with the *balai* of the Malay, or the *balay* in Visayan and *bahay* in Tagalog languages in the Philippines. Firstly, in the case of floating houses in Sabang Gibong riverine community, the technology was developed by people of Manobo and Visayan descent as a combination of external influence and indigenous ingenuity. As presented and discussed in the referred publications (Sitcharon & Malaque III, 2023a, 2023b) and shown in Fig. 12.1, there are four typologies of floating houses found in Agusan Marsh, referred in local dialect as *hapa*, *hanger*, *hapa-hanger*, and *dos andanas*. *Hapa* is considered the safest type, which has been tested against strong winds and typhoons. *Hanger* as the second type, this provides better ventilation and easier access in cleaning the floaters (bundles of bamboo poles), and other structural members under the main floor called *batangan* and *sakbat*.

Batangan are the round logs on top of the floaters which carry the weight of the house and ensure that the upper layer is not reached by water. On top of the *batangan* is *sakbat* which could be round logs or lumber to serve as base of the main house. To enjoy the benefits of the first two types, the third type is a combination of *hapa* and *hanger* types. For few households who could afford, *dos andanas* type provides bigger spaces, but transferring from one location to another during the seasonal flooding and safety against strong winds become challenges. In terms of settlement patterns, the rivers as defining factors influenced the linear arrangement of floating houses. These houses are either within or adjacent to the rivers, depending on the water level and current (see Fig. 2, left). During the dry season, half of the floating houses are on the rivers anchored adjacent to the bank, while others are setting on dryland. However, during the flooding season, all these houses afloat and would be moved to safer locations, in most cases anchoring together in clusters to be more resilient against strong water currents (see Fig. 2, right).

In another study area referred to as La Flora riverine community, cases of floating houses, stilt houses, and houses combining stilt and floating technologies were presented and discussed in another referred publication (Malaque III & Sitcharon, 2024). As shown in Fig. 12.3 (left), typologies of floating houses identified in the case of Sabang Gibong were also found in La Flora situated along the main Agusan

Fig. 12.2 Left: Sabang Gibong during the dry season in 2019 with Gibong River at the foreground and main Agusan River at the background (adopted from Sitcharon & Malaque III, 2023a, 2023b); and right: cluster of floating houses anchoring together at the junction of the two rivers during the height of a flooding season (photo taken by the author during a site visit and immersion in January 2024)

River and the tributary Mayat River which is a small river that empties into the greater Agusan River. While these houses can float all throughout the year, during the dry season, some houses remain afloat and moored by ropes to adjacent trees along the riverbanks, and some sit on land near or at the riverbanks. In the case of La Flora, stilt houses (Fig. 12.3, right) are the most common type, of which, in few cases, they are situated at the riverbank but in most cases, they are situated in the inland part of the community.

This house type has a stilt height of 12 feet (3.60 metres) or higher since the water level during annual flood recurrence may reach as high as 10 feet (3.00 metres). The stilt house design has a unique variation by having a first floor, which is an open structure. Referred as *permirong andana* elevated from the ground at approximately four feet (1.20 meters), when flood water is not very high, households utilise the open area as a social space that also serves as the living room, kitchen, and comfort room of the households. On the occasion that floodwater level is higher reaching the first floor, the main floor referred as *ikaduhang andana* or the second floor, serves as sleeping quarters and storage for household items, keeping the inhabitants safe

Fig. 12.3 Left: landscape of La Flora riverine community during the dry season and its floating and stilt houses; and right: a stilt house situated in the inland part of La Flora (photos taken by the author during fieldwork in April 2024)

from floodwater. Moreover, the third type is a combination of both floating and stilt house technologies. Typically located along the riverfront during dry season or at the banks of Agusan River, this type acts as a stilt house during the dry season, but it will float with bamboo floaters inserted below the main floor during the flooding season, which will be moored to an adjacent tree to secure it..

4 Discussion

Geographic factors serve as a primary determinant in shaping settlement patterns and housing typologies across the two riverine communities under comparison. In the case of Sabang Gibong, the use of floating houses would be the best option being located at the intersection of two major rivers and considering regular flooding and strength of the water current in the area. For permanently situated house types such as bungalows or houses on stilts, there would be no guarantee of safety because, at some point during the flooding season, plants and other floating debris accumulate in the area. With this environmental condition, the community found it to be an unsafe option since floating debris may get stuck on the stilts, which will eventually damage the structure.

Moreover, hardwoods are not locally available, necessary to be used as stilt posts. While softwood for the house structure and bamboo for the floaters are readily available, thus, 100 percent of households in Sabang Gibong inhabited floating houses. In the case of La Flora, those households who prefer to live with, or along, the river would choose floating houses if bamboo were available. Due to the limited availability of bamboo and the relative abundance of timber in the area, majority of households in La Flora opt to construct and inhabit stilt houses. Unlike in Sabang Gibong, the physical geography of La Flora has an inland area which is connected to adjacent dryland communities and to the town centre through a road, making it more accessible especially during dry season. Considering its landscape features, among other factors, La Flora situated various housing types to include floating and stilt houses, plus the combination of floating and stilt house technology, as shown in Figs. 12.4, 12.5, and 12.6.

As shown in Fig. 12.4, floating houses during dry season will remain afloat in the river and be moored by ropes to adjacent trees, or in other cases, sitting on land near, or at, the riverbanks. Dry season is typically from April to November, but in other narratives, it is also from March to September. During flooding season, as shown in Fig. 12.5, all floating houses afloat and moored to adjacent trees or clusters to be more resilient against strong water currents. Flooding season is typically is from December to March, but in other narratives, it is from October to February. At the height of flooding season, as shown in Fig. 12.6, house type combining stilt and floating house technology will float with bamboo floaters inserted below the main floor, and moored to a tree to secure it. For the stilt houses located inland, the second floor will serve as sleeping quarters and storage of household items when the floodwater level is higher, reaching the first floor that is typically designed as an open type. Stilt houses

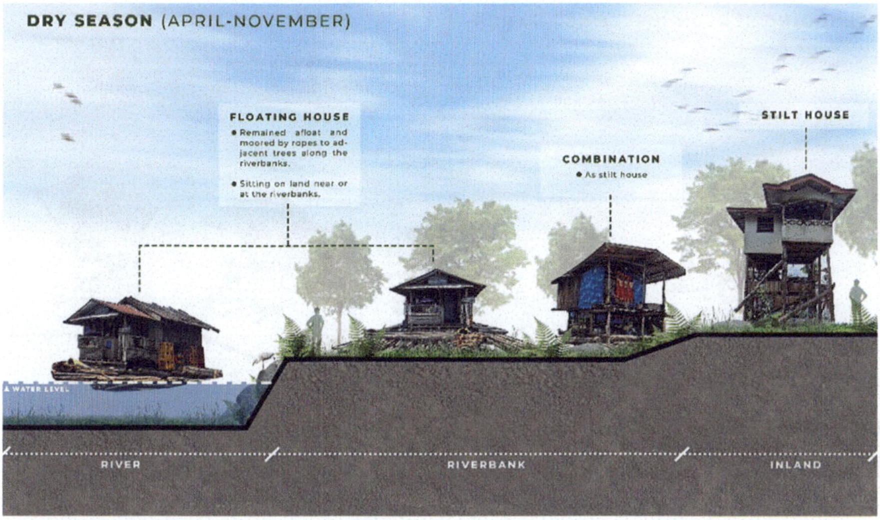

Fig. 12.4 Location of houses during dry season typically from April to November

Fig. 12.5 Location of houses during flooding season, typically from December to March

were designed by the community where they can also monitor floodwater levels so they will have enough time to prepare for and avoid impending damage brought by flooding as a natural hazard in the marsh environment.

The community played an important role from the construction and to the sustainability of floating and stilt houses. In the case of Sabang Gibong (Sitcharon & Malaque III, 2023a, 2023b), where all inhabitants live in floating houses, for example,

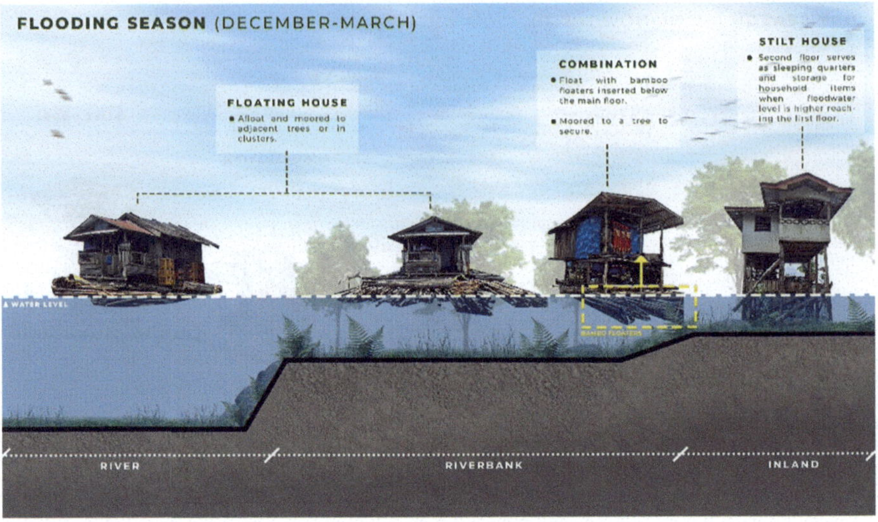

Fig. 12.6 Location of houses at the height of flooding season

one necessary consideration in the practice of floating houses is the financial capacity of households. While most of the building materials are locally and freely available, payment for labour is still required when needing the expertise of local carpenters..

However, the community has a traditional practice locally called *lugoy* or *lugoy-lugoyan*, that literally means mutual assistance wherein people help each other with no expected monetary return. In the practice of floating houses, this is demonstrated in the annual replacement of floaters. The same mutual assistance is also done in the construction of stilt houses in La Flora (Malaque III & Sitcharon, 2024). While floaters can be replaced by four or more people, wooden stilt installation requires 15 to 20 people. Thus, when one may ask for help from others in the community, depending on availability, kindness will be returned when the helper requires the same or another kind of assistance in the near future. With no monetary payments, snacks or meals will be served, even including alcoholic drinks such as the locally known rum brand *Tanduay*, as celebration for the success of the communal undertaking.

The sustainable dimensions—encompassing environmental, economic, and social factors as outlined by Moon (2014)—can be meaningfully examined through the lens of vernacular architecture, particularly in the context of floating and stilt houses. Firstly, with adaptation to water level change as one example for environment as a dimension, floating houses adapt to water level change, though dependent on quality of floater and strength of anchor. In terms of flood resilience, in Sabang Gibong (Sitcharon & Malaque III, 2023b), for example, research informants inferred that there is no flooding in their area because the floodwaters do not enter inside their houses. In the case of stilt houses in La Flora (Malaque III & Sitcharon, 2024), flood resilience is also demonstrated by having an open *permirong andana* or the first floor at the stilts below the main second floor. At the height of flooding season, the *ikaduhang andana* or the main second floor serves as the living quarters, making the inhabitants safe from floodwaters..

Moreover, with the use of local materials, conservation of environment to sustain the availability of local resources is assured by the community since the practice requires a regular supply of timber and bamboo. Secondly, in terms of long-term usage as one example of the economic dimension, houses made of softwood can last from 10 to 15 years, while houses made of hardwood may remain up to 20–30 years. Moreover, due to limited access to resources and to avoid financial strain, houses are incrementally constructed.

Thirdly, in terms of social support and a high sense of community as factors for social dimension, the community practice of *lugoy-lugoyan* as collaborative effort is demonstrated in the transferring or towing their houses, in the maintenance and replacement of *batangan* or floaters for the floating houses, and in the setting up of main stilts for the stilt houses. Moreover, both floating and stilt house technologies have proven their safety against natural calamities brought by flooding and have demonstrated community resilience. The community believe that they can survive economic constraint, natural calamity, or any hazard because they support each other. For example, during flooding season, the floating houses in Sabang Gibong (shown in Fig. 2, right) are anchoring together in clusters for better protection against strong currents, and the stilt houses in La Flora are designed so that the community can monitor floodwater levels for better preparedness. Accordingly, the sustainability of floating and stilt houses—serving as an expression of community resilience—is evidenced through their integration of environmental, economic, and social dimensions.

The architecture of floating and stilt houses, as a manifestation of indigenous knowledge systems and practices, has demonstrated its viability in harmonising with the ecological conditions of the marshland landscape. However, considering the transformations and challenges introduced by twenty-first century trends and technological achievements, the vernacular architecture and spatial configurations of riverine communities in Agusan Marsh, Philippines—examined in this chapter—may need further scholarly attention and critical assessment.

One theoretical lens, as cited in this chapter, is the socio-ecological model, first developed by Bronfenbrenner (1979), which is recently cited and further discussed by Sadownik (2023). Although the writings of Bronfenbrenner (1979) and Sadownik (2023) focused on the ecology of child development, the same ecological principle can be seen in the practices of indigenous communities in Agusan Marsh reflected in their vernacular architecture and demonstrated with their community resilience in the context of the marsh environment. Bronfenbrenner's (1979) theory explains human development through layers of systems that affect people directly or indirectly. The series of social systems, to include microsystem(s), mesosystem, exosystem, macrosystem, and chronosystem (Bronfenbrenner, 1979, as cited by Sadownik, 2023, pp. 84–85) does not operate in a vacuum; instead, they are interconnected with each other.

Highlighting on macrosystem, generally as the '*socio-cultural context that includes the legal framework, cultural values, customs, and principles*' (Sadownik,

2023, p. 85), in the case of the communities in Agusan Marsh, their indigenous knowledge system and practices contextualise the interconnectedness of other systems, as well as the relationship between people and other ecological elements operating in both social and natural environment systems. From the point of view of chronosystem in a wider perspective including the course of time and historical changes in the socio-cultural environment, the twenty-first century trends and influences cannot be denied in the formation of the current housing and settlements despite their local and vernacular setting. Evident in the present construction of floating and stilt houses, while local building materials such as timber and bamboo are dominantly used, house construction also uses synthetic modern materials such as galvanised iron sheets, glass doors and windows, and the like. Although twenty-first century technologies have significantly shaped contemporary housing design and construction, it is equally important to note that indigenous knowledge, such as those demonstrating community resilience, offers valuable insights and potential solutions to pressing global challenges, including climate change adaptation, resilience-building, and disaster risk reduction and management.

5 Conclusions

The diversity of housing typologies observed in the riverine communities of Sabang Gibong and La Flora exemplifies a form of vernacular architecture rooted in indigenous knowledge systems, which demonstrates adaptive responses to the ecological conditions of the marshland environment. While the Sabang Gibong community was able to develop purely floating architecture due to its own geography, the community in La Flora with different geographical characteristics, was also able to develop other variations such as stilt houses and houses combining stilt and floating technologies. This reflects how an indigenous community was able to work with the dynamic environmental conditions, which resulted in a healthy and continued human-environment relation shaping the built environment of the community, and in return shaped the lifeways of the inhabitants. In light of contemporary challenges, particularly those related to climate change adaptation and disaster risk reduction, this chapter highlights the relevance of indigenous architectural practices as a source of insight for promoting resilience, environmental sustainability, and the preservation of vernacular traditions and knowledge systems.

Acknowledgements Sincere gratitude is hereby expressed to the residents of Sabang Gibong and La Flora riverine communities in the Municipality of Talacogon, Province of Agusan del Sur, Philippines for allowing the conduct of the case study and for sharing their invaluable indigenous knowledge system and practices.

References

Asquith, L. (2006). Lessons from the vernacular: Integrated approaches and new methods for housing research. In L. Asquith & M. Vellinga (Eds.), *Vernacular architecture in the twenty-first century: Theory, education and practice* (pp. 128–144). Taylor & Francis.

Aung MS, Pretty A (2022) Creating resilience through empowering self-build strategies in a Myanmar refugee camp. In: Izadpanahi P, Perugia F (Eds.) Architectural Science and User Experience: How can Design Enhance the Quality of Life, 55th International Conference of the Architectural Science Association, pp. 189–198.

Bronfenbrenner, U. (1979). *The ecology of human development: Experiments by nature and design.* Harvard University Press.

Goyal M, Silva MM (2023) The Socio-environmental importance of indigenous landscape in the case of Dharavi Slum. In: Proceedings of the 1st Southeast Asian Urbanisms Seminar, pp. 88–95.

Haas, S., Gianoli, A., & Van Eerd, M. (2021). The roles of community resilience and risk appraisal in climate change adaptation: The experience of the Kannagi Nagar resettlement in Chennai. *Environment & Urbanization, 33*(2), 560–578.

Malaque III IR, Golimlim AP (2019) Socio-spatial processes in the construction of low-income housing towards climate change adaptation and disaster risk reduction. In: Agrawal A, Gupta R (Eds.) Revisiting the Role of Architecture for Surviving Development, 53rd International Conference of the Architectural Science Association, pp. 441–450.

Malaque III IR, Sitcharon RMP (2024) Floating and stilt houses in Agusan Marsh, Philippines: Vernacular architecture and water urbanism in the 21st century, In: Zhang F, Yu R, Bischeri C, Liu T, Khoshbakht M (Eds.) Harmony in Architectural Science and Design: Sustaining the Future, 57th International Conference of the Architectural Science Association 2024, pp. 216–223.

Montresor F (2023) Craft villages of the Red River Delta, Vietnam: socio-ecological evolutions falling outside of the urban-rural dichotomy. In: Proceedings of the 1st Southeast Asian Urbanisms Seminar, pp. 96–103.

Moon C (2014) Applications of sustainable factors in floating architecture. In: World Sustainable Building 2014 Barcelona Conference-Conference Proceedings 2014(4).

Ostrom, E. (1990). *Governing the commons: The evolution of institutions for collective action.* Cambridge University Press.

Rapoport, A. (1988). Spontaneous settlement as vernacular architecture. In C. J. Patton (Ed.), *Spontaneous shelter: International perspectives and prospects* (pp. 51–77). Temple University Press.

Rapoport, A. (2006). Vernacular design as a model system. In L. Asquith & M. Vellinga (Eds.), *Vernacular architecture in the twenty-first century: Theory, education and practice* (pp. 179–198). Taylor & Francis.

Rudofsky, B. (1964). *Architecture without architects: A short introduction to non-pedigreed architecture.* The Museum of Modern Art.

Sadownik AR (2023) Bronfenbrenner: ecology of human development in ecology of collaboration, In: Sadownik AR, Višnjić Jevtić A (Eds.) (Re)theorising More-than-parental Involvement in Early Childhood Education and Care, International Perspectives on Early Childhood Education and Development 40, pp. 83–95.

Sitcharon RMP, Malaque III IR (2023a) Building with nature: the case of floating houses in Agusan Marsh, Philippines, In: Dewsbury M, Tanton D, Al-shamma Z, Tan J (Eds.) Architectural Science and User Experience: Sustainability and Health: the nexus of carbon neutral architecture and well-being, 56th International Conference of the Architectural Science Association 2023, pp. 184–197.

Sitcharon RMP, Malaque III IR (2023b) Living afloat: management regimes and sustainable dimensions of floating houses in Agusan Marsh. In: Proceedings of the 1st Southeast Asian Urbanisms Seminar, pp. 35–43.

Stassi ST (2023) Rivers in the making: placemaking practices in modified riverine neighborhoods. In: Proceedings of the 1st Southeast Asian Urbanisms Seminar, pp. 113–121.

Trang KM, Thu NHD, Thang PN (2023) Water integrated in development of infrastructure space for the expansion of urban area in Ho Chi Minh City. In: Proceedings of the 1st Southeast Asian Urbanisms Seminar, pp. 08–16.
Udomcharoenchaikit P (2023) Water oriented development: water urbanism and placemaking strategies as Bangkok city climate adaptation strategies. In: Proceedings of the 1st Southeast Asian Urbanisms Seminar, pp. 17–25.
United Nations (1987) Our Common Future: Report of the World Commission on Environment and Development. https://digitallibrary.un.org/record/139811?ln=en&v=pdf

From Vision to Practice: Advancing Nature-Based Urbanism

Sara Alidoust, Wendy Y. Chen, Richard Fuller, and Ali Cheshmehzangi

Abstract This book advocates for a fundamental paradigm shift in decision-making, city shaping, and urban living—one that repositions nature as a central actor rather than a passive backdrop. It calls for the integration of ecological systems and recognition of nature's agency alongside inclusive community participation to create resilient, equitable, and thriving urban environments. Such an approach is essential for cities to adopt transformative strategies that simultaneously address pressing societal challenges—such as climate change, social equity, and public health—while enhancing biodiversity.

Nature-based urbanism transcends geographical, political, and cultural boundaries. It is not confined to specific contexts but offers a universal framework for reimagining urban futures. By embedding nature as an active agent of change within planning and governance, this book demonstrates how cities can move beyond conventional sustainability narratives toward regenerative models that foster symbiotic relationships between people and ecosystems.

Keywords Nature-based urbanism · Nature-based solution · Governance · Place-making · Climate resilience · Sustainable development · Multispecies planning · Integrated planning · Regenerative planning

S. Alidoust (✉) · A. Cheshmehzangi
School of Architecture, Design and Planning, The University of Queensland, Brisbane, Australia
e-mail: s.alidoust@uq.edu.au

W. Y. Chen
Department of Geography, The University of Hong Kong, Pok Fu Lam, Hong Kong

R. Fuller
School of the Environment, The University of Queensland, Brisbane, Australia

Centre for Biodiversity and Conservation Science, The University of Queensland, Brisbane, Australia

A. Cheshmehzangi
School of Architecture and Urban Planning, Hunan University, Changsha, China

A. Cheshmehzangi et al. (eds.), *Nature-based Urbanism*, Urban Sustainability,
https://doi.org/10.1007/978-981-95-6864-2_13

1 Reflecting on the Journey: Nature-Based Urbanism

In today's rapidly urbanising world, marked by intensifying climate risks and deepening social inequities, the issue of our relationship with nature is highly important. Nature-based urbanism offers a critical opportunity for a paradigm shift. It is defined as actions inspired by, supported by, or copied from nature, aiming to advance not only environmental but also societal and economic challenges (European Commission, 2015). This approach positions nature not as a passive element but as an active agent for change, restoring ecosystems, improving climate resilience, and promoting socio-ecological health. By strategically integrating natural processes and elements into urban planning, nature-based urbanism delivers multiple co-benefits, strengthening both environmental integrity and societal well-being.

This book is organised into three interconnected sections. The first section, *Rethinking Urban Theory through Nature*, starts with reinterpreting Kevin Lynch's seminal theory of urban form to frame the 'image of nature-based urbanism'. Moving beyond visual interpretations of cities, it proposes a systematic understanding of urban environments as dynamic socio-ecological constructs. The section calls for prioritising multisensory, embodied experiences rather than abstract or purely technical design—creating spaces that can be felt, heard, and inhabited. It embraces movement and changes as inherent qualities of nature-based place-making and aligns planning with natural rhythms and cycles, such as seasonal variations and weather patterns. These principles are translated into adaptive design and planning strategies that respond to evolving human needs while supporting local ecosystems.

The second section of the book on *Governance, Participation and Multispecies Planning* demonstrates the importance of prioritising nature in urban governance to enhance liveability for all. It advocates mainstreaming nature-based strategies into core planning instruments, including master plans, zoning regulations, building codes, and municipal budgets. This cannot happen without capacity building within and collaborations between different stakeholder groups. Breaking down knowledge silos through interdisciplinary teams is essential for integrated decision-making.

The section highlights a persistent gap between official planning narratives and real-world practices, where multi-species considerations remain largely marginalised. To bridge this gap, the emphasis must shift toward evidence-based, multi-criteria decision-making, guided by standardised metrics and supported by robust monitoring systems. These tools are essential for evaluating the effectiveness of nature-based interventions and informing adaptive management strategies. Equally important is the establishment of long-term, stable governance structures to ensure continuity and accountability in implementation.

Section three of this book focuses on *Nature as Everyday Practice and Place-Making*. It discusses the loss of human-nature interaction that accelerates ecological degradation and highlights the role of eco-centric urbanism in restoring the balance between urban development and environmental considerations. Indigenous knowledge emerges as a vital resource, offering adaptive strategies rooted in vernacular architecture—such as the floating houses of Agusan Marsh in Mindanao, Philippines, which exemplify resilience and ecological sensitivity. Living with water, the vernacular architecture reflects Indigenous ingenuity that fosters community resilience in the face of climate change.

The section also emphasises the cultural dimension of nature in everyday life, showing how stories, images, and experiences can inspire sustainable behaviours—from children's daily mobility patterns to community-led place-making. It invites a more holistic view and deeper appreciation of the benefits of nature in cities, which extends to inspire a more respectful approach to day-to-day life. Attention is drawn to the stark contrasts between nature's rhythms and resource-intensive human routines, urging a more respectful approach to urban living.

Across theory, governance, and practice, several common threads emerge in this book. Central to nature-based urbanism is the need for systems thinking and integrated approaches that transcend disciplinary silos. Realising the full potential of nature-based urbanism demands sustained interdisciplinary research—bridging ecological science, design and planning, social justice, and technology—to develop robust, context-specific models of practice. Equally important is the recognition of nature's agency alongside human participation, acknowledging that ecological systems are active co-constituents of urban life rather than passive backdrops. The book also emphasises the role of cultural and Indigenous knowledge as a foundation for resilience, offering insights that bridge ecological integrity with social and cultural continuity. Finally, a strong commitment to equity and inclusivity underpins the vision for shaping socio-ecological futures that are just, adaptive, and sustainable. The journey of this book—through rethinking urban theory, governance, and everyday practices and place-making—marks a step toward cities that are ecologically grounded, socially just, and environmentally resilient.

2 Key Insights and Takeaways: What We Have Learned

Nature-based urbanism is not only about achieving outcomes, but it also requires a shift toward prioritising the process of decision-making, planning, and city shaping. This involves redefining priorities and values for all stakeholder groups and developing early visions of how nature can act as an agent for change. Implementing nature-based urbanism needs a cultural shift in urban decision-making. Major barriers include silo thinking, fragmented governance, limited technical capacity, socio-spatial inequalities, and the absence of robust monitoring and financing mechanisms. Urban planning and city shaping need systems thinking and holistic approaches, enabled through inter and multidisciplinary collaboration. Cities must be understood as dynamic socio-ecological ecosystems rather than static built environments.

Nature-based urbanism needs to be integrated in decision-making and city shaping across all levels as discussed in this book, from building and architectural design and site planning to precinct and neighbourhood planning, metropolitan strategies, and regional frameworks. This integration requires not only practical implementation but also a fundamental shift in mindset. Some scholars have even called for a more transformative 'turn', moving beyond nature-based solutions toward 'nature-based thinking' (Randrup et al., 2020). This integration applies to both new urban policies and spatial plans as well as to the retrofitting and revitalisation of existing urban spaces and communities.

To move beyond the pressure of delivering purely technical solutions, adopting place-based and participatory approaches can open constructive pathways forward. These approaches reveal possibilities that extend beyond rational, top-down methods. Redefining priorities in community-led projects and place-based strategies can create meaningful and lasting impacts that extend beyond their immediate contexts.

The concepts of participation and agency are interconnected yet distinct in their application to urban spatial studies. Beyond active stakeholder participation, agency—the capacity to act and influence change—is equally critical. Agency often fosters advocacy and can reshape social dynamics, driving transformative outcomes. However, conventional design and planning strategies frequently overlook the agency of nature's biotic diversity, focusing instead on the human benefits and reinforcing a transactional relationship with the environment. Biotic entities are not passive objects for management or aesthetic contemplation; they are co-constituents of urban ecosystems, possessing autonomous capacities for movement, adaptation, and resilience. Indigenous knowledge and values offer profound insights for advancing nature-based urbanism. They emphasise the preservation of vernacular traditions and knowledge systems while promoting environmental sustainability and community resilience. These perspectives offer alternative frameworks for designing cities that respect ecological processes and cultural heritage.

Finally, it is important to note that nature-based urbanism is not exclusive to Global North. In fact, many models developed in the North are inadequate or inapplicable to the Global South, where urbanisation follows different trajectories and faces distinct challenges. These include complex trade-offs between economic growth, environmental sustainability, socio-cultural priorities, and human health. Addressing these realities requires context-specific, participatory approaches that move beyond one-size-fits-all solutions. This calls for a fundamental shift in development logic and mindset, from residual greening to integrated ecological systems; from rigid engineered control to ecological collaboration, and from top-down decision-making to inclusive participation. Nature-based urbanism represents a paradigm shift that must be adapted across diverse geographies, political contexts, and communities with varying values, priorities, and constraints.

3 Moving Forward: Designing Nature into Cities

This book suggests that the way cities are planned and built must undergo a fundamental transformation—a shift toward integrating natural processes and elements into urban planning and city making. Building on the foundations laid in this volume, the forthcoming book, *Designing Nature into Cities*, will advance the discourse on nature-based urbanism by examining the integration of green and blue infrastructure across city planning, city making and city living. The volume will explore strategies for nature-based urbanism at different scales and through different lenses—from micro-landscape, including living walls and vegetated building envelopes to citywide ecological networks.

The book calls for place-sensitive planning and design that moves beyond green branding toward grounded accountability and support for local material ecosystems. A key emphasis is on native species, not only as ecologically appropriate but as vital elements to support local biodiversity and reinforce cultural identity, highlighting their role in enhancing urban resilience. It explores the dialectical relationship between humans and nature as a foundation for sustainable place-making, recognising the dynamic interplay between ecological systems and urban life. A socio-spatial lens will guide investigations into the structures and dynamics of natural spaces and their role in shaping urban identity, culture, and everyday life.

The volume addresses critical challenges such as ecological vulnerability, economic dependence, and fragmented governance, offering a strategic framework that integrates ecological restoration, economic transition, and community participation. By positioning nature-based solutions as both a climate mitigation tool and a catalyst for inclusive urban development, the book argues for integrated approaches that link environmental recovery with climate justice and urban equity.

4 Concluding Reflections

Nature-based urbanism is not a new jargon for sustainable development in academia or practice, nor is it a design method or planning approach. It represents a fundamental shift in mindset, a rethinking of cities as complex socio-ecological ecosystems. This transformation calls for placing nature, greenery, and landscape at the heart of every design and planning process, ensuring they are considered as essential structural elements rather than secondary features. Nature-based urbanism draws inspiration from, is supported by, and copies from nature. It needs a renewed vision and definition of decision-making in cities and communities that is multi and interdisciplinary, evolutionary and dynamic through feedback loops and continuous monitoring and evaluation, and strong cross-sector partnerships for implementation.

However, implementation remains a critical challenge. Too often, nature-based solutions are approached without fully acknowledging the socio-political, institutional, and governance contexts that shape urban decisions and actions. Their success depends not only on technical design but also on addressing the material and discursive dimensions that influence long-term outcomes (Kremer et al., 2016).

Finally, as Oliveira (2025) argues, the very definition of nature remains contested and ambiguous, creating uncertainty about what constitutes 'naturalness' within nature-based solutions. Recognising plural interpretations of nature and integrating diverse perspectives on human-nature relationships are therefore essential steps toward advancing the effectiveness, inclusivity, and legitimacy of nature-based solutions in urban planning theory and practice.

References

European Commission. (2015). *Towards an EU Research and Innovation Policy Agenda for Nature-Based Solutions & Re-Naturing Cities-Final Report of the Horizon 2020 Expert Group on 'Nature Based Solutions and Re-Naturing Cities'*. Publications Office of the European Union.

Kremer, P., Hamstead, Z., Haase, D., McPhearson, T., Frantzeskaki, N., Andersson, E., Kabisch, N., Larondelle, N., Rall, E. L., Voigt, A., & Baró, F. (2016). Key Insights for the Future of Urban Ecosystem Services Research. *Ecology and Society, 21*(2).

Oliveira, F. L. (2025). Nature in Nature-based Solutions in Urban Planning. *Landscape and Urban Planning, 256*, 105282.

Randrup, T. B., Buijs, A., Konijnendijk, C. C., & Wild, T. (2020). Moving beyond the Nature-based Solutions Discourse: Introducing Nature-based Thinking. *Urban Ecosystems, 23*(4), 919–926.

The manufacturer's authorised representative in the EU is Springer Nature Customer Service Centre GmbH, Europaplatz 3, 69115 Heidelberg, Germany. If you have any concerns regarding our products, please contact ProductSafety@springernature.com

Printed and bound by CPI Group (UK) Ltd, Croydon, CR0 4YY
21/09/2026
02220012-0008